Experimental Chemistry

Fourth Edition

James F. Hall
University of Massachusetts at Lowell

Houghton Mifflin Company **Boston** **New York**

Senior Sponsoring Editor:	**Richard Stratton**
Assistant Editor:	**Marianne Stepanian**
Director of Manufacturing:	**Michael O'Dea**
Executive Marketing Manager:	**Karen Natale**

Cover Design:	Harold Burch Design, New York
Cover Image:	Pete Turner/The Image Bank

Printed in the U.S.A.

ISBN: 0-669-41800-5

6789-VG-05 04 03 02 01

Contents

Preface

Introduction: Why We Study Chemistry

On the simplest level, we say that chemistry is "the study of matter and the transformations it undergoes." On a more intellectual level, we realize that chemistry can explain (or try to explain) such things as the myriad chemical reactions in the living cell, the transmission of energy by superconductors, the working of transistors, and even how the oven and drain cleaners we use in our homes function.

The study of chemistry is required for students in many other fields because it is a major unifying force among these other subjects. Chemistry is the study of matter itself. In other disciplines, particular *aspects* of matter or its applications are studied, but the basis for such study rests in a firm foundation in chemistry.

Laboratory study is required in chemistry courses for several reasons. In some cases, laboratory work may serve as an *introduction* to further, more difficult laboratory work (this is true of the first few experiments in this laboratory manual). Before you can perform "meaningful" or "relevant" experiments, you must learn to use the basic tools of the trade.

Sometimes laboratory experiments in chemistry are used to *demonstrate* the topics covered in course lectures. For example, every chemistry course includes a lengthy exposition of the properties of gases and a discussion of the gas laws. A laboratory demonstration of these laws may clarify them.

Laboratory work can teach you the various standard *technique*s used by scientists in chemistry and in most other fields of science. For example, pipets are used in many biological, health, and engineering disciplines when a precisely measured volume of liquid is needed. Chemistry lab is an excellent place to learn techniques correctly; you can learn by trial and error in general chemistry lab rather than on the job later.

Finally, your instructor may use the laboratory as a means of judging how much you have learned and understood from your lectures. Sometimes students can read and understand their textbook and lecture notes and do reasonably well on examinations without gaining much practical knowledge of the subject. The lab serves as a place where classroom knowledge can be synthesized and applied to realistic situations.

What Will Chemistry Lab Be Like?

General chemistry laboratories are generally 2–3 hours long, and most commonly meet once each week. The laboratory is generally conducted by a teaching assistant—a graduate student with several years of experience in chemistry laboratory. Smaller colleges, which may not have a graduate program in chemistry, may employ instructors whose specific duties consist of running the general chemistry labs or upperclass student assistants who work under the course professor's supervision. Your professor undoubtedly will visit the laboratory frequently to make sure that everything is running smoothly.

The first meeting of a laboratory section is often quite hectic. This is perfectly normal, so do not judge your laboratory by this first meeting. Part of the first lab meeting will be devoted to checking in to the laboratory. You will be assigned a workspace and locker and will be asked to go through the locker to make sure that it contains all the equipment you will need. Your instructor will then conduct a brief orientation to the laboratory, giving you the specifics of how the laboratory period will be operated. Be sure to ask any questions you may have about the operation of the laboratory.

Your instructor will discuss with you the importance of safety in the laboratory and will point out the emergency equipment. *Pay close attention to this discussion so that you will be ready for any eventuality.*

Finally, your instructor will discuss the experiment you will be performing after check-in. Although we have tried to make this laboratory manual as explicit and complete as possible, your instructor will certainly offer tips or advice based on his or her own experience and expertise. Take advantage of this information.

How Should You Prepare for Lab?

Laboratory should be one of the most enjoyable parts of your study of chemistry. Instead of listening passively to your instructor, you have a chance to witness chemistry in action. However, if you have not suitably *prepared* for lab, you will spend most of your time wondering what is going on.

This manual has been written and revised to help you prepare for laboratory. First of all, each experiment has a clearly stated *Objective*. This is a very short summary of the lab's purpose. This Objective will suggest what sections of your textbook or lecture notes might be appropriate for review before lab. Each experiment also contains an *Introduction* that reviews explicitly some of the theory and methods to be used in the experiment. You should read through this Introduction *before* the laboratory period and make cross-references to your textbook while reading. Make certain that you understand fully what an experiment is supposed to demonstrate before you attempt that experiment.

The lab manual contains a detailed *Procedure* for each experiment. The Procedures have been revised in this edition to be as clear as possible. Do not wait until you are actually performing the experiment to read this material. Study the Procedure before the lab period, and question the instructor before the lab period on anything you do not understand. Sometimes it is helpful to write up a summary or overview of a long Procedure. This can prevent major errors when you are actually performing the experiment. Another technique students use in getting ready for lab is to prepare a *flow chart,* indicating the major procedural steps, and any potential pitfalls, in the experiment. You can keep the flow chart handy while performing the experiment to help in avoiding any gross errors as to "what comes next." A flow chart is especially useful for planning your time most effectively: for example, if a procedure calls for heating something for an hour, a flow chart can help you plan what else you can get done during that hour.

This lab manual contains a *Laboratory Report* for each Choice of each experiment. Each Laboratory Report is preceded by a set of *pre-laboratory questions*, which your instructor will probably ask you to complete before coming to lab. Sometimes these pre-laboratory questions will include numerical problems similar to those you will encounter in processing the data you will be collecting in the experiment. Obviously, if you review a calculation before lab, things will be that much easier when you are actually in the lab.

In other cases, the pre-laboratory questions may ask you to do some literature research in your school's science library. The study of chemistry is based not only on current and future experiments, but also on a careful review of what has come before. It is very important to develop an acquaintance with the major reference works of chemical literature.

New in the Fourth Edition

Several new experiments have been added to this edition to expand its appeal: a simple demonstration of "counting by weighing"; an analysis of calcium in antacid tablets with EDTA; and two new experiments with a biochemical orientation. Additionally, several previous experiments have been rearranged into more rational units that should better fit the available time in the lab period.

As always, the author encourages those who use this manual—both students and instructors—to assist him in improving it for the future: Any comments or criticism will be gratefully appreciated.

James F. Hall
hallj@woods.uml.edu

Laboratory Glassware and Other Apparatus

Introduction

When you open your laboratory locker for the first time, you are likely to be confronted with a bewildering array of various sizes and shapes of glassware and other apparatus. Glass is used more than any other material for the manufacture of laboratory apparatus because it is relatively cheap, usually easy to clean (and keep clean), and can be heated to fairly high temperatures or cooled to quite low temperatures. Most importantly, glass is used because it is impervious to and nonreactive with most reagents encountered in the beginning chemistry laboratory. Most glassware used in chemistry laboratories is made of borosilicate glass, which is relatively sturdy and safe to use at most temperatures. Such glass is sold under such trade names as Pyrex® (Corning Glass Co.) and Kimax® (Kimble Glass Co.). If any of the glassware in your locker does not have either of these trade names marked on it, consult with your instructor before using it at temperatures above room temperature.

Many pieces of laboratory glassware and apparatus have special names that have evolved over the centuries chemistry has been studied. You should learn the names of the most common pieces of laboratory apparatus to make certain that you use the correct equipment for the experiments in this manual. Study the drawings of common pieces of laboratory apparatus that are shown on the next few pages. Compare the equipment in your locker with these drawings, and identify all the pieces of apparatus that have been provided to you. It might be helpful for you to label the equipment in your locker, at least for the first few weeks of the term.

While your locker may not contain all the apparatus in the drawings, be sure to ask your instructor for help if there is equipment in your locker that is not described in the drawings. While examining your locker equipment, watch for chipped, cracked, or otherwise imperfect glassware. *Imperfect glassware is a major safety hazard, and must be discarded.* Don't think that because a beaker has just a little crack, it can still be used. *Replace all glassware that has any cracks, chips, star fractures, or any other deformity.* Most college and university laboratories will replace imperfect glassware free of charge during the first laboratory meeting of the term, but may assess charges if breakage is discovered later in the term.

You may wish to clean the set of glassware in your locker during the first meeting of your laboratory section. This is certainly admirable, but be warned that lab glassware has an annoying habit of becoming dirty again without apparent human intervention. Glassware always looks clean when wet, but tends to dry with water spots (and show minor imperfections in the glass that are not visible when wet). It is recommended that you thoroughly clean all the locker glassware at the start, and then rinse the glassware out before its first use. In the future, clean glassware before leaving lab for the day, and rinse before using. Instructions for the proper cleaning of laboratory glassware are provided after the diagrams of apparatus.

Laboratory Glassware and Other Apparatus

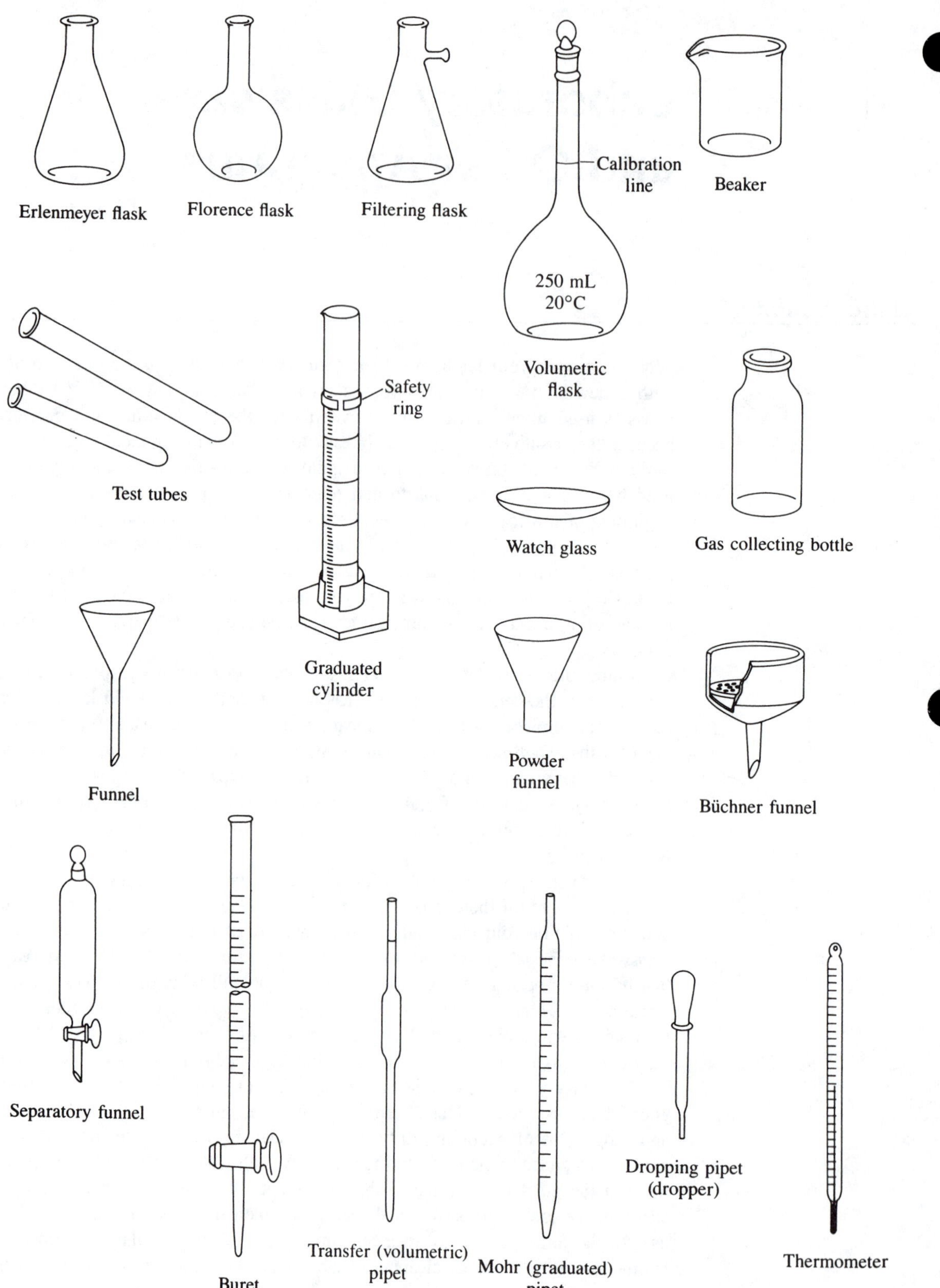

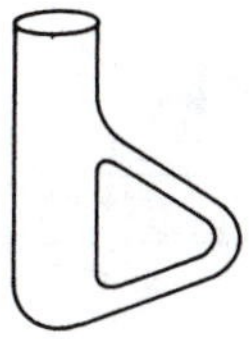

Thiele tube

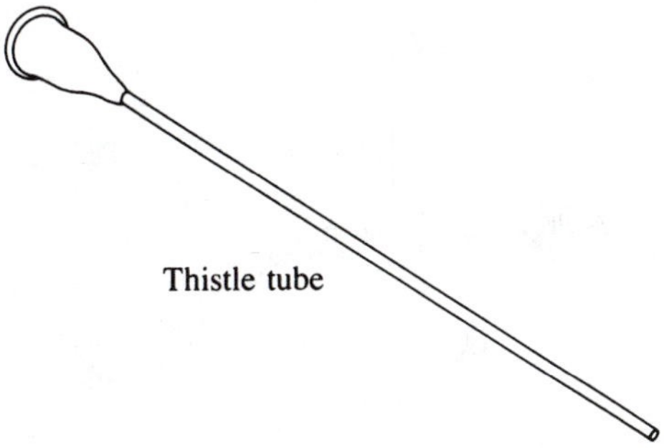

Thistle tube

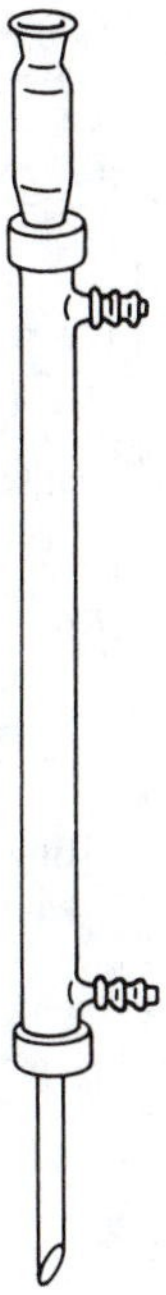

Liebig condenser
(water-cooled)

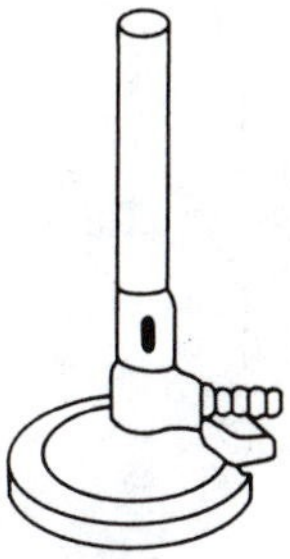

Bunsen
burner

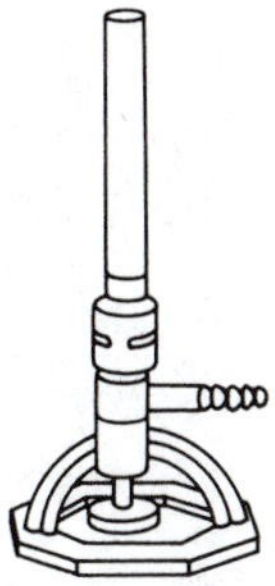

Tirrill burner

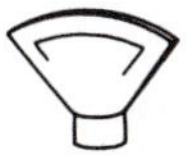

Wing top
(flame spreader)

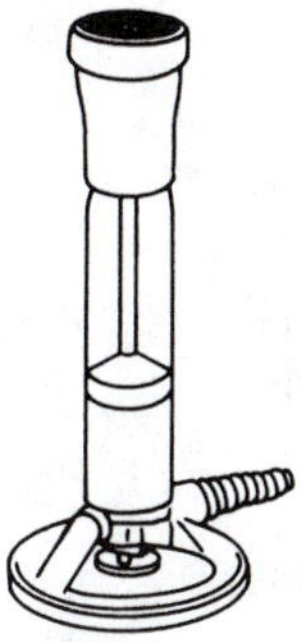

Meker burner

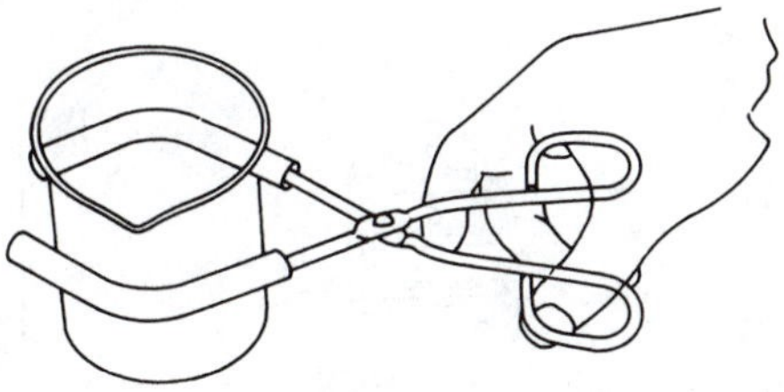

Beaker tongs

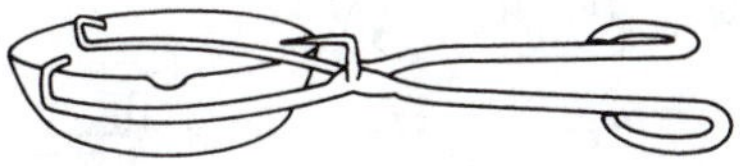

Dish tongs

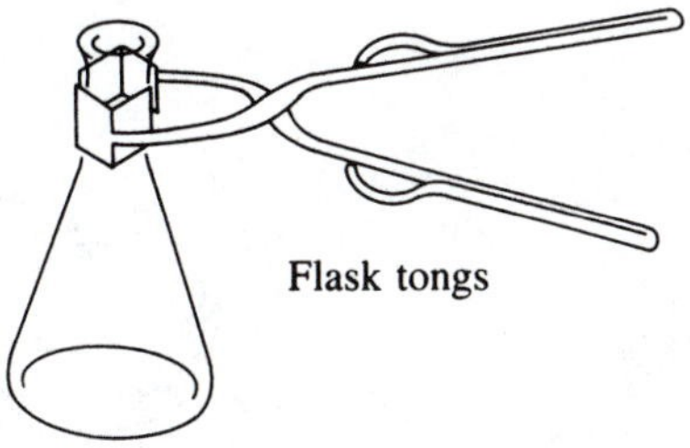

Flask tongs

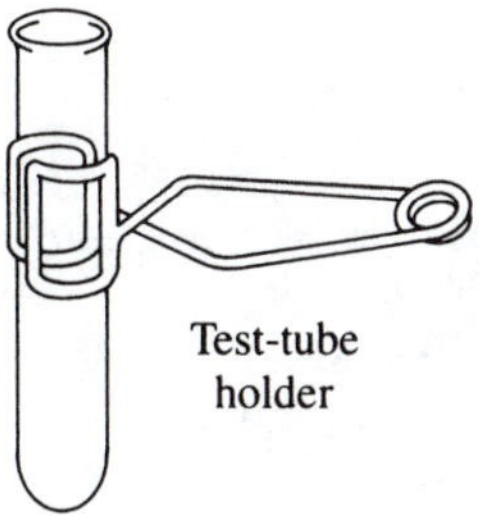

Test-tube
holder

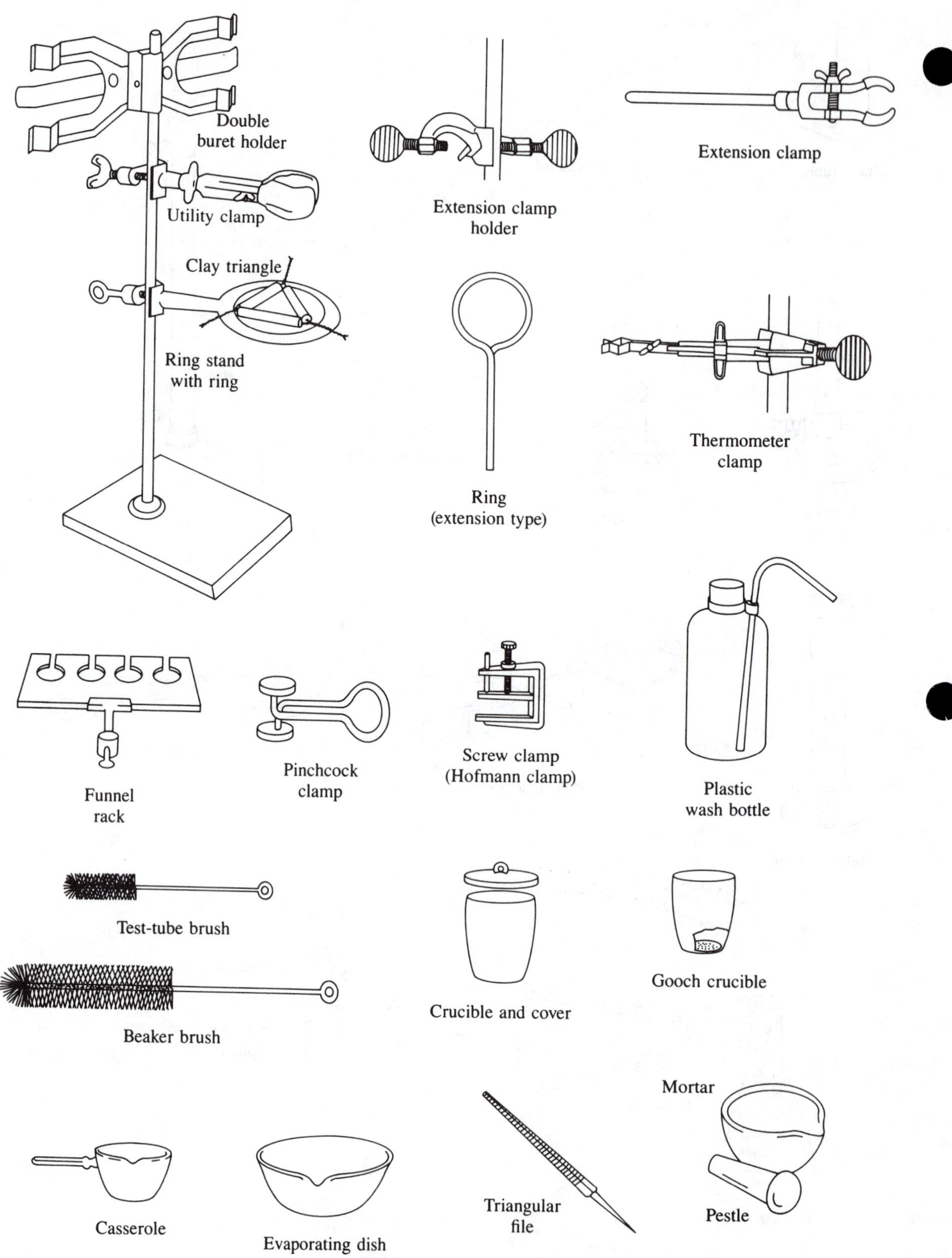
Double
buret holder
Utility clamp
Clay triangle
Ring stand
with ring
Extension clamp
holder
Extension clamp
Ring
(extension type)
Thermometer
clamp
Funnel
rack
Pinchcock
clamp
Screw clamp
(Hofmann clamp)
Plastic
wash bottle
Test-tube brush
Beaker brush
Crucible and cover
Gooch crucible
Mortar
Pestle
Casserole
Evaporating dish
Triangular
file

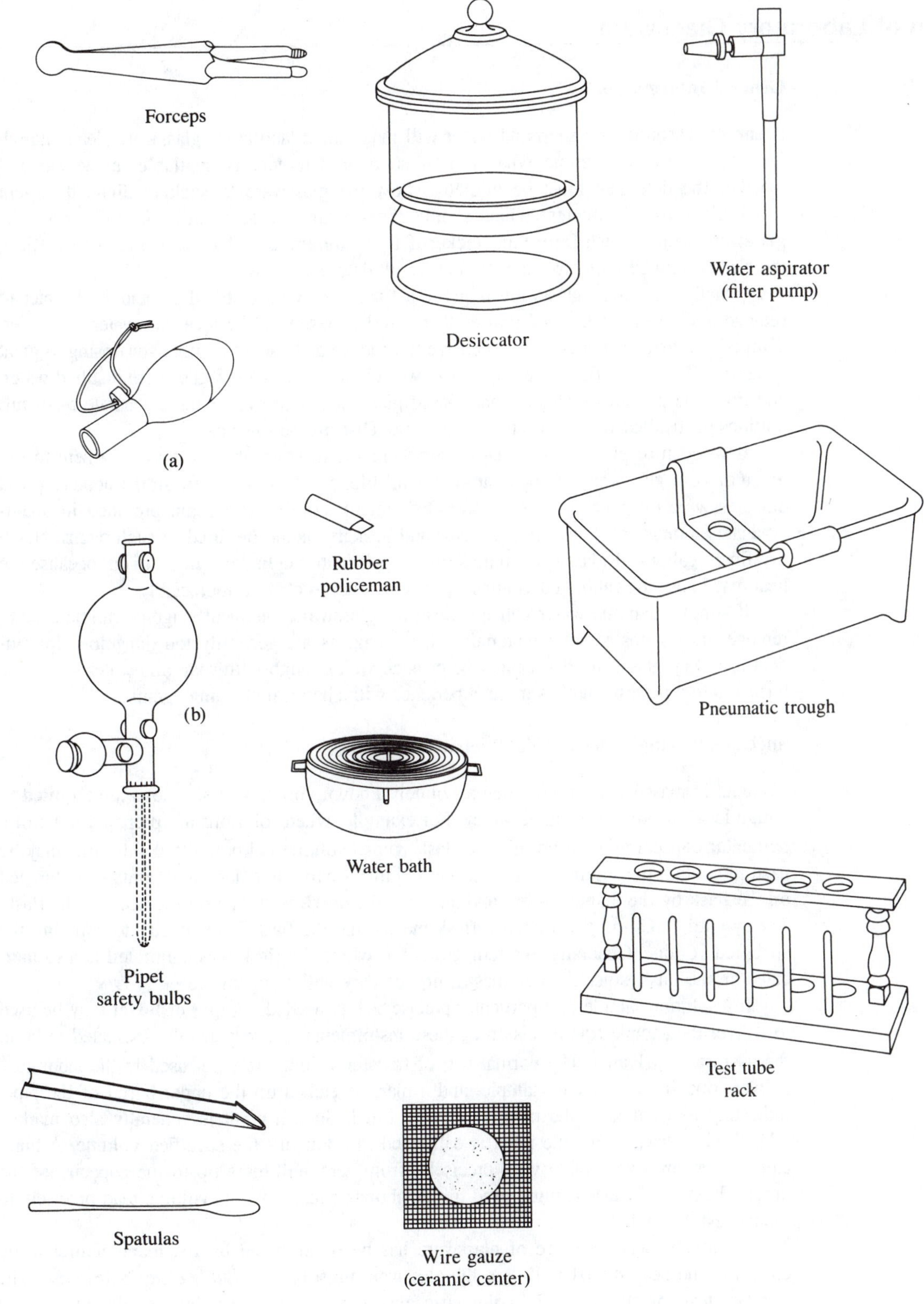
Forceps
Desiccator
Water aspirator
(filter pump)
(a)
Rubber
policeman
(b)
Pneumatic trough
Water bath
Pipet
safety bulbs
Test tube
rack
Spatulas
Wire gauze
(ceramic center)

Cleaning of Laboratory Glassware

General Information

A simple washing with soap and water will make most laboratory glassware clean enough for general use. Determine what sort of soap or detergent is available in the lab and whether the detergent must be diluted. Allow the glassware to soak in dilute detergent solution for 10–15 minutes (which should remove any grease or oil), and then scrub the glassware with a brush from your locker if there are any caked solids on the glass. Rinse the glassware well with tap water to remove all detergent.

Usually laboratory glassware is given a final rinse with distilled or deionized water to remove any contaminating substances that may be present in the local tap water. However, distilled/deionized water is very expensive to produce and cannot be used for rinsing in great quantities. Therefore, fill a plastic squeeze wash bottle from your locker with distilled water, and rinse the previously cleaned and rinsed glassware with two or three separate 5–10 mL portions of distilled water from the wash bottle. Discard the rinsings.

If wooden or plastic drying hooks are available in your lab, you may use them to dry much of your glassware. If hooks are not available, or items do not fit on the hooks, spread out glassware on paper towels to dry. Occasionally, drying ovens are provided in undergraduate laboratories. Only simple flasks and beakers should be dried in such ovens. Never dry finely calibrated glassware (burets, pipets, graduated cylinders) in an oven, because the heat may cause the calibrated volume of the container to change appreciably.

If simple soap and water will not clean the glassware, chemical reagents can be used to remove most stains or solid materials. These reagents are generally too dangerous for student use. Any glassware that cannot be cleaned well enough with soap and water should be turned in to a person who has more experience with chemical cleaning agents.

Special Cleaning Notes for Volumetric Glassware

Volumetric glassware is used when absolutely known volumes of solutions are required to a high level of precision and accuracy. For example, when solutions are prepared to be of a particular concentration, a volumetric flask whose volume is known to ± 0.01 mL may be used to contain the solution. The absolute volume a particular flask will contain is stamped on the flask by the manufacturer, and an exact **fill mark** is etched on the neck of the flask. The symbol "TC" on a volumetric flask means that the flask is intended "to contain" the specified volume. Generally, the temperature at which the flask was calibrated is also indicated on the flask, since the volumes of liquids vary with temperature.

If a solution sample of a particular precise size is needed, a **pipet** or **buret** may be used to deliver the sample (the precision of these instruments is also generally indicated to be to the nearest ± 0.01 mL). The normal sort of transfer volumetric pipet used in the laboratory delivers one specific size of sample, and a mark is etched on the upper barrel of the pipet indicating to what level the pipet should be filled. Such a pipet is generally also marked "TD," which means that the pipet is calibrated "to deliver" the specified volume. A buret can deliver any size sample very precisely, from zero milliliters up to the capacity of the buret. The normal Class A buret used in the laboratory can have its volume read precisely to the nearest 0.01 mL.

Obviously, when a piece of glassware has been calibrated by the manufacturer to be correct to the nearest 0.01 milliliter, the glassware must be *absolutely clean* before use. The standard test for cleanliness of volumetric glassware involves watching a film of distilled water run down the interior sides of the glassware. Water should flow in sheets (a continuous film) down the inside of volumetric glassware, without beading up anywhere on the inside surface.

If water beads up anywhere on the interior of volumetric glassware, the glassware must be soaked in dilute detergent solution, scrubbed with a brush (except for pipets), and rinsed thoroughly (both with tap water and distilled water). The process must be repeated until the glassware is absolutely clean. Narrow bristled brushes are available for reaching into the narrow necks of volumetric flasks, and special long-handled brushes are available for scrubbing burets. Since brushes cannot be fitted into the barrel of pipets, if it is not possible to clean the pipet completely on two or three attempts, the pipet should be exchanged for a new one (special pipet washers are probably available in the stockroom for cleaning pipets that have been turned in by students as uncleanable).

Volumetric glassware is generally used wet. Rather than drying the glassware (possibly allowing the glassware to become water-spotted, which may cause incorrect volumes during use), the user rinses the glassware with whatever is going to be used ultimately in the glassware. For example, if a buret is to be filled with standard acid solution, the buret still wet from cleaning is rinsed with small portions of the same acid (with the rinsings being discarded). Rinsing with the solution that is going to be used with the glassware insures that no excess water will dilute the solution to be measured. *Volumetric glassware is never heated in an oven,* since the heat may destroy the integrity of the glassware's calibration. If space in your locker permits, you might wish to leave volumetric glassware, especially burets, filled with distilled water between laboratory periods. A buret that has been left filled with water will require much less time to clean on subsequent use (consult with the instructor to see if this is possible).

Safety in the Chemistry Laboratory

Introduction

Chemistry is an experimental science. You cannot learn chemistry without getting your hands dirty. Any beginning chemistry student faces the prospect of laboratory work with some apprehension. It would be untruthful to say that there is no element of risk in a chemistry lab. *Chemicals can be dangerous.* The more you study chemistry, the larger the risk will become. However, if you approach your laboratory work calmly and studiously, you will minimize the risks.

During the first laboratory meeting, you should ask your instructor for a brief tour of the laboratory room. Ask him or her to point out for you the locations of the various pieces of emergency apparatus provided by your college or university. At your bench, construct a map of the laboratory, noting the location of the exits from the laboratory and the location of all safety equipment. Close your eyes, and test whether you can locate the exits and safety equipment from memory. A brief discussion of the major safety apparatus and safety procedures follows. A Safety Quiz is provided at the end of this section to test your comprehension and appreciation of this material. For additional information on safety in your particular laboratory, consult with your laboratory instructor or course professor.

Protection for the Eyes

Government regulations, as well as common sense, demand the wearing of protective eyeware while you are in the laboratory. Such eyeware must be worn even if you personally are not working on an experiment. Figure 1 shows one common form of plastic **safety goggle.**

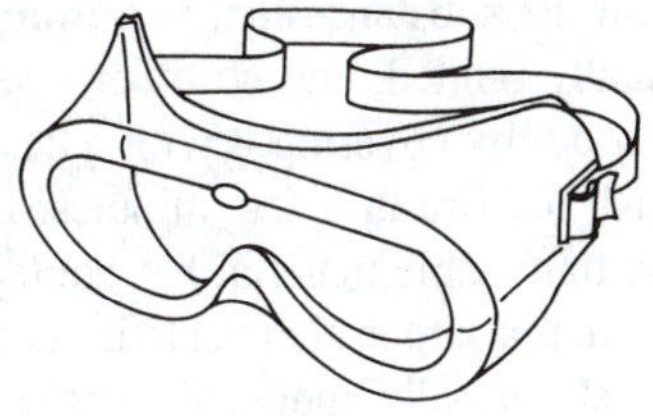

Figure 1. A typical plastic student safety goggle.

Although you may not use the particular type of goggle shown in the figure, your eyeware must include shatterproof lenses and side shields that will protect you from splashes. *Safety glasses must be worn at all times while you are in the laboratory, whether or not you are working with chemicals.* Failure to wear safety glasses may result in your being failed or withdrawn from your chemistry course, or in some other disciplinary action.

In addition to protective goggles, an **eyewash fountain** provides eye protection in the laboratory. Should a chemical splash near your eyes, you should use the eyewash fountain before the material has a chance to run in behind your safety glasses. If the eyewash is not near your bench, wash your eyes quickly but thoroughly with water from the nearest source, and then use the eyewash. A typical eyewash fountain is indicated in Figure 2:

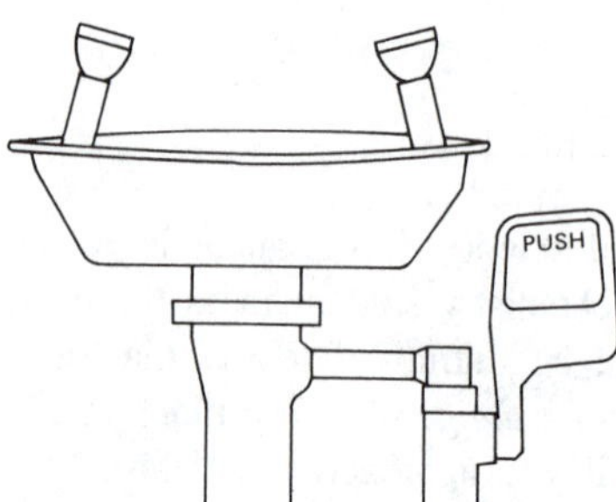

Figure 2. Laboratory emergency eyewash station.

The eyewash has a panic bar that enables the eyewash to be activated easily in an emergency. If you need the eyewash, don't be modest—*use it immediately.* It is critical that you protect your eyes properly.

Protection from Fire

The danger of uncontrolled fire in the chemistry laboratory is very real, since the lab typically has a fairly large number of flammable liquids in it, and open-flame gas burners are generally used for heating (see pages xxii–xxiv for proper use of the gas burner). With careful attention, however, the danger of fire can be reduced considerably.

Always check around the lab before lighting a gas burner to ensure that no one is pouring or using any flammable liquids. Be especially aware that the vapors of most flammable liquids are heavier than air and tend to concentrate in sinks (where they may have been poured) and at floor level. Since your laboratory may be used by other classes, always check with your instructor before beginning to use gas burners.

The method used to fight fires that occur in spite of precautions being taken depends on the size of the fire and on the substance that is burning. If only a few drops of flammable liquid have been accidentally ignited, and no other reservoir of flammable liquid is nearby, the fire can usually be put out by covering it with a beaker. This deprives the fire of oxygen and will usually extinguish the fire in a few minutes. Leave the beaker in place for several minutes to ensure that the flammable material has cooled and will not flare up again.

In the unlikely event that a larger chemical fire occurs, carbon dioxide **fire extinguishers** are available in the lab (usually mounted near one of the exits from the room). An example of a typical carbon dioxide fire extinguisher is shown in Figure 3. Before activating the extinguisher, pull the metal safety ring from the handle. Direct the output from the extinguisher at the base of the flames. The carbon dioxide not only smothers the flames, it also cools the flammable material quickly. If it becomes necessary to use the fire extinguisher, be sure afterward to turn the extinguisher in at the stockroom so that it can be refilled immediately. If the carbon dioxide extinguisher does not immediately put down the fire, evacuate the laboratory and call the fire department. Carbon dioxide fire extinguishers must *not* be used on fires involving magnesium or certain other reactive metals, since carbon dioxide may react vigorously with the burning metal and make the fire worse.

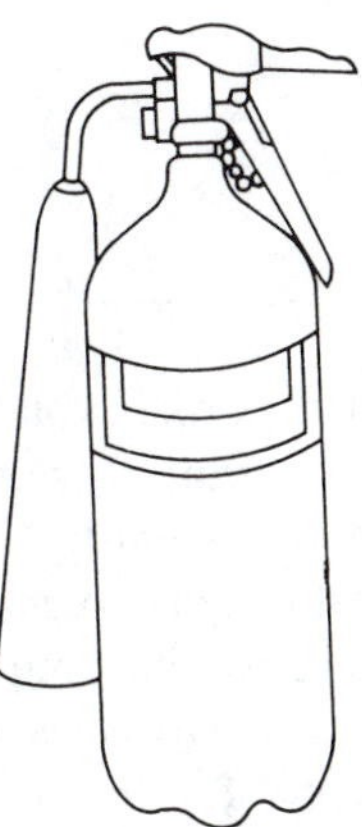

Figure 3. A typical carbon dioxide fire extinguisher. Pull the metal ring to activate the extinguisher.

One of the most frightening and potentially tragic accidents is the igniting of a person's clothing. For this reason, certain types of clothing are *not appropriate* for the laboratory and must not be worn. Since sleeves are most likely to come in closest proximity to flames, any garment that has bulky or loose sleeves should not be worn in the laboratory. Certain fabrics should also be avoided; such substances as silk and certain synthetic materials may be highly flammable. Ideally, students should wear laboratory coats with tightly fitting sleeves made specifically for the chemistry laboratory. This clothing may be required by your particular college or university. Long hair also presents a clear danger if it is allowed to hang loosely in the vicinity of the flame. Long hair must be pinned back or held with a rubber band.

In the unlikely event a student's clothing or hair is ignited, his or her neighbors must take prompt action to prevent severe burns. Most laboratories have two options for extinguishing such fires: the **water shower** and the **fire blanket**. Figure 4 shows a typical laboratory emergency water shower:

Figure 4. Laboratory emergency shower. Use the shower to extinguish clothing fires and in the event of a large-scale chemical spill.

Showers such as this generally are mounted near the exits from the laboratory. In the event someone's clothing or hair is on fire, *immediately* push or drag the person to the shower and pull the metal ring of the shower. Safety showers generally dump 40–50 gallons of water, which should extinguish the flames. Be aware that the showers cannot be shut off once the metal ring has been pulled. For this reason, the shower cannot be demonstrated. (But note that the showers are checked for correct operation on a regular basis.)

Protection from Chemical Burns

Most acids, alkalis, and oxidizing and reducing agents must be assumed to be corrosive to skin. It is impossible to avoid these substances completely, since many of them form the backbone of the study of chemistry. Usually, a material's corrosiveness is proportional to its concentration. Most of the experiments in this manual have been set up to use as dilute a mixture of reagents as possible, but this does not entirely remove the problem. Make it a personal rule to *wash your hands regularly* after using any chemical substance and to wash immediately, with plenty of water, if any chemical substance is spilled on the skin.

After working with a substance known to be particularly corrosive, you should wash your hands immediately even if you did not spill the substance. Someone else using the bottle of reagent may have spilled the substance on the side of the bottle. It is also good practice to hold a bottle of corrosive substance with a paper towel, or to wear plastic gloves, during pouring. Do not make the mistake of thinking that because an experiment calls for dilute acid, this acid cannot burn your skin. Some of the acids used in the laboratory are not volatile, and as water evaporates from a spill, the acid becomes concentrated enough to damage skin. Whenever a corrosive substance is spilled on the skin, you should inform the instructor immediately. If there is any sign whatsoever of damage to the skin, you will be sent to your college's health services for evaluation by a physician.

In the event of a major chemical spill, in which substantial portions of the body or clothing are affected, you must use the emergency water shower. Forget about modesty, and get under the shower immediately.

Protection from Toxic Fumes

Many volatile substances may have toxic vapors. A rule of thumb for the chemistry lab is, "If you can smell it, it can probably hurt you." Some toxic fumes (such as those of ammonia) can overpower you immediately, whereas other toxic fumes are more insidious. The substance may not have that bad an odor, but its fumes can do severe damage to the respiratory system. There is absolutely no need to expose yourself to toxic fumes. All chemistry laboratories are equipped with **fume exhaust hoods**. A typical hood is indicated in Figure 5.

Figure 5. A common type of laboratory fume exhaust hood. Use the hood whenever a reaction involves toxic fumes. Keep the glass door of the hood partially closed to provide for rapid flow of air.

The exhaust hood has fans that draw vapors out of the hood and away from the user. The hood is also used when flammable solvents are required for a given procedure, since the hood will remove the vapors of such solvents from the laboratory and reduce the hazard of fire. The hood is equipped with a safety-glass window that can be used as a shield from reactions that could become too vigorous. Naturally, the number of exhaust hoods available in a particular laboratory is limited, but *never neglect to use the hood if it is called for*, merely to save a few minutes of waiting time. Finally, reagents are sometimes stored in a hood, especially if the reagents evolve toxic fumes. Be sure to return such reagents to the designated hood after use.

Protection from Cuts and Simple Burns

Perhaps the most common injuries to students in the beginning chemistry laboratory are simple cuts and burns. Glass tubing and glass thermometers are used in nearly every experiment and are often not prepared or used properly. Most glass cuts occur when the glass (or thermometer) is being inserted through rubber stoppers in the construction of an apparatus. Use glycerine as a lubricant when inserting glass through rubber (several bottles of glycerine will be provided in your lab). Glycerine is a natural product of human and animal metabolism; it may be applied liberally to any piece of glass. Glycerine is water soluble, and though it is somewhat messy, it washes off easily. You should always remove glycerine before using an apparatus since the glycerine may react with the reagents to be used.

Most simple burns in the laboratory occur when a student forgets that an apparatus may be hot and touches it. Never touch an apparatus that has been heated until it has cooled for at least five minutes, or unless specific tongs for the apparatus are available.

Report any cuts or burns, no matter how apparently minor, to the instructor immediately. If there is any visible damage to the skin, you will be sent to your college's health services for immediate evaluation by a physician. What may seem like a scratch may be adversely affected by chemical reagents or may become infected; therefore, it must be attended to by trained personnel.

Proper Use of the Laboratory Burner

The laboratory burner is the most commonly used apparatus in the general chemistry laboratory, and may pose a major hazard if not used correctly and efficiently.

The typical laboratory burner is correctly called a **Tirrel burner** (though the term Bunsen burner is often used in a generic sense). A representation of a Tirrel burner is indicated in Figure 6. Compare the burner you will be using with the burner shown in the figure, and consult with your instructor if there seems to be any difference in construction. Burners from different manufacturers may differ slightly in appearance and operation from the one shown in the illustration.

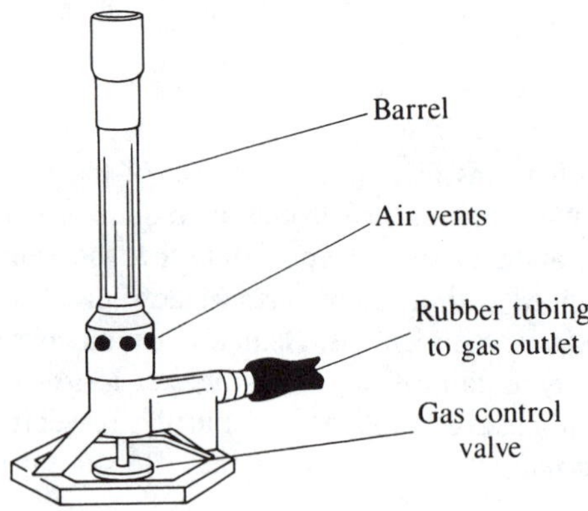

Figure 6. A Tirrel burner of the sort most commonly found in student laboratories. Compare this burner with the one you will use for any differences in construction or operation.

Most laboratories are supplied with natural gas, which consists primarily of the hydrocarbon methane (CH_4). If your college or university is some distance from natural gas lines, your laboratory may be equipped with bottled gas, which consists mostly of the hydrocarbon propane (C_3H_8). In this case, your burner may have modifications to allow the efficient burning of propane.

A length of thin-walled rubber tubing should attach your burner to the gas main jet. If your burner has a screw-valve on the bottom for controlling the flow of gas, the valve should be closed (by turning in a right-hand direction) before you light the burner. The barrel of the burner should be rotated to close the air vents (or slide the air vent cover over the vent holes if your burner has this construction).

To light the burner, turn the gas main jet to the open position. If your burner has a screw-valve on the bottom, open this valve until you hear the hiss of gas. Without delay, use your striker or a match to light the burner. If the burner does not light on the first attempt, shut off the gas main jet, and make sure that all rubber tubing connections are tight. Then reattempt to light the burner. You may have to increase the flow of gas, using the screw-valve on the bottom of the burner.

After lighting the burner, the flame is likely to be yellow and very unstable (easily blown about by any drafts in the lab). The flame at this point is relatively cool (the gas is barely burning) and would be very inefficient in heating. To make the flame hotter and more stable, open the air vents on the barrel of the burner slowly to allow oxygen to mix with the gas before combustion. This should cause the flame's size to decrease and its color to change. A proper mixture of air and gas gives a pale blue flame, with a bright blue cone-shaped region in the center. The hottest part of the flame is directly above the tip of the bright blue cone. Whenever an item is to be heated strongly, it should be positioned directly above this bright blue cone. You should practice adjusting the flame to get the ideal pale blue flame with its blue inner cone, using the control valve on the bottom of the burner or the gas main jet.

Protection from Apparatus Accidents

An improperly constructed apparatus can create a major hazard in the laboratory, not only to the person using the apparatus, but to his or her neighbors as well. Any apparatus you set up should be constructed exactly as indicated in this manual. If you have any question as to whether you have set up the apparatus correctly, ask your instructor to check the apparatus before you begin the experiment.

Perhaps the most common apparatus accident in the lab is the tipping over of a flask or beaker while it is being heated or otherwise manipulated. All flasks should be *clamped securely* with an adjustable clamp to a ring support. This is indicated in Figure 7. Be aware that a vacuum flask is almost guaranteed to tip over during suction filtration if it is not clamped. This will result in loss of the crystals being filtered and will require that you begin the experiment again.

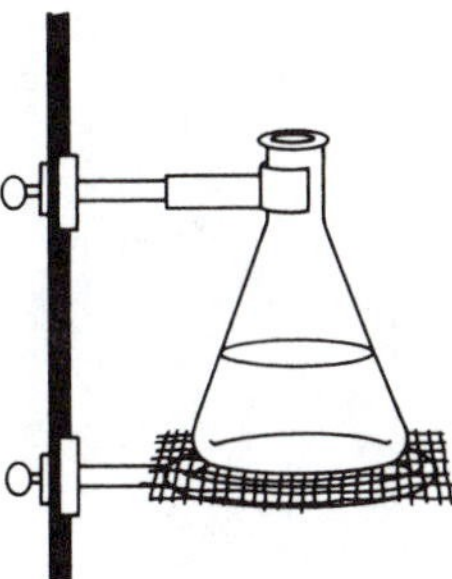

Figure 7. One method of supporting a flask. Be sure to clamp all glassware securely to a ringstand before using.

An all-too-common lab accident occurs when a liquid must be heated to boiling in a test tube. Test tubes hold only a few milliliters of liquid and require only a few seconds of heating to reach the boiling point. A test tube cannot be heated strongly in the direct full heat of the burner flame. The contents of the test tube will super-heat and blow out of the test tube like a bullet from a gun. Ideally, when a test tube requires heating, a **boiling water**

bath in a beaker should be used. If this is not possible, then hold the test tube at a 45° angle a few inches above the flame and heat only *briefly*, keeping the test tube moving constantly (from top to bottom, and from side to side) through the flame during the heating. *Aim the mouth of the test tube away from yourself and your neighbors.* See Figure 8.

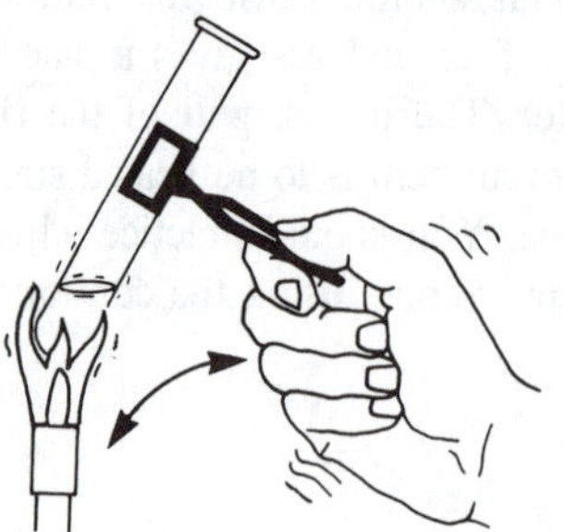

Figure 8. Method for heating a test tube containing a small quantity of liquid. Heat only for a few seconds, keep the test tube moving, and aim the mouth of the test tube away from you and your neighbors. Do not apply the flame to the very bottom of the test tube: heat evenly along the entire length of the liquid sample

Safety Regulations

- Wear safety glasses at all times while in the laboratory.
- Do not wear short skirts, shorts, and bare-midriff shirts in the laboratory.
- Do not wear scarves and neckties in the lab, because they may be ignited accidentally in the burner flame.
- Men who have long beards must secure them away from the burner flame.
- Open-toed shoes and sandals, as well as thin canvas sneakers, are not permitted in the laboratory.
- Never leave Bunsen burners unattended when lighted.
- Never heat solutions to dryness unless this is done in an evaporating dish on a hotplate or over a boiling water bath.
- Never heat a "closed system" such as a stoppered flask.
- Never smoke, chew gum, eat, or drink in the laboratory, since you may inadvertently ingest some chemical substance.
- Always use the smallest amount of substance required for an experiment; more is never better in chemistry. Never return unused portions of a reagent to their original bottle.
- Never store chemicals in your laboratory locker unless you are specifically directed to do so by the instructor.
- Never remove any chemical substance from the laboratory. In many colleges, removal of chemicals from the laboratory is grounds for expulsion or other severe disciplinary action.
- Keep your work area clean, and help keep the common areas of the laboratory clean. If you spill something in a common area, remember that this substance may injure someone else.
- Never fully inhale the vapors of any substance. Waft a tiny amount of vapor toward your nose.
- To heat liquids, always add 2–3 boiling stones to make the boiling action smoother.
- Never add water to a concentrated reagent when diluting the reagent. Always add the reagent to the water. If water is added to a concentrated reagent, local heating and density effects may cause the water to be splashed back.
- Never work in the laboratory unless the instructor is present. If no instructor is present during your assigned work time, report this to the senior faculty member in charge of your course.
- Never perform any experiment that is not specifically authorized by your instructor. Do not play games with chemicals!
- Dispose of all reaction products as directed by the instructor. In particular, observe the special disposal techniques necessary for flammable or toxic substances.
- Dispose of all glass products in the special container provided.

Information on Hazardous Substances and Procedures

Many of the pre-laboratory assignments in this manual require that you use several chemical handbooks available in any scientific library to look up various data for the substances you will be working with. You should also use such handbooks as a source of information about the hazards associated with the substances and procedures you will be using. A particularly useful general reference is the booklet *Safety in Academic Laboratories*, published by the American Chemical Society (your course professor or laboratory instructor may have copies).

Each of the experiments in this manual includes a section entitled *Safety Precautions,* which gives important information about expected hazards. This manual also lists some of the hazards associated with common chemical substances in Appendix J. Such problems as flammability and toxicity are described in terms of a low/medium/high rating scale. If you have any questions about this material, consult with your instructor before the laboratory period.

You may now be even more hesitant about facing your chemistry laboratory experience, now that you have read all these warnings. Be cautious . . . be careful . . . be thoughtful . . . but do not be afraid. Every precaution will be taken by your instructor and your college or university for your protection. Realize, however, that *you* bear ultimate responsibility for safety in the laboratory.

Safety Quiz

1. Sketch below a representation of your laboratory, showing clearly the location of the following: exits, fire extinguishers, safety shower, eyewash fountain, and fume hoods.

2. Why must you wear safety glasses or goggles at all times while you are in the laboratory, even when you are not personally working on an experiment?

____

____

3. Suppose your neighbor in the lab were to spill several milliliters of a flammable substance near her burner, and the substance then ignited. How should this relatively small fire be extinguished?

____

____

____

4. What steps should be taken if a larger scale fire were to break out in the laboratory?

____

____

____

5. What measures should be taken if a student's clothing were to be ignited?

6. List five types of clothing or footwear that are *not* acceptable in the laboratory, and explain why they are not acceptable.

7. Suppose you were pouring concentrated nitric acid from a bottle into your reaction flask, and you spilled the acid down the front of your shirt or blouse. What should you do?

8. What steps should you have taken if you had spilled the nitric acid on your hand rather than on your clothing?

9. What are the most common student injuries in the chemistry lab, and how can they be prevented?

10. Where should reactions involving the evolution of toxic gases be performed?

11. Why should you immediately clean up any chemical spills in the laboratory?

12. Why should you never deviate from the published procedure for an experiment?

13. Why should apparatus always be clamped securely to a metal ring stand before starting a chemical reaction?

14. Why should you never eat, drink, or smoke while you are in the laboratory?

15. Suppose the experiment you are to perform in a given lab period involves substances that are toxic or corrosive. What should you do *before* the laboratory to prepare yourself for using these substances safely?

The Laboratory Notebook and Laboratory Reports

Keeping a Lab Notebook

The practicing scientist must keep current, accurate records of his or her work. As observations are made during a chemical reaction, or as numerical data are recorded and processed, a complete, up-to-date record must be made of the observations and data. For this purpose, chemists and other scientists keep a working notebook with them at all times. All data, observations, hypotheses, and preliminary conclusions are recorded directly in the notebook during the course of experimental work. Scientists *never* record data on scrap paper, since such papers may be lost or transcribed incorrectly.

Keeping a research-style notebook is not an easy task for a beginning science student—it takes practice. The notebook should report both successful and unsuccessful experiments. If you make a mistake, write down what happened so that you will not make the same mistake again. All calculations you have performed before, during, or after an experiment should be entered, both to help you understand your results and to help your instructor find any errors you may have made. A poor grade on a lab report is oftentimes due to a simple arithmetic error made in processing data; if you can find the error by reviewing your notebook calculations, your instructor may be only too happy to regrade your results.

Although the notebook is your personal record, it must be *neat* and *organized.* Random jottings are little help in writing up reports on your experiments or in pre-examination reviews. Numerical data should be *tabulated* whenever possible for ready reference and should always be clearly *labeled* with the units of the data measurement. Oftentimes a graphical treatment of data is appropriate; many research notebooks consist of square-ruled paper, which makes the plotting of at least rough graphs possible. Observations written in the notebook should be complete enough that someone who has at least your level of scientific training could understand what you observed or measured.

Various types of notebooks are sold in college bookstores. Your instructor may ask you to purchase a specific type. In general, research-type notebooks must have sewn bindings to prevent the addition or removal of pages. Loose-leaf or spiral bound notebooks are not generally acceptable, since pages may be removed or added without notice. When you first purchase your notebook, immediately number the pages consecutively from front cover to last page, and write your name on the cover of the notebook.

Specific rules for keeping a general chemistry notebook will differ from instructor to instructor, and from school to school. The following are some guidelines that most research scientists follow:

1. All entries to the notebook should be made in ink as they occur; pencil is never permitted in a notebook.
2. Your instructor may not permit you to remove your notebook from the laboratory. Also, your instructor may collect your notebook for grading from time to time. Therefore carbon copies of all notebook entries should be made on plain paper that you can remove from the lab. This will enable you to write up an experiment even if your instructor has kept your notebook.
3. The first three to four leaves of the notebook should be used as a table of contents. As you begin an experiment, you should make an entry in this table that will permit you and your instructor to find information easily.
4. Ordinarily only the right-hand pages of the notebook should be used for information. Start each experiment on a new right-hand page. On the first page of a new experiment, list the title of the experiment, the date on which it was performed, the name of any partner with whom you worked, a short statement of the purpose or objective of the experiment, and *all pre-laboratory data and calculations* (e.g., formula weights of the reagents to be used in the experiment). Subsequent right-hand pages should be used for recording all data and observations that arise during the experiment and for a discussion of the *procedure* used. Data should be planned for. Read the experiment carefully before coming to the laboratory, and construct a table before lab into which all data can be entered. Observations should also be planned for as much as possible. This manual is meant to help you identify appropriate observations, but you should enter whatever observations or comments you consider appropriate. Also use right-hand pages for presenting your final results and any conclusions you may make from these results. With your results and conclusions, you should also include a brief discussion of *sources of error* in the experiment.
5. The left-hand pages of the notebook are used primarily for *quick calculations* and for writing *brief notes* to yourself (for example, any special announcements about the experiment). Detailed calculations are written on additional right-hand pages, however. Calculations or other data treatment should never be limited to numbers; they should show the method used for the calculation, either in words or in a dimensional analysis set-up or formula. If there are several repeat determinations in a given experiment, the method of calculation only has to be shown once with only the actual results of repeat determinations indicated. All data and all calculated results should be given to the level of precision permitted by the experiment.
6. A laboratory notebook should contain a description of the experimental procedure. However, a detailed, step-by-step list of manipulations is not needed. Generally, at the introductory level, laboratory manuals give procedures in greater detail than would be necessary for the professional scientist. Try to understand the basics of what procedural method is used in an experiment, and summarize this in a few sentences. Often, if a published procedure is being followed, a reference to that procedure is sufficient (giving the book or journal article from which the procedure was taken, the authors' names, page numbers, and publication data). If you vary from a published procedure, either on your instructor's direction or on your own initiative, you should discuss how your actual procedure deviated from the published procedure.
7. Mistakes do happen—even to Nobel-prize-winning scientists! No one expects your experiment to be perfect. Mistakes in recording of data are especially common. However, mistakes in the lab notebook must be treated in a special way. *No data may ever be erased or obliterated from a notebook, or the en-*

tire notebook is invalidated. If you make a simple mistake in a datum, draw a single clean line through the incorrect datum, and write in the new datum legibly. Make certain that the incorrect datum is still clearly readable, since incorrect data often turn out to be correct after all. Your instructor may ask you to initial errata and to include a short explanation as to why the datum is incorrect.

Writing Laboratory Reports

The exact type of laboratory report you will be asked to submit will depend on your instructor's preference or on the policy of your particular college or university. Two types of laboratory reports are common in general chemistry labs.

The first type is like the report pages at the back of each experiment in this manual. Proper, thoughtful completion of these lab report pages may constitute a complete report in some situations. Each set of lab report pages contains a set of *pre-laboratory questions*, space for writing significant data and observations (which are keyed to the Procedure for the experiment), and a series of questions to answer after you have completed the laboratory and have processed your data. The pre-laboratory questions sometimes present a numerical problem similar to the problems you will encounter in processing your experimental data; at other times, the questions demand library research. The pre-laboratory questions are designed to get you thinking about the experiment you will be performing, so make good use of them. The questions at the end of the lab report suggest a broader context for the experiment and provide ideas for further reading as well.

Some instructors may prefer more formal reports than those provided for in this manual. A longer report more closely resembles those done by professional scientists working in industrial and academic research facilities. Typically, a formal laboratory report should be typewritten and should include the following: a title page, an introduction (giving the objectives of the experiment and a short review of the theory behind the experiment), a discussion of the procedure used for the experiment (or a reference to a published procedure), a table listing all recorded data and a separate table listing all calculated or derived results, a discussion of the results (including an analysis of the likely sources of error), any conclusions you may have made based on what you observed or measured, and a list of references you have consulted during the course of preparing for the experiment or in writing up the results. Your instructor may ask you to include additional information with your report or may modify the components of the report as listed here.

Whichever type of report your instructor requests, realize that a report should represent *your own work.* If you have questions about your data or calculations, it is best to consult with the instructor during his or her conference hours, rather than asking a friend for help.

Experiment

1 The Laboratory Balance: Mass Determinations

Objective

Familiarity with the various instruments used for making physical measurements in the laboratory is essential to the study of experimental chemistry. In this experiment, you will investigate the uses and limits of the various types of laboratory balances.

Introduction

The accurate determination of *mass* is one of the most fundamental techniques for students of experimental chemistry. **Mass** is a direct measure of the *amount of matter* in a sample of some substance. That is, the mass of a sample is a direct indication of the number of atoms or molecules the sample contains. Since chemical reactions occur in proportion to the number of atoms or molecules of reactant present, it is essential that the mass of reactant used in a process be accurately and precisely known.

Various types of balances are available in the typical general chemistry laboratory. Such balances differ in their construction, appearance, operation, and in the level of precision (number of significant figures) they permit in mass determinations. Three of the most common types of laboratory balance are indicated in Figures 1-1, 1-2, and 1-3. Determine which sort of balance your laboratory is equipped with, and ask your instructor for a demonstration of the use of the balance if you are not familiar with its operation.

The method of operation differs for the three types of balances, and you should ask your instructor for a demonstration of the proper procedure to use. There are, however, some general points to keep in mind when using any balance:

1. Always make sure that the balance gives a reading of 0.000 grams when nothing is present on the balance pan. Adjust the tare or zero knob if necessary. If the balance cannot be set to zero, ask the instructor for help.
2. All balances, but especially electrical/electronic balances, are damaged by moisture. Do not pour liquids in the immediate vicinity of the balance. Clean up any spills immediately from the balance area.
3. No reagent chemical substance should ever be weighed directly on the pan of the balance. Ideally, reagents should be weighed directly into the beaker or flask in which they are to be used. Plastic weighing boats may also be used if several reagents are required for an experiment. Pieces of filter paper or weighing paper should ordinarily *not* be used for weighing of reagents.

4. Procedures in this manual are generally written in such terms as "weigh 0.5 grams of substance (to the nearest milligram)." This does not mean that exactly 0.500 grams of substance is needed. Rather, the statement means to obtain an amount of substance between 0.450 and 0.550 gram, but to record the actual amount of substance taken (e.g., 0.496 grams). Unless a procedure states explicitly to weigh out an exact amount (e.g., "weigh out exactly 5.00 grams of NaCl"), you should not waste time trying to obtain an exact amount. However, always record the amount actually taken to the precision of the balance used.
5. For accurate mass determinations, the object to be weighed must be at room temperature. If a hot or warm object is placed on the pan of the balance, such an object causes the air around it to become heated. Warm air rises, and the motion of such warm air may be detected by the balance, giving mass determinations that are significantly less than the true value.
6. For many types of balances, there are likely to be small errors in the absolute masses of objects determined with the balance, particularly if the balance has not been properly calibrated or has been abused. For this reason, most weighings in the laboratory should be performed by a difference method: an empty container is weighed on the balance, and then the reagent or object whose mass is to be determined is added to the container. The resulting difference in mass is the mass of the reagent or object. Because of possible calibration errors, the same balance should be used throughout a procedure.

Safety Precautions

- Wear safety glasses at all times while in the laboratory.

Apparatus/Reagents Required

Unknown mass samples provided by the instructor, rubber stoppers, small beakers

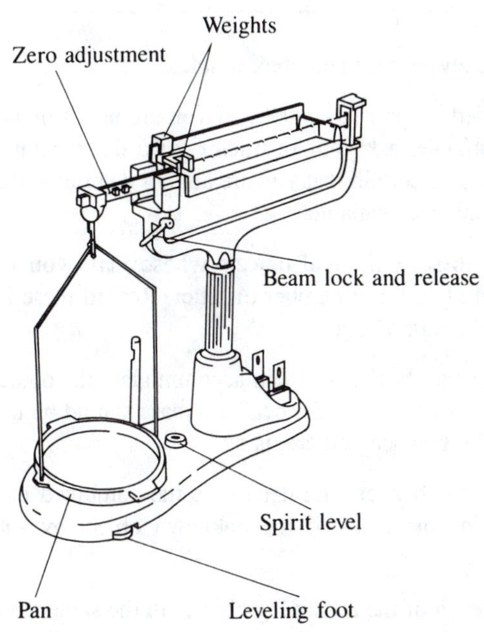

Figure 1-1. Triple beam balance. Never weigh a chemical directly on the balance, since it is more difficult to zero this type of balance if the balance becomes dirty.

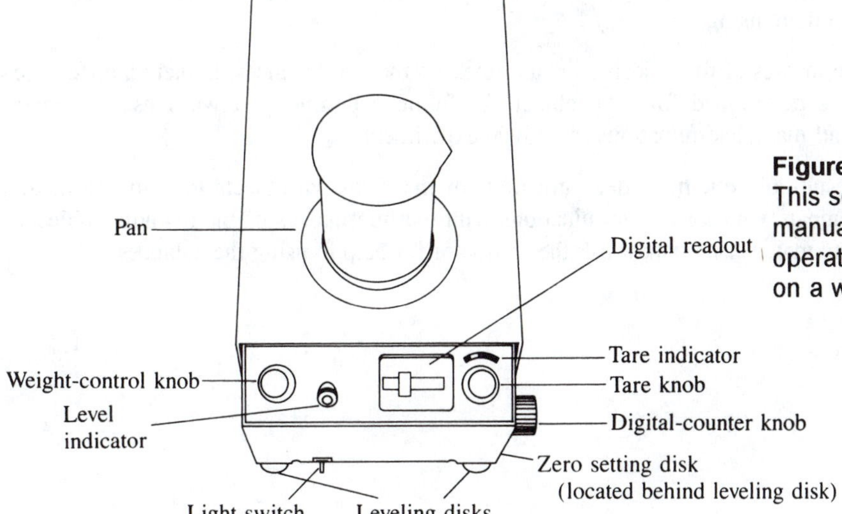

Figure 1-2. Top-loading electric balance. This sort of balance is operated manually: the digits are dialed in by the operator until the pointer comes to rest on a whole number.

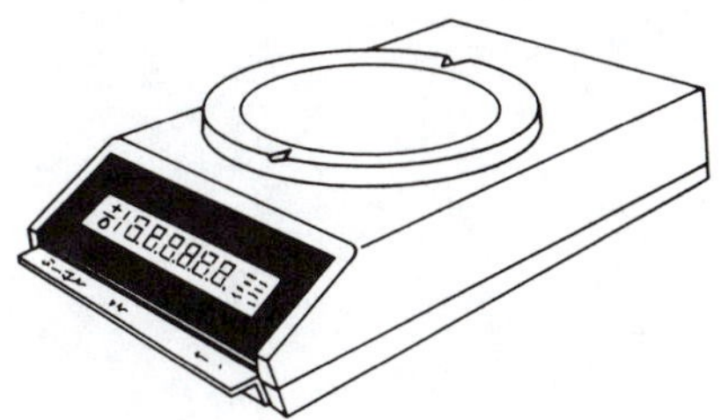

Figure 1-3. Digital electronic balance. The balance directly gives the mass when an object to be weighed is placed on the pan

Procedure

Record all data and observations directly in your notebook in ink.

Examine the balances that are provided in your laboratory. If you are not familiar with the operation of the type of balance available, ask your instructor for a demonstration of the appropriate technique. In particular, make certain that you are able to determine the level of precision permitted by each type of balance available.

Your instructor will provide you with several small objects whose mass you will determine. The objects are coded with an identifying number or letter. Record these identification codes in your notebook and on the report page.

Determine and record the mass of a small beaker that can accommodate the objects whose masses are to be determined. The determination of the beaker's mass should be to the level of precision permitted by the particular balance you are using.

Transfer the first unknown object to the beaker, and determine the combined mass of the beaker and object. Record. Determine the mass of the unknown object by subtraction. Record.

Determine and record the masses of each of the remaining objects in the same manner.

Use a different balance from that used earlier, and determine the masses of each of the unknown objects on the second balance in the same manner already described. The determination of the beaker's mass should be to the level of precision permitted by the particular balance you are using.

Compare the masses of the objects as determined on the two balances. Is there a difference in the masses determined for each object? In future experiments, always use the *same balance* for all mass determinations in a given experiment.

Show the results of your mass determinations of the unknown objects to your instructor, who will compare your mass determinations with the true masses of the unknown objects. If there is any major discrepancy, ask the instructor for help in using the balances.

Name ______________________ Section ______________________

Lab Instructor ______________________ Date ______________________

EXPERIMENT 1 The Laboratory Balance: Mass Determinations

PRE-LABORATORY QUESTIONS

1. What are the differences in meaning between *mass* and *weight*? In the laboratory, do we determine the mass or the weight of objects?

____

____

____

____

2. Explain the following: Weigh approximately 5 grams of NaCl to the nearest milligram.

____

____

____

____

Name ______________________ Section ______________________

Lab Instructor ______________________ Date ______________________

EXPERIMENT 1 The Laboratory Balance: Mass Determinations

RESULTS/OBSERVATIONS

1. First balance

Mass of empty beaker ______________________

ID number of object	Mass of beaker plus object	Mass of object itself

2. Second balance

Mass of empty beaker ______________________

ID number of object	Mass of beaker plus object	Mass of object itself

3. Difference in masses determined on the two balances

ID number of object	Difference in mass

QUESTIONS

1. Why is it important always to use the *same balance* during the course of an experiment?

2. What *error* is introduced in a mass determination if the object being weighed is *warm*? Why?

3. Why should reagent chemicals *never* be weighed directly on the pan of the balance?

EXPERIMENT

2 The Use of Volumetric Glassware

Objective

Familiarity with the various instruments used for making physical measurements in the laboratory is essential to the study of experimental chemistry. In this experiment, you will investigate the uses and limits of the various types of volumetric glassware.

Introduction

Most of the glassware in your laboratory locker has been marked by the manufacturer to indicate the volume contained by the glassware when filled to a certain level. The graduations etched or painted onto the glassware by the manufacturer differ greatly in the precision they indicate, depending on the type of glassware and its intended use. For example, beakers and Erlenmeyer flasks are marked with very approximate volumes, which serve merely as a rough guide to the volume of liquid in the container. Other pieces of glassware, notably burets, pipets, and graduated cylinders, are marked much more precisely by the manufacturer to indicate volumes. It is important to distinguish when a *precise* volume determination is necessary and appropriate for an experiment and when only a *rough* determination of volume is needed

Glassware that is intended to contain or to deliver specific precise volumes is generally marked by the manufacturer with the letters "TC" (to contain) or "TD" (to deliver). For example, a flask that has been calibrated by the manufacturer to contain exactly 500 mL of liquid at 20°C would have the legend "TC 20 500 mL" stamped on the flask. A pipet that is intended to deliver a precise 10.00 mL sample of liquid at 20°C would be stamped with "TD 20 10 mL." It is important not to confuse "TC" and "TD" glassware: such glassware may not be used interchangeably. The temperature (usually 20°C) is specified with volumetric glassware since the volume of a liquid changes with temperature, which causes the density of the liquid to change. While a given pipet will contain or deliver the same *volume* at any temperature, the *mass* (amount of substance present in that volume) will vary with temperature.

A. Graduated Cylinders

The most common apparatus for routine determination of liquid volumes is the graduated cylinder. Although a graduated cylinder does not permit as precise a determination of volume as do other volumetric devices, for many applications the precision of the graduated cylinder is sufficient. Figures 2-1 and 2-2 show typical graduated cylinders. In Figure 2-2, notice the plastic safety ring, which helps to keep the graduated cylinder from breaking if it is tipped over. In Figure 2-1, compare the difference in graduations shown for the 10-mL and 100-mL cylinders. Examine the graduated cylinders in your lab locker, and determine the smallest graduation of volume that can be determined with each cylinder.

Figure 2-1. Expanded view of 10-mL and 100-mL cylinders. Greater precision is possible with the 10-mL cylinder, since each numbered scale division represents 1 mL.

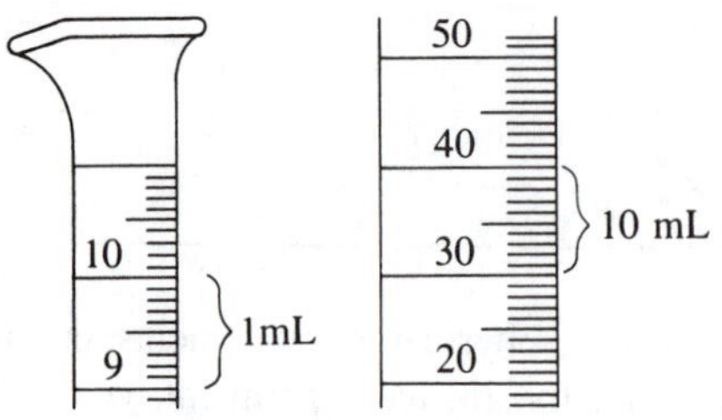

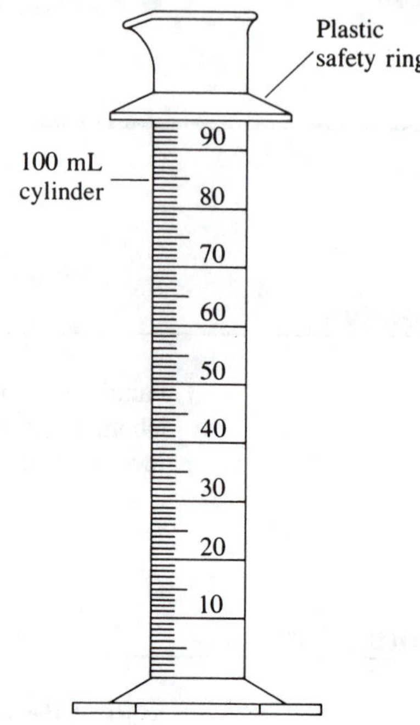

Figure 2-2. A 100-mL graduated cylinder with a plastic safety ring

When water (or an aqueous solution) is contained in a narrow glass container such as a graduated cylinder, the liquid surface is not flat as might be expected. Rather, the liquid surface is curved downward (see Figure 2-3). This curved surface is called a **meniscus**, and is caused by an interaction between the water molecules and the molecules of the glass container wall. When reading the volume of a liquid that makes a meniscus, hold the graduated cylinder so that the meniscus is at eye level, and read the liquid level at the *bottom* of the curved surface.

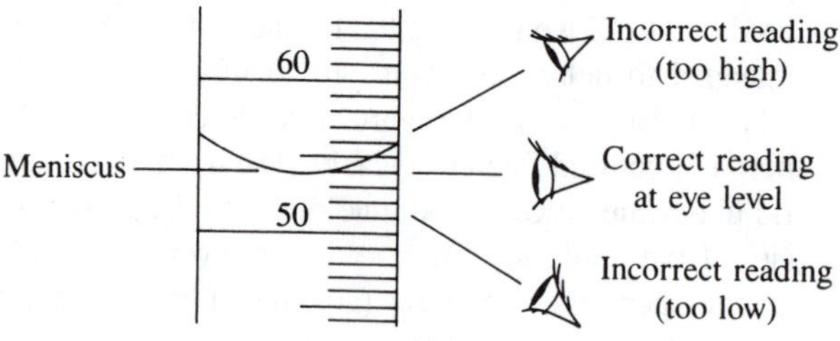

Figure 2-3. Reading a meniscus. Read the bottom of the meniscus while holding at eye level.

B. Pipets

When a more precise determination of liquid volume is needed than can be provided by a graduated cylinder, a transfer pipet may be used. Pipets are especially useful if several measurements of the same volume are needed (such as in preparing similar-sized samples of a liquid substance). Two types of pipet are commonly available, as indicated in Figure 2-4. The Mohr pipet is calibrated at least at each milliliter and can be used to deliver any size sample (up to the capacity of the pipet). The volumetric transfer pipet can deliver only one size sample (as stamped on the barrel of the pipet), but generally it is easier to use and is more reproducible.

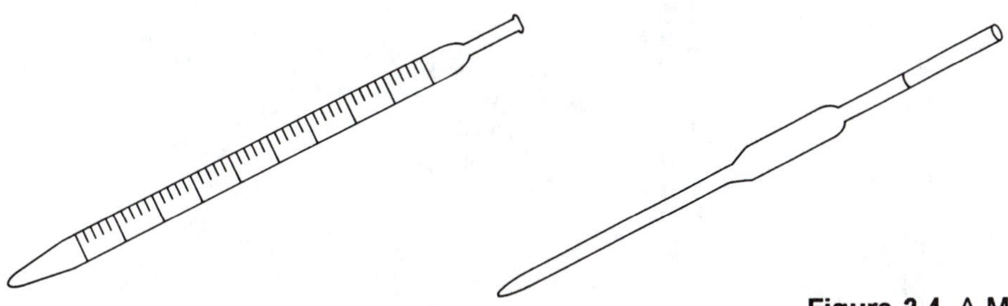

Figure 2-4. A Mohr pipet (left) and a volumetric pipet (right).

Pipets are filled using a rubber safety bulb to supply the suction needed to draw liquid into the pipet. *It is absolutely forbidden to pipet by mouth in the chemistry laboratory.* Two common types of rubber safety bulb are shown in Figure 2-5.

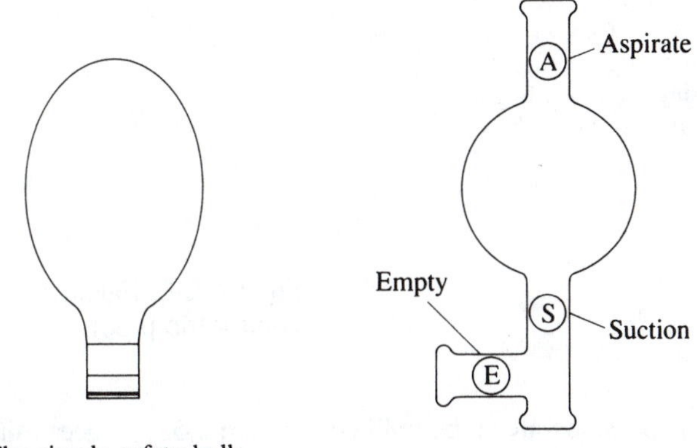

Figure 2-5. Pipet safety bulbs. Never pipet by mouth.

The simple bulb should *not* actually be placed onto the barrel of the pipet. This would most likely cause the liquid being measured to be sucked into the bulb itself. Rather, squeeze the bulb, and just *gently* press the opening of the bulb against the opening in the barrel of the pipet to apply the suction force, keeping the tip of the pipet under the surface of the liquid being sampled. Do not force the pipet into the plastic tip of the safety bulb, or the pipet may break. Allow the suction to draw liquid into the pipet until the liquid level is 1 or 2 inches above the calibration mark on the barrel of the pipet. At this point, quickly place your index finger over the opening at the top of the pipet to prevent the liquid level from falling. By gently releasing the pressure of your index finger, the liquid level can be allowed to fall until it reaches the calibration mark of the pipet. The tip of the pipet may then be inserted into the container that is to receive the sample and the pressure of the finger removed to allow the liquid to flow from the barrel of the pipet. (See Figure 2-6.)

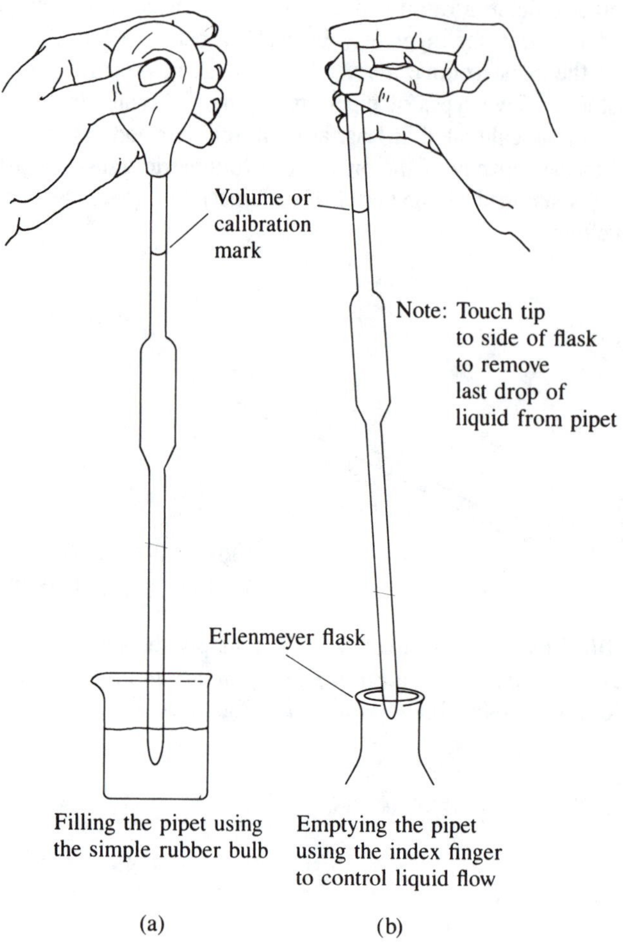

Figure 2-6. Filling technique for volumetric pipet.

To use the more expensive valve-type bulb (see Figure 2-5b), squeeze the valve of the bulb marked *A*, and simultaneously squeeze the large portion of the rubber bulb itself to expel air from the bulb. Press valve *A* a second time, release the pressure on the bulb, and attach the bulb to the top of the pipet. Insert the tip of the pipet under the surface of the liquid to be measured, and squeeze the valve marked *S* on the bulb, which will cause liquid to begin to be sucked into the pipet. When the liquid level has risen to an inch or two above the calibration mark of the pipet, stop squeezing valve *S* to stop the suction. Transfer the pipet to the vessel to receive the liquid, and press valve *E* to empty the pipet. The use of this sort of bulb generally requires considerable practice to develop proficiency.

When using either type of pipet, observe the following rules:

1. The pipet must be scrupulously clean before use: wash with soap and water, rinse with tap water, and then with distilled water. If the pipet is clean enough for use, water will not bead up anywhere on the inside of the barrel.
2. To remove rinse water from the pipet (to prevent dilution of the solution to be measured), rinse the pipet with several small portions of the solution to be

measured, discarding the rinsings in a waste beaker for disposal. It is not necessary to completely fill the pipet with the solution for rinsing.

3. The tip of the pipet must be kept under the surface of the liquid being measured out during the entire time suction is being applied, or air will be sucked into the pipet.
4. Allow the pipet to drain for at least a minute when emptying to make certain the full capacity of the pipet has been delivered. Remove any droplets of liquid adhering to the tip of the pipet by touching the tip of the pipet to the side of the vessel that is receiving the sample.
5. If you are using the same pipet to measure out several different liquids, you should rinse the pipet with distilled water between liquids, and follow with a rinse of several small portions of the next liquid to be measured.

C. Burets

When samples of various sizes must be dispensed or measured precisely, a buret may be used. The buret consists of a tall, narrow calibrated glass tube, fitted at the bottom with a valve for controlling the flow of liquid. The valve is more commonly called a **stopcock**. (See Figure 2-7.)

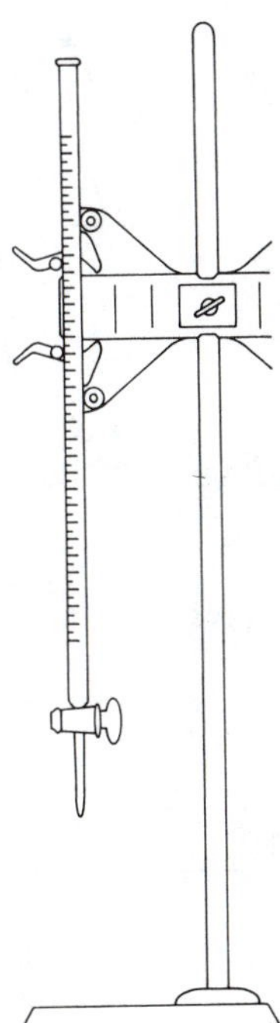

Figure 2-7. A volumetric buret. Typically, 50-mL burets are used in chemistry labs.

Like a pipet, a buret must be scrupulously clean before use. The precision permitted in reading a buret is 0.02 mL, but if the buret is not completely clean, this level of precision is not attainable. To clean the buret, first use soap and water, using a special long-handled buret brush to scrub the interior of the glass. Then rinse the buret with tap water, followed by several rinsings with distilled water.

Before use, the buret should be rinsed with several small portions of the solution to be used in the buret. The buret should be tilted and rotated during the rinsings, to make sure that all rinse water is washed from it. Discard the rinsings. After use, the buret should again be rinsed with distilled water. Many of the reagent solutions used in burets may attack the glass of the buret if they are not removed. This would destroy the calibration. To speed up the cleaning of a buret in future experiments, the buret may be left filled with distilled water during storage between experiments (if your locker is large enough to permit this).

A common mistake made by beginning students is to fill the buret with the reagent solution to be dispensed to exactly the 0.00 mark. This is not necessary or desirable in most experiments, and wastes time. The buret should be filled to a level that is comfortable for you to read (based on your height). The precise initial liquid level reading of the buret should be taken before the solution is dispensed and again after the liquid is dispensed. The readings should be made to the nearest 0.02 mL. The volume of liquid dispensed is then obtained by simple subtraction of the two volume readings.

Safety Precautions

- Wear safety glasses at all times while in the laboratory.
- When using a pipet, use a rubber safety bulb to apply the suction force. ***Never pipet by mouth.***
- Rinse the buret carefully. Do not attempt to admit water directly to the buret from the cold water tap. Fill a beaker with tap water, and pour from the beaker into the buret.

Apparatus/Reagents Required

Graduated cylinders, pipets and safety bulb, buret and clamp, beakers, distilled water

Procedure

Record all data and observations directly in your notebook in ink.

A. The Graduated Cylinder

Your instructor will set up a display of several graduated cylinders filled with different amounts of colored water. Several sizes of cylinder are available (10-mL, 25-mL, 50-mL, 100-mL). Examine each cylinder, paying particular attention to the marked scale divisions on the cylinder. For each graduated cylinder, to what fractional unit of volume does the smallest scale mark correspond?

Read the volume of liquid contained in each graduated cylinder and record. Make your readings to the level of precision permitted by each of the cylinders.

Check your readings of the liquid levels with the instructor before proceeding, and ask for assistance if your readings differ from those provided by the instructor.

Clean and wipe dry your 25-mL graduated cylinder and a 50-mL beaker (a rolled-up paper towel will enable you to dry the interior of the graduate). Weigh the graduated cylinder and beaker and record the mass of each to the nearest milligram (0.001 gram).

Obtain about 100 mL of distilled water in a clean Erlenmeyer flask. Determine and record the temperature of the distilled water.

Fill the graduated cylinder with distilled water so that the meniscus of the water level lines up with the 25-mL calibration mark of the cylinder. Place distilled water in the 50-mL beaker up to the 25-mL mark.

Weigh the graduated cylinder and the 50-mL beaker to the nearest milligram (0.001 gram) and calculate the mass of water each contains.

Using the Density of Water table from Appendix H to this manual, calculate the volume of water present in the graduated cylinder and beaker from the exact mass of water present in each.

Compare the *calculated* volume of water (based on the mass of water) to the observed volumes of water determined from the calibration marks on the cylinder and beaker. Calculate the percentage difference between the calculated volume and the observed volume from the calibration marks. Why are the calibration marks on laboratory beakers taken only to be an approximate guide to volume?

B. The Pipet

Obtain a 25-mL pipet and rubber safety bulb. Clean the pipet with soap and water. Rinse the pipet with tap water, and then with small portions of distilled water. Practice filling and dispensing distilled water from the pipet until you feel comfortable with the technique. Ask your instructor for assistance if you have any difficulties in the manipulation.

Clean and wipe dry a 150-mL beaker. Weigh the beaker to the nearest milligram (0.001 gram) and record.

Obtain about 100 mL of distilled water in a clean Erlenmeyer flask. Determine and record the temperature of the water.

Pipet 25 mL of the distilled water from the flask into the clean beaker you have weighed. Reweigh the beaker containing the 25 mL of water. Determine the weight of water transferred by the pipet.

Using the Density of Water table from Appendix H to this manual, calculate the *volume* of water transferred by the pipet from the *mass* of water transferred. Compare this calculated volume to the volume of the pipet as specified by the manufacturer. Any significant difference in these two volumes is an indication that you need additional practice in pipeting. Consult with your instructor for help.

How does the volume dispensed by the pipet compare to the volumes as determined in Part A using a graduated cylinder or beaker?

C. The Buret

Obtain a buret and set it up in a clamp on your lab bench.

Fill the buret with tap water, and check to make sure that there are no leaks from the stopcock before proceeding. If the stopcock leaks, have the instructor examine the stopcock to make sure that all the appropriate washers are present. If the stopcock cannot be made leakproof, replace the buret.

Clean the buret with soap and water, using a long-handled buret brush to scrub the inner surface of the buret. Rinse all soap from the buret with tap water, being sure to flush water through the stopcock as well. Rinse the buret with several small portions of distilled water, then fill the buret to above the zero mark with distilled water.

Open the stopcock of the buret and allow the distilled water to run from the buret into a beaker or flask. Examine the buret while the water is running from it. If the buret is clean enough for use, water will flow in sheets down the inside surface of the buret without beading up anywhere. If the buret is not clean, repeat the scrubbing with soap and water.

Once the buret is clean, refill it with distilled water to a point somewhat below the zero mark. Determine the precise liquid level in the buret to the nearest 0.02 mL.

With a paper towel, clean and wipe dry a 150-mL beaker. Weigh the beaker to the nearest milligram (0.001 g).

Place the weighed beaker beneath the stopcock of the buret. Open the stopcock of the buret and run water into the beaker until approximately 25 mL of water have been dispensed. Determine the precise liquid level in the buret to the nearest 0.02 mL. Calculate the volume of water that has been dispensed from the buret by subtraction of the two liquid levels.

Reweigh the beaker containing the water dispensed from the buret to the nearest milligram, and determine the mass of water transferred to the beaker from the buret.

Use the Density of Water table from Appendix H to calculate the volume of water transferred from the mass of the water. Compare the volume of water transferred (as determined by reading the buret) with the calculated volume of water (from the mass determinations). If there is any significant difference between the two volumes, most likely you need additional practice in the operation and reading of the buret.

How does the volume dispensed by the buret compare to the volumes as determined in Part A using a graduated cylinder or beaker? How does the volume dispensed by the buret compare to that dispensed using a pipet in Part B?

Name ______________________________ Section ______________________

Lab Instructor ______________________________ Date ______________________

EXPERIMENT 2 The Use of Volumetric Glassware

PRE-LABORATORY QUESTIONS

1. Pipets used for the transfer of samples of aqueous solutions are always *rinsed* with a small portion of the solution to be used before the sample is taken. Calculate the percentage error arising in an experiment if 1-mL, 5-mL, and 10-mL pipets are used for transfer and each pipet contains 5 drops of water adhering to the inside of the barrel. A single drop of water has an approximate volume of 0.05 mL.

____

____

____

____

____

2. It is important to make certain that there is no air bubble in the tip of the buret below the stopcock *before* the initial reading of the liquid level in the buret is taken. If a 0.5-mL air bubble is present in the tip of a buret, what percent error in 10-mL, 20-mL, and 40-mL samples will result if the air bubble is dislodged during the dispensing of the samples?

____

____

____

____

____

____

Name ______________________ Section ______________________

Lab Instructor ______________________ Date ______________________

EXPERIMENT 2 The Use of Volumetric Glassware

RESULTS/OBSERVATIONS

A. The Graduated Cylinder

Identifying color of cylinder water	Volume contained in cylinder
______________________	______________________
______________________	______________________
______________________	______________________

Mass of *empty* 25-mL graduated cylinder ______________________

Mass of cylinder plus water sample ______________________

Mass of water in cylinder ______________________

Temperature of water sample ______________________

Density of water at this temperature ______________________

Calculated volume of water transferred ______________________

Difference between observed volume and calculated volume ______________________

Mass of empty 50-mL beaker ______________________

Mass of beaker plus water sample ______________________

Mass of water in beaker ______________________

Temperature of water sample ______________________

Density of water at this temperature ______________________

Calculated volume of water transferred ______________________

Difference between observed volume and calculated volume ______________________

B. The Pipet

Mass of empty 150-mL beaker ______________________________

Mass of beaker plus water sample ______________________________

Mass of water transferred ______________________________

Temperature of water sample ______________________________

Density of water at this temperature ______________________________

Calculated volume of water transferred ______________________________

Difference between actual and calculated volumes transferred ______________________________

C. The Buret

Initial liquid level in buret ______________________________

Final liquid level in buret ______________________________

Volume of liquid transferred ______________________________

Mass of empty beaker ______________________________

Mass of beaker plus water ______________________________

Mass of water transferred ______________________________

Temperature of water ______________________________

Density at this temperature ______________________________

Calculated volume of water transferred ______________________________

Difference between actual and calculated volumes transferred ______________________________

QUESTION

Based on your experience in this experiment, briefly discuss the relative precision permitted by a graduated cylinder, a pipet, and a buret. Give several circumstances under which you would use each instrument, in preference to the other two.

____

____

____

____

EXPERIMENT

3 Density Determinations

Objective

Density is an important property of matter, and may be used as a method of identification. In this experiment, you will determine the densities of regularly and irregularly shaped solids as well as of pure liquids and solutions.

Introduction

The density of a sample of matter represents the mass contained within a unit volume of space in the sample. For most samples, a unit volume means 1.0 mL. The units of density, therefore, are quoted in terms of grams per milliliter (g/mL) or grams per cubic centimeter (g/cm^3) for most solid and liquid samples of matter.

Since we seldom deal with exactly 1.0 mL of substance in the general chemistry laboratory, we usually say that the density of a sample represents the mass of the specific sample divided by its particular volume.

$$\text{density} = \frac{mass}{volume}$$

Because the density does in fact represent a *ratio*, the mass of any size sample, divided by the volume of that sample, gives the mass that 1.0 mL of the same sample would possess.

Densities are usually determined and reported at 20°C (around room temperature) because the volume of a sample, and hence the density, will often vary with temperature. This is especially true for gases, with smaller (but still often significant) changes for liquids and solids. References (such as the various chemical handbooks) always specify the temperature at which a density was measured.

Density is often used as a point of identification in the determination of an unknown substance. In later experiments, you will study several other physical properties of substances that are used in the identification of unknown substances. For example, the boiling and melting points of a given substance are characteristic of that substance and are used routinely in identification of unknown substances. Suppose an unknown's boiling and melting points have been determined, but on consulting the literature, it is found that more than one substance has these boiling and melting points. The *density* of the unknown might then be used to distinguish the unknown. It is very unlikely that two substances would have the same boiling point, melting point, and density.

Density can also be used to determine the concentration of solutions in certain instances. When a solute is dissolved in a solvent, the density of the *solution* will be different from that of the *pure solvent* itself. Handbooks list detailed information about the densities of solutions as a function of their composition (typically, in terms of the *percent solute* in the solution). If a sample is known to contain only a single solute, the density of the solution could be measured experimentally, and then the handbook could be consulted to determine what concentration of the solute gives rise to the measured solution density.

The determination of the density of certain physiological liquids is often an important screening tool in medical diagnosis. For example, if the density of urine differs from normal values, this may indicate a problem with the kidneys secreting substances which should not be lost from the body. The determination of density (specific gravity) is almost always performed during a urinalysis.

Several techniques are used for the determination of density. The method used will depend on the *type of sample* and on the level of *precision* desired for the measurement. For example, devices have been constructed for determinations of the density of urine, that permit a quick, reliable, routine determination. In general, a density determination will involve the determination of the mass of the sample with a balance, but the method used to determine the volume of the sample will differ from situation to situation. Several methods of volume determination are explored in this experiment.

For *solid* samples, different methods may be needed for the determination of the volume, depending on whether or not the solid is regularly shaped. If a solid has a *regular shape* (e.g., cubic, rectangular, cylindrical), the volume of the solid may be determined by geometry:

$$\text{For a cubic solid, volume} = (\text{edge})^3$$
$$\text{For a rectangular solid, volume} = \text{length} \times \text{width} \times \text{height}$$
$$\text{For a cylindrical solid, volume} = \pi \times (\text{radius})^2 \times \text{height}$$

If a solid does *not* have a regular shape, it may be possible to determine the volume of the solid from Archimedes' principle, which states that an insoluble, nonreactive solid will *displace* a volume of liquid equal to its own volume. Typically, an irregularly shaped solid is added to a liquid in a volumetric container (such as a graduated cylinder) and the *change* in liquid level determined.

For liquids, very precise values of density may be determined by pipeting an exact volume of liquid into a sealable weighing bottle (this is especially useful for highly volatile liquids) and then determining the mass of liquid that was pipeted. A more convenient method for routine density determinations for liquids is to weigh a particular volume of liquid as contained in a graduated cylinder.

Safety Precautions

- Wear safety glasses at all times while in the laboratory.
- The unknown liquids may be flammable, and their fumes may be toxic. Keep the unknown liquids away from open flames, and do not inhale their vapors. Dispose of the unknown liquids as directed by the instructor.
- Dispose of the metal samples only in the container designated for their collection.

Apparatus/Reagents Required

Unknown liquid sample, unknown metal sample, sodium chloride

Procedure

Record all data and observations directly in your notebook in ink.

A. Determination of the Density of Solids

Obtain a regularly shaped solid, and record its identification number. With a ruler, determine the physical dimensions (e.g., length, width, height, radius) of the solid to the nearest 0.2 mm. From the physical dimensions, calculate the volume of the solid.

Determine the mass of the regularly shaped solid to at least the nearest mg (0.001 g). From the mass and volume, calculate the density of the solid.

Obtain a sample of metal pellets (shot) and record its identification code number. Weigh a sample of the metal of approximately 50 g, but record the actual mass of metal taken to the nearest mg (0.001 g).

Add water to your 100-mL graduated cylinder to approximately the 50-mL mark. Record the exact volume of water in the cylinder to the precision permitted by the calibration marks of the cylinder.

Pour the metal sample into the graduated cylinder, making sure that none of the pellets sticks to the walls of the cylinder above the water level. Stir/shake the cylinder to make certain that no air bubbles have been trapped among the metal pellets. (See Figure 3-1.)

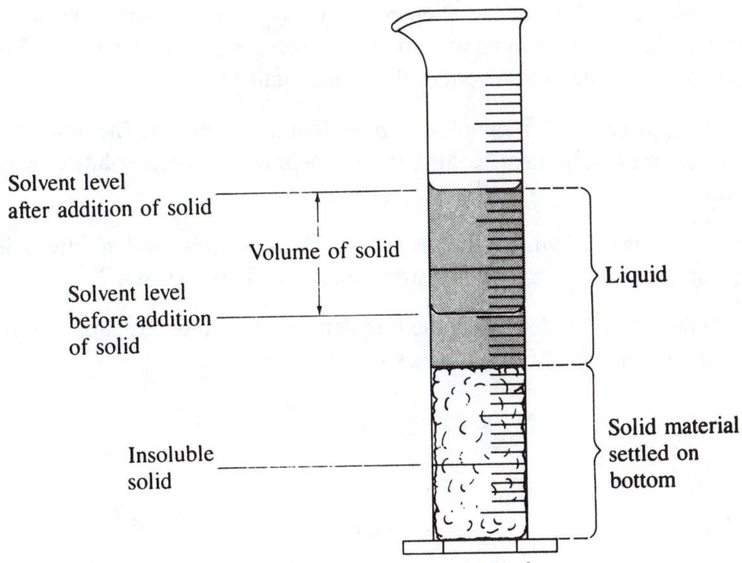

Figure 3-1. Measurement of volume by displacement. A nonsoluble object displaces a volume of liquid equal to its own volume.

Read the level of the water in the graduated cylinder, again making your determination to the precision permitted by the calibration marks of the cylinder. Assuming that the metal sample does not dissolve in or react with water, the change in water levels represents the volume of the metal pellets.

Calculate the density of the unknown metal pellets.

After blotting them dry with a paper towel, turn in the metal pellets to your instructor (*do not discard*).

B. Density of Pure Liquids

Clean and dry a 25-mL graduated cylinder (a rolled-up paper towel should be used). Weigh the dry graduated cylinder to the nearest mg (0.001 g).

Add distilled water to the cylinder so that the water level is above the 20-mL mark but below the 25-mL mark. Determine the temperature of the water in the cylinder.

Reweigh the cylinder to the nearest milligram.

Record the exact volume of water in the cylinder, to the level of precision permitted by the calibration marks on the barrel of the cylinder.

Calculate the density of the water. Compare the measured density of the water with the value listed in the handbook for the temperature of your experiment.

Clean and dry the graduated cylinder.

Obtain an unknown liquid and record its identification number. Determine the density of the unknown liquid, using the method just described for water.

C. Density of Solutions

The concentration of solutions is often conveniently described in terms of the solutions' *percentage composition* on a weight basis. For example, a 5% sodium chloride solution contains 5 g of sodium chloride in every 100 g of solution (which corresponds to 5 g of sodium chloride for every 95 g of water present).

Prepare solutions of sodium chloride in distilled water consisting of the following percentages by weight: 5%, 10%, 15%, 20%, 25%. Prepare 25 mL of each solution (you do *not* have to prepare 100 g of each solution to be able to use the percentage composition). Make the weight determinations of solute and solvent to the nearest milligram.

Using the method described earlier for samples of *pure* liquids, determine the density of each of your sodium chloride solutions. Record the temperature of each solution while determining its density.

Construct a graph of the *density* of your solutions versus the *percentage of NaCl* the solution contains. What sort of relationship exists between density and composition?

Use a handbook of chemical data to determine the true density of each of the solutions you prepared. Calculate the error in each of the densities you determined.

Name ______________________________ Section ______________________

Lab Instructor ______________________________ Date ______________________

EXPERIMENT 3 Density Determinations

PRE-LABORATORY QUESTIONS

1. A metal sphere weighing 15.45 g is added to 21.27 mL of water in a graduated cylinder. The water level rises to 24.78 mL. Calculate the *density* of the metal.

2 An empty beaker weighs 32.4257 g. A 10-mL pipet sample of an unknown liquid is transferred to the beaker. The beaker weighs 40.1825 g when weighed with the liquid in it. Calculate the *density* of the unknown liquid.

3. A term that is easily confused with density is *specific gravity*. For example, urinalysis reports commonly give the specific gravity of the sample rather than the density. What is meant by specific gravity? What are the units of specific gravity?

Name ______________________ Section ______________

Lab Instructor ______________________ Date ______________

EXPERIMENT 3 Density Determinations

RESULTS/OBSERVATIONS

A. Density of Solids

ID number of regular solid ______________ Shape ______________

Dimensions of solid ______________________________

Calculated volume of solid ______________________________

Mass of solid ______________ Density ______________

ID number of metal pellets ______________ Mass ______________

Initial water level ______________ Final water level ______________

Calculated density of metal ______________________________

B. Density of Pure Liquids

Mass of empty graduated cylinder ______________________________

Mass of cylinder plus water ______________________________

Volume of water ______________ Density ______________

Temperature ______________ Handbook density ______________

ID number of unknown liquid ______________________________

Mass of cylinder plus liquid ______________________________

Volume of liquid ______________ Density ______________

C. Density of Solutions

% NaCl	*Density measured*	*Temperature*	*Density handbook value*	*% error*
5%	______	______	______	______
10%	______	______	______	______
15%	______	______	______	______
20%	______	______	______	______
25%	______	______	______	______

QUESTIONS

1 What error would be introduced into the determination of the density of the regularly shaped solid if the solid were *hollow*? Would the density be too high or too low?

____

____

____

____

2. What error would be introduced into the determination of the density of the irregularly shaped metal pellets if you had not stirred/shaken the pellets to remove adhering *air bubbles*? Would the density be too high or too low?

____

____

____

____

3. An alternative procedure for determining the density of a liquid would be to pipet a sample of the liquid into a weighed flask; then reweigh the flask to determine the mass of liquid transferred. Would this alternative procedure be likely to give greater precision in the density determination? Why? Why do you suppose this alternative procedure was not used?

____

____

____

____

4. Your data for the density of sodium chloride solutions should have produced a straight line when plotted. How could this plot be used to determine the density of any concentration of sodium chloride solution?

____

____

____

____

____

5. Using a handbook of chemical data, look up the densities of the sodium chloride solutions listed below. Give the solution temperature listed in the reference work.

5% NaCl solution Density ____________________ g/mL

10% NaCl solution Density ____________________ g/mL

15% NaCl solution Density ____________________ g/mL

20% NaCl solution Density ____________________ g/mL

25% NaCl solution Density ____________________ g/mL

Temperature for the above densities of NaCl solutions ________________ ^{3}C

Reference ____________________________________

EXPERIMENT

4 The Determination of Boiling Point

Objective

In this experiment, you will first check your thermometer for errors by determining the temperature of two stable equilibrium systems. You will then use your calibrated thermometer in determining the boiling point of an unknown substance.

Introduction

The most common laboratory device for the measurement of temperature is, of course, the thermometer. The typical thermometer used in the general chemistry laboratory permits the determination of temperatures from –20° to 120°C. Most laboratory thermometers are constructed of glass, and so they are very fragile. Most general chemistry lab thermometers contain a red-colored organic liquid as the temperature-sensing fluid. This liquid is flammable and may be toxic: If a thermometer is broken, you should shut off any flames in the vicinity and consult with the instructor. Until recently, many lab thermometers contained elemental mercury metal as the temperature-sensing fluid. Mercury, especially its vapor, is very toxic, however, and most labs have replaced mercury thermometers with the red-liquid type.

The typical laboratory thermometer contains a bulb (reservoir) of temperature-sensing liquid at the bottom; it is this portion of the thermometer which actually senses the temperature. The glass barrel of the thermometer above the liquid bulb contains a fine capillary opening in its center, into which the liquid rises as it expands in volume when heated. The capillary tube in the barrel of the thermometer has been manufactured to very strict tolerances, and it is very regular in cross-section along its length. This ensures that the rise in the level of liquid in the capillary tube as the thermometer is heated will be directly related to the temperature of the thermometer's surroundings.

Although the laboratory thermometer may appear similar to the sort of clinical thermometer used for determination of body temperature, the laboratory thermometer does *not* have to be shaken before use. Medical thermometers are manufactured with a constriction in the capillary tube that is intended to prevent the liquid level from changing once it has risen. The liquid level of a laboratory thermometer, however, changes immediately when removed from the substance whose temperature is being measured. For this reason, temperature readings with the laboratory thermometer must be made while the bulb of the thermometer is actually present in the material being determined.

Because the laboratory thermometer is so fragile, it is helpful to check that the thermometer provides reliable readings before any important determinations are made with it. Often, thermometers develop nearly invisible hairline cracks along the barrel, making them unsuitable for further use. This happens especially if you are not careful in opening and closing your laboratory locker.

To check whether or not your thermometer is operating correctly, you will *calibrate* the thermometer. To do this, you will determine the reading given by your thermometer in two systems whose temperature is known with certainty. If the readings given by your thermometer differ by more than one degree from the true temperatures of the systems measured, you should exchange your thermometer, and then calibrate the new thermometer. A mixture of ice and water has an equilibrium temperature of 0°C, and will be used as the first calibration system. A boiling water bath, whose exact temperature can be determined from the day's barometric pressure, will be used as the second calibration system in this experiment.

Once your thermometer has been calibrated, you will use the thermometer to determine the *boiling point* of an unknown liquid as a means of identiifying the liquid. The boiling point of a pure substance is important because it is *characteristic* for a given substance (at a particular barometric pressure). That is, under the same laboratory conditions, a given substance will always have the *same* boiling point. Characteristic physical properties (such as the boiling point of a pure substance) are of immense help in the *identification of unknown substances.* Such properties are routinely reported in scientific papers when new substances are isolated or synthesized, and are compiled in tables in the various handbooks of chemical data that are available in science libraries. When an unknown liquid substance is isolated from a chemical system, its boiling point may be measured (along with certain other characteristic properties) and then compared with previously tabulated data. If the experimentally determined physical properties of the unknown match those found in the literature, you can typically assume that you have identified the unknown substance.

The **boiling point** of a liquid is defined as the temperature at which the vapor escaping from the surface of the liquid has a pressure equal to the pressure existing above the liquid. In the most common situation of a liquid boiling in a container open to the atmosphere, the pressure above the liquid will be the day's barometric pressure. In other situations, the pressure above a liquid may be reduced by means of a vacuum pump or aspirator, which enables the liquid to be boiled at a much lower temperature in an open container (this is especially useful in chemistry when a liquid is unstable, possibly decomposing if it were heated to its normal boiling point under atmospheric pressure). When boiling points are tabulated in the chemical literature, the pressure at which the boiling point determinations were made are also listed. The method to be used for the determination of boiling point is a semimicro method that requires only a few drops of liquid. Since boiling point determinations are so common, they will be used as an aid in identification of substances in several later experiments in this manual.

The apparatus used for heating samples in this experiment is called a **Thiele tube**. The Thiele tube contains oil (typically mineral oil) as a fluid, which permits the determination of temperatures up to about 200°C. The Thiele tube is constructed in such a way that when the side arm is heated, the warm oil will rise and enter the main chamber of the tube, which provides for circulation of the oil and for a more uniform temperature. Samples to be placed in the Thiele tube are ordinarily positioned so that the sample is aligned with the top branch of the side arm. When using a Thiele tube, ***remember that it contains hot oil,*** which can be dangerous if caution is not exercised.

Safety Precautions

- Wear safety glasses at all times while in the laboratory.
- Thermometers are often fitted with rubber stoppers as an aid in supporting the thermometer with a clamp. Inserting a thermometer through a stopper must be done carefully to prevent breaking of the thermometer, which would release mercury and possibly cut you. Your instructor will demonstrate the proper technique for inserting your thermometer through the hole of a rubber stopper. Glycerine is used to lubricate the thermometer and stopper. Protect your hands with a towel during this procedure.
- The red liquid used as the temperature-sensing liquid in lab thermometers is flammable. If a red-liquid thermometer breaks, extinguish all flames in the vicinity. Mercury is poisonous and is absorbed through the skin. Its vapor is toxic. If mercury is spilled from a broken thermometer, inform the instructor immediately so that the mercury can be removed. Do not attempt to handle spilled mercury.
- The liquids used in this experiment are flammable. Although only small samples of the liquids are used, the danger of fire is not completely eliminated. Keep all liquids away from open flames.
- The liquids used in this experiment are toxic if inhaled or absorbed through the skin. The liquids should be disposed of in the manner indicated by the instructor.
- *Caution*: Oil is used as the heating fluid in the Thiele tube used for the boiling/melting point determinations that follow. Hot oil may spatter if it is heated too strongly, especially if any moisture is introduced into the oil from glassware that is not completely dry. The oil may smoke or ignite if heated above 200°C.

Apparatus/Reagents Required

Thermometer and clamp, several beakers, Thiele tube, melting point capillaries, glass tubing (5–6-mm-diameter), burner and rubber tubing, file, scissors, medicine dropper, unknown sample for boiling point determination, ice

Procedure

Record all data and observations directly in your notebook in ink.

A. Calibration of the Thermometer

Fill a 400-mL beaker with ice, and add tap water to the beaker until the ice is covered with water. Stir the mixture with a stirring rod for 30 seconds.

Clamp the thermometer to a ringstand so that the bottom 2–3 inches of the thermometer is dipping into the ice bath. Make certain that the thermometer is suspended freely in the ice bath and is not touching either the walls or the bottom of the beaker.

Allow the thermometer to stand in the ice bath for 2 minutes, and then read the temperature indicated by the thermometer to the nearest 0.2 degree. Remember that the thermometer must be read while still in the ice bath. If the reading indicated by the thermometer differs from 0°C by more than one degree, replace the thermometer and repeat the ice bath calibration.

Allow the thermometer to warm to room temperature by resting it in a safe place on the laboratory bench.

Set up an apparatus for boiling as indicated in Figure 4-1, using a 100-mL beaker containing approximately 75 mL of water. Add 2–3 boiling chips to the water, and heat the water to boiling.

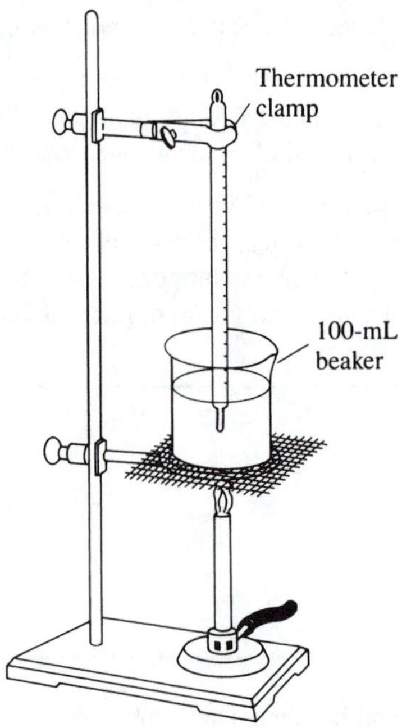

Figure 4-1. Apparatus for boiling. Make certain that the thermometer is freely suspended in the water and is not touching the walls of the beaker.

Using a clamp, suspend the thermometer so that it is dipping halfway into the boiling water bath. Make certain that the thermometer is not touching the walls or bottom of the beaker. Allow the thermometer to stand in the boiling water for 2 minutes; then record the thermometer reading to the nearest 0.2°C.

A boiling water bath has a temperature near 100°C, but the actual temperature of boiling water is dependent on the barometric pressure and changes with the weather from day to day. Your instructor will list the current barometric pressure on the chalkboard. Using a handbook of chemical data, look up the actual boiling point of water for this barometric pressure and record.

If your measured boiling point differs from the handbook value for the provided barometric pressure by more than one degree, exchange your thermometer at the stockroom, and repeat the calibration of the thermometer in both the ice bath and the boiling water bath.

B. Determination of Boiling Point

Set up a Thiele tube filled with oil as indicated in Figure 4-2. Examine the oil to make certain that it is not cloudy or contaminated with any substance. If the oil appears cloudy or contaminated, consult with the instructor about replacing the oil. The oil bath will be used to heat the boiling point sample evenly. Remember that hot oil can spatter or ignite if heated too quickly or too strongly.

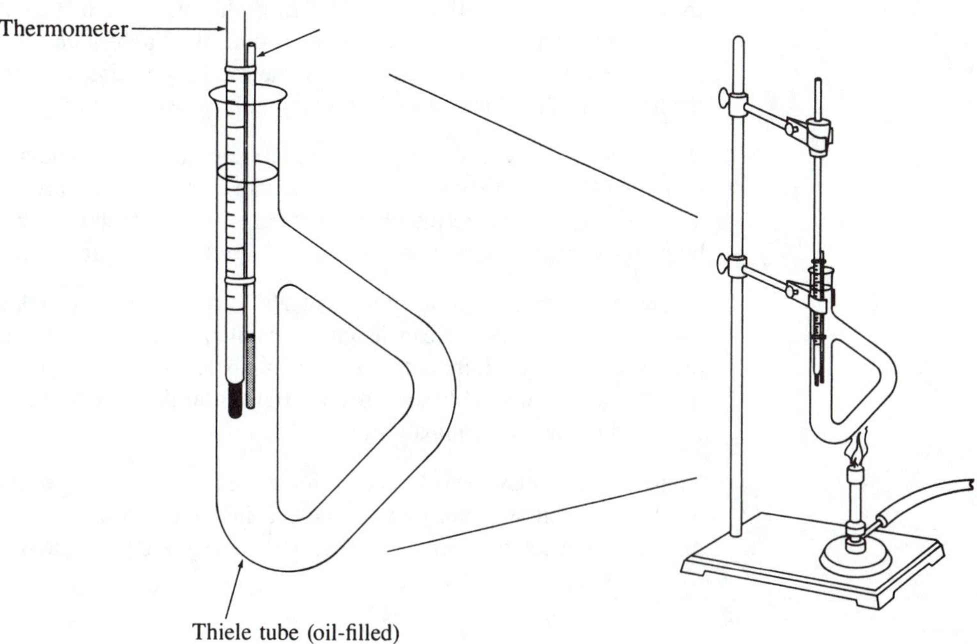

Figure 4-2. Thiele tube oil bath for boiling/melting determinations. Exercise caution when dealing with heated oil.

Obtain a clean 10×75-mm semimicro test tube, which will contain the boiling point sample. Attach the test tube to the lower end of your thermometer with two small rubber bands.

Obtain an unknown liquid for boiling point determination from your instructor, and record the identification code number of the unknown in your notebook and on the lab report page. Transfer part of the unknown sample to the small test tube, until the test tube is approximately half full.

Obtain a melting point capillary. If the capillary tube is open at both ends, heat one end briefly in a flame to seal it off. Using a glass file, carefully cut the capillary about 1 inch from the sealed end. Do *not* fire-polish the cut end of the capillary. Place the small portion of capillary *sealed end up* into the boiling point sample in the test tube.

The capillary has a rough edge at the cut end, which serves as a surface at which bubbles can form during boiling. The capillary is filled with air when inserted sealed end up into the liquid, and the presence of this air can be used to judge when the vapor pressure of the unknown liquid reaches atmospheric pressure.

Lower the thermometer/sample into the oil bath, so that the bulb of the thermometer and the sample are level with the upper branch of the Thiele tube's side arm. The Thiele tube is constructed in such a manner that when the side arm is heated, oil will circulate from the side arm into the main chamber. This makes it unnecessary to stir the oil bath.

Begin heating the side arm of the Thiele tube with a low flame, so that the temperature rises by one or two degrees per minute. Watch the small capillary tube in the unknown sample while heating. As the sample is heated, air in the capillary tube will begin to bubble from the capillary. As the air bubbles from the capillary, it is gradually replaced by vapor of the unknown. As the liquid begins to boil, the bubbles coming from the capillary will form a continuous, rapid stream. When the capillary begins to bubble continuously, remove the heat from the Thiele tube. The liquid will continue to boil.

Watch the capillary in the unknown sample carefully after removing the heat. The bubbling will slow down and stop after a few minutes, and the capillary will begin to fill with the unknown liquid. Record the temperature at which the bubbling stops. At this point (where bubbling stops), the vapor pressure of the liquid is just equal to the barometric pressure.

Allow the oil bath to cool by at least 20°C. Add additional unknown liquid to the small test tube, as well as a fresh length of capillary tube (it is not necessary to remove the previous capillary). Repeat the determination of the boiling point of the unknown. If the repeat determination of boiling point differs from the first determination by more than one degree, do a third determination.

Your instructor may provide you with the identity of your boiling point sample. If so, look up the true boiling point of your sample in a handbook of chemical data. Calculate the percent difference between your measured boiling point and the literature value.

Name ______________________ Section ______________________

Lab Instructor ______________________ Date ______________________

EXPERIMENT 4 The Determination of Boiling Point

PRE-LABORATORY QUESTIONS

1. Why is a *mixture* of ice and water, rather than ice alone, used in calibrating a thermometer?

____

____

____

____

____

2. Why does the boiling point of a liquid vary with the *barometric pressure*?

____

____

____

____

____

3. Using a handbook of chemical data, find the boiling points of water at the pressures indicated below. Give the reference for the data you list.

75 mm Hg ______________________

375 mm Hg ______________________

760 mm Hg ______________________

799 mm Hg ______________________

Reference ______________________

Name ______________________ Section ______________________

Lab Instructor ______________________ Date ______________________

EXPERIMENT 4 The Determination of Boiling Point

RESULTS/OBSERVATIONS

A. Calibration of the Thermometer

Temperature in ice/water bath ______________________

Thermometer error at ice/water temperature ______________________

Temperature in boiling water ______________________

Barometric pressure ______________________

True boiling point of water at this pressure ______________________

Thermometer error in boiling point of water ______________________

B. Determination of Boiling Point

Identification number of unknown liquid sample ______________________

First determination of boiling point ______________________

Second determination of boiling point ______________________

Mean value for unknown boiling point ______________________

QUESTIONS

1. What is meant by the *normal* boiling point of a substance?

____

____

2. Food products such as cake mixes often list special directions for cooking the products in high-altitude areas. Why are special directions needed for such situations? Would a food take a longer or shorter time period to cook under such conditions? Why?

____

____

____

EXPERIMENT

5 The Determination of Melting Point

Objective

The melting point of a pure substance is a characteristic property of the substance. In this experiment you will determine the melting point of an unknown sample, and will then confirm the identity of the uknown by a "mixed" melting point determination.

Introduction

The melting point of a pure substance is a *characteristic* property for a given substance. That is, under the same laboratory conditions, a given substance will always have the *same* melting point. Characteristic physical properties (such as the melting point of a pure substance) are of immense help in the identification of unknown substances. Such properties are routinely reported in scientific papers when new substances are isolated or synthesized, and are compiled in tables in the various handbooks of chemical data that are available in science libraries. When an unknown substance is isolated from a chemical system, its melting point may be measured (along with certain other characteristic properties) and then *compared* with tabulated data. If the experimentally determined physical properties of the unknown *match* those found in the literature, you can typically assume that you have identified the unknown substance.

When a pure solid substance melts during heating, the melting usually occurs quickly at one specific, characteristic temperature. For certain substances, especially more complicated organic substances or biological substances that tend to decompose slightly when heated, the melting may occur over a span of a few degrees, called the **melting range**. Melting ranges are also commonly observed if the substance being determined is not completely pure. The presence of an impurity will *broaden* the melting point of the major component and will also *lower* the temperature at which melting begins. Melting points of solid substances are routinely reported in the scientific literature and are tabulated in handbooks for use in identification of unknown substances. Melting point determinations are very common and will be used as an aid in identification of substances in several later experiments in this manual.

The apparatus used for heating samples in this experiment is called a **Thiele tube**. The Thiele tube contains oil (typically mineral oil) as a fluid, which permits the determination of temperatures up to about 200°C. The Thiele tube is constructed in such a way that when the side arm is heated, the warm oil will rise and enter the main chamber of the tube, which provides for circulation of the oil and for a more uniform temperature. Samples to be placed in the Thiele tube are ordinarily positioned so that the sample is aligned with the top branch of the side arm. When using a Thiele tube, remember that it contains hot oil, which can be dangerous if caution is not exercised.

In today's experiment, you will first make duplicate determinations of the melting point of an unknown pure substance provided by your instructor. Your instructor will then list for you on the chalkboard the names of the substances fromwhich the unknowns were prepared, along with the literature values for the melting points of these substances. Based on your experimental data, you will choose one of the substances listed on the chalkboard as the most likely candidate for the identity of your unknown. You will then perform a "mixed" melting point determination of a combination of a small amount of your unknown substance and an authentic sample of the substance you believe the unknown to be. If the two substances (unknown and authentic sample) are the *same substance*, the mixture should melt at the same temperature as in your previous melting point determinations of the unknown alone. If you choose the identity of your unknown *incorrectly,* then the mixture of your unknown and the authentic sample will not be a pure substance, and will demonstrate a much lower and broader melting range than did the pure unknown alone.

Safety Precautions

- Wear safety glasses at all times while in the laboratory.
- Thermometers are often fitted with rubber stoppers as an aid in supporting the thermometer with a clamp. Inserting a thermometer through a stopper must be done carefully to prevent breaking of the thermometer, which would release mercury and possibly cut you. Your instructor will demonstrate the proper technique for inserting your thermometer through the hole of a rubber stopper. Glycerine is used to lubricate the thermometer and stopper. Protect your hands with a towel during this procedure.
- The red liquid used as the temperature-sensing liquid in some lab thermometers is flammable. If a red-liquid thermometer breaks, extinguish all flames in the vicinity. Mercury may be used as the temperature-sensing fluid in other types of thermometers. Mercury is poisonous and is absorbed through the skin. Its vapor is toxic. If mercury is spilled from a broken thermometer, inform the instructor immediately so that the mercury can be removed. Do not attempt to handle spilled mercury.
- The solid samples used in this experiment may be toxic if ingested or if absorbed through the skin. Wash immediately if spilled on skin and consult with the instructor.
- ***Caution:*** Oil is used as the heating fluid in the Thiele tube used for the boiling/melting point determinations that follow. Hot oil may spatter if it is heated too strongly, especially if any moisture is introduced into the oil from glassware that is not completely dry. The oil may smoke or ignite if heated above 200°C.

Apparatus/Reagents Required

Thermometer and clamp, several beakers, Thiele tube, melting point capillaries, 5–6-mm-diameter glass tubing, burner and rubber tubing, file, scissors, medicine dropper, unknown sample for melting point determination

Procedure

Record all data and observations directly in your notebook in ink.

A. Melting Point of Unkown

Obtain a solid unknown sample for melting point determination from your instructor. Record the identification code number in your notebook and on the lab report form. If the solid is not finely powdered, grind some of the crystals on a watchglass or flat glass plate with the bottom of a clean beaker. Set aside part of this unknown sample for use in Part B.

Pick up a few crystals of the unknown solid in the open mouth of a melting point capillary tube. Tap the sealed end of the capillary tube on the lab bench to pack the crystals into a tight column at the sealed end of the tube. Repeat this process until you have a column of crystals approximately 2-3 mm high at the sealed end of the capillary.

Set up the Thiele tube apparatus as indicated in Figure 5-1. Attach the capillary tube containing the crystals to the thermometer with one or two small rubber bands, and position the capillary so that the crystals are *next to* the temperature-sensing bulb of the thermometer.

Lower the thermometer into the oil bath, and begin heating the side arm of the Thiele tube with a very small flame. Adjust the flame as necessary so that the temperature rises by one or two degrees per minute.

Watch the crystals in the capillary tube, and record the exact temperature at which the crystals first *begin* to melt, and the exact temperature at which the last portion *finishes* melting. Record these two temperatures as the melting range of the unknown.

Allow the oil bath to cool by *at least* 20°C.

Prepare another sample of the unknown crystals in a fresh capillary tube, and repeat the determination of the melting point. If this second determination differs significantly from the first determination, repeat the experiment a third time. Calculate the average value of your melting point determinations.

B. Mixed Melting Point

Your instructor will have listed on the chalkboard the names and melting points of the pure substances from which the unknown samples were prepared. By comparing the experimental value for the melting point of your unknown with the melting points of the pure substances listed on the chalkboard, choose the most likely candidate for the identity of your unknown. Obtain an authentic sample of this substance from your instructor or the stockroom.

Mix a small amout of the authentic sample with a small amount of the unknown sample (saved from Part A). Use approximately equal amounts of the two materials to give a total volume about the size of a pea. Use a stirring rod to mix together the two components of the mixture.

Pick up a few crystals of the mixture in the open mouth of a melting point capillary tube. Tap the sealed end of the capillary tube on the lab bench to pack the crystals into a tight column at the sealed end of the tube. Repeat this process until you have a column of crystals approximately 1 cm high at the sealed end of the capillary.

Attach the capillary tube containing the crystals to the thermometer of the Thiele tube apparatus with one or two small rubber bands, and position the capillary so that the crystals are *next to* the temperature-sensing bulb of the thermometer.

Lower the thermometer into the oil bath, and begin heating the side arm of the Thiele tube with a very small flame. Adjust the flame as necessary so that the temperature rises by only one or two degrees per minute.

Watch the crystals in the capillary tube, and record the exact temperature at which the crystals first *begin* to melt, and the exact temperature at which the last portion *finishes* melting. Record these two temperatures as the melting range of the mixture.

Allow the oil bath to cool by *at least* 20°C.

Prepare another sample of the crystals in a fresh capillary tube, and repeat the determination of the melting point. If this second determination differs significantly from the first determination, repeat the experiment a third time. Calculate the average value of your melting point determinations.

If the average melting temperature range for the *mixture* of unknown and authentic sample does not differ from the melting point of the unknown *itself* (Part A) by more than 1°C, you may assume that you have identified the unknown correctly.

If the melting range of the mixture differs significantly from the melting point of the unknown itself, consult with the instructor about performing additional melting point determinations of the unknown.

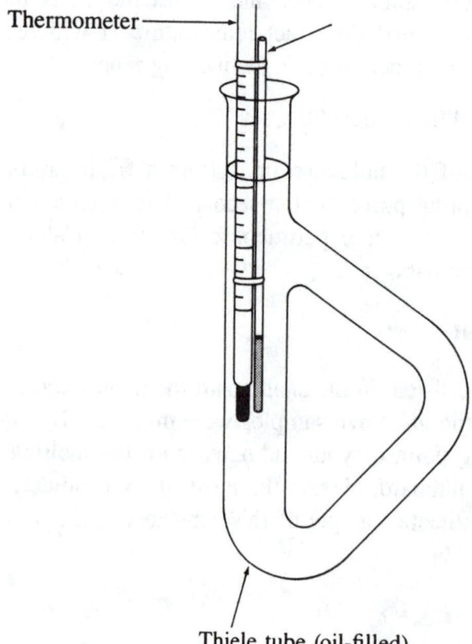

Figure 5-1. Thiele tube oil bath for boiling/melting determinations. Exercise caution when dealing with heated oil.

Name ______________________________ Section ______________________

Lab Instructor ______________________________ Date ______________________

EXPERIMENT 5 The Determination of Melting Point

PRE-LABORATORY QUESTIONS

1. What does it mean to say that the melting point of a pure substance is a *characteristic* property of the substance?

2. What effect does the presence of an *impurity* have on the melting point of a substance?

3. What is meant by a *mixed melting point* determination?

Name ______________________ Section ______________________

Lab Instructor ______________________ Date ______________________

EXPERIMENT 5 The Determination of Melting Point

RESULTS/OBSERVATIONS

A. Melting Point of Unknown

Identification number of unknown solid sample ______________________

First determination of melting point ______________________

Second determination of melting point ______________________

Mean value for melting point of unknown ______________________

B. Mixed Melting Point

Substance taken for mixed melting point determination ______________________

First determination of melting point ______________________

Second determination of melting point ______________________

Mean value for melting point of mixture ______________________

QUESTIONS

1. Suggest a *reason* why the melting point of a binary mixture is lowered and broadened, compared to the melting point of either component of the mixture.

____

____

____

____

____

2. Using a handbook, look up the normal melting points of each of the following substances:

NaCl ______________________

Biphenyl ______________________

Naphthalene ______________________

EXPERIMENT

6 The Solubility of a Salt

Objective

In this experiment, you will determine the solubility of a given salt at various temperatures. Also you will prepare the solubility curve for your salt.

Introduction

The term **solubility** in chemistry has both general and specific meanings. In everyday situations, we might say that a salt is *soluble*, meaning that experimentally, we were able to dissolve a sample of the salt in a particular solvent.

In a specific sense, however, the *solubility* of a salt refers to a definite numerical quantity. Typically, the solubility of a substance is indicated as the *number of grams of the substance that will dissolve in 100 g of the solvent.* More often than not the solvent is water. In that case the solubility could also be indicated as the number of grams of solute that dissolve in 100 mL (the density of water is near to 1.00 g/mL under ordinary conditions).

Since solubility refers to a specific, experimentally determined amount of substance, it is not surprising that the various handbooks of chemical data contain extensive lists of solubilities of various substances. In looking at such data in a handbook, you will notice that the *temperature* at which the solubility was measured is always given. Solubility *changes* with temperature. For example, if you like your tea extra sweet, you have undoubtedly noticed that it is easier to dissolve two teaspoons of sugar in hot tea than in iced tea. For many substances, the solubility increases with increasing temperature. For a number of other substances, however, the solubility decreases with increasing temperature.

For convenience, **graphs** of solubilities are often used rather than lists of solubility data from a handbook. A graph of the solubility of a substance versus the temperature will clearly indicate whether or not the solubility increases or decreases as the temperature is raised. If the graph is carefully prepared, the specific numerical solubility may be read from the graph.

It is important to distinguish experimentally between *whether* a substance is soluble in a given solvent, and *how fast* or *how easily* the substance will dissolve. Sometimes an experimenter may wrongly conclude that a salt is not soluble in a solvent, when actually the solute is merely dissolving at a very slow rate. The *speed* at which a solvent dissolves has nothing to do with the final *maximum quantity* of solute that can enter a given amount of solvent. In practice, we use various techniques to speed up the dissolving process, such as grinding the solute to a fine powder or stirring/shaking the mixture. Such techniques will *not* affect the final amount of solute that ultimately dissolves, however.

The solubility of a salt in water represents the amount of solute necessary to reach a state of *equilibrium* between saturated solution and undissolved additional solute. This number is a *constant* for a given solute/solvent combination at a constant temperature.

Safety Precautions

- Wear safety glasses at all times while in the laboratory.
- Use glycerine when inserting the thermometer through the stopper. Protect your hands with a towel.
- Some of the salts used in this experiment may be toxic. Wash your hands after use. Dispose of the salts as directed by the instructor.

Apparatus/Reagents Required

8-inch test tube fitted with 2-hole cork or slotted stopper, copper wire for stirring, thermometer, 50-mL buret, salt for solubility determination

Procedure

Record all data and observations directly in your notebook in ink.

Obtain a salt for the solubility determination. If the salt is presented as an *unknown*, record the code number in your notebook (otherwise, record the formula and name of the salt). If the salt is not finely powdered, grind it to a fine powder in a mortar.

Fit an 8-inch test tube with a 2-hole stopper (either cork or slotted rubber). Using glycerine, insert your thermometer (*Caution!*) in one of the holes of the stopper in such a way that the thermometer can still be read from 0° to 100°C. This may involve making a slot in the stopper through which the thermometer can be read.

Obtain a length of heavy-gauge copper wire for use in stirring the salt in the test tube. If the copper wire has not been prepared for you, form a loop in the copper wire in such a way that the loop can be placed around the thermometer when in the test tube. Fit the copper wire through the second hole in the stopper, making sure that the hole in the rubber stopper is big enough that the wire can be easily agitated in the test tube. (See Figure 6-1.)

Place about 300 mL of water in a 400-mL beaker, and heat the water to boiling.

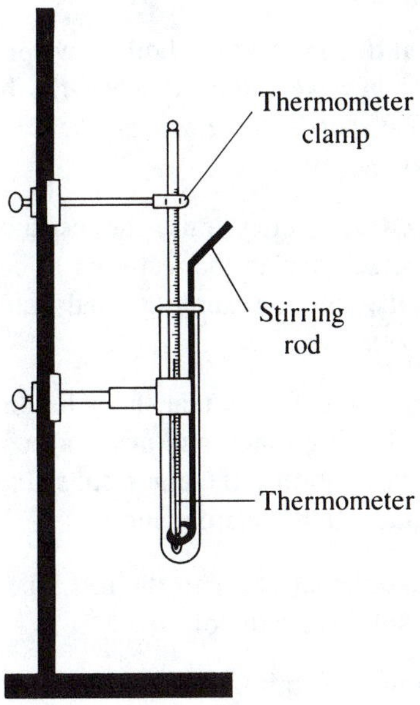

Figure 6-1. Apparatus for stirring a soluble salt. Be ceratin the thermometer bulb dips into the solution being measured.

While the water is heating, weigh the empty clean, dry 6-inch test tube (*without* the stopper/thermometer/stirring assembly). Make the weight determination to the nearest milligram (0.001 g).

Add approximately 5 g of your salt for the solubility determination to the test tube, and reweigh the test tube and its contents. Again, make the weight determination to the nearest milligram.

Clean the buret with soap and water, and then rinse the buret with tap water, followed by several rinses with distilled water. Fill the buret with distilled water. Make sure that water flows freely from the stopcock of the buret, but that the stopcock does not leak.

Record the reading of the initial water level in the buret to the nearest 0.02 mL. (Recall that water makes a meniscus. Read the water level across the *bottom* of the meniscus.)

In the following procedure, record in your notebook *each time* a portion of water is added from the buret. It is essential to know the amount of water used in the determination.

Add 3.00 ± 0.01 mL of water from the buret to the salt in the test tube. Record the precise amount of water used.

Attach the stopper with thermometer and stirrer, and clamp the test tube vertically in the boiling water bath. Adjust the thermometer so that the bulb of the thermometer will be *immersed* in the solution in the test tube. The test tube should be set up so that the contents of the test tube are immersed fully in the boiling water. See Figure 6-1. Using the stirring wire, gently stir the salt in the test tube until it dissolves.

If the salt does not dissolve completely after several minutes of stirring in the boiling water bath, remove the test tube and add 1.00 ± 0.01 mL additional water from the buret. Record. Return the test tube to the boiling water bath and stir.

If the salt is still not completely dissolved at this point, add 1.00 ± 0.01 mL water portions (one at a time) until the salt just dissolves. Record.

When all the salt has been dissolved at the boiling water temperature, the solution will be *nearly* saturated, and will *become* saturated when the heating is stopped. Minimize the amount of time the test tube spends in the boiling water bath to restrict any possible loss of water from the test tube by evaporation.

After the salt has dissolved completely, raise the test tube out of the boiling water. With constant stirring, allow the solution in the test tube to cool spontaneously in the air. Observe the temperature of the solution carefully, and note the temperature where the *first crystals of salt begin to form.*

The first formation of crystals indicates that the solution is saturated at that temperature. Reheat the test tube in the boiling water, and make a second determination of the temperature at which the first crystal forms. If your results disagree by more than one degree, reheat the solution and make a third determination.

Add 1.00 ± 0.01 mL of additional water to the test tube. Record. Reheat the test tube in boiling water until all the solid has redissolved.

Remove the test tube from the boiling water and allow it to cool again spontaneously. Make a determination of the saturation temperature for solution in the same manner as indicated earlier. Repeat the determination of the new saturation temperature as a check on your measurement.

Repeat the addition of 1.00-mL water samples, with determination of the saturation temperatures, until you have at least six sets of data. Keep accurate records as to how much water has been added from the buret at each determination.

If the saturation temperature drops sharply on the addition of the 1.00-mL samples, consult with the instructor about reducing subsequent additions of water to only 0.50-mL increments. If the saturation temperature does not change enough on the addition of 1.00-mL samples, increase the size of the water samples added to 2.00-mL increments. Keep accurate records of how much water is added.

From your data at each of the saturation temperatures, calculate the *mass of salt that would dissolve in 100 g of water* at that temperature. Assume that the density of water is exactly 1.00 g/mL, so that your buret additions in milliliters will be equivalent to the weight of water being added.

On a piece of graph paper, plot the solubility curve for your salt, using saturation temperatures on the horizontal axis and solubilities per 100 g of water on the vertical axis. Attach the graph to your laboratory report.

Name ______________________ Section ______________________

Lab Instructor ______________________ Date ______________________

EXPERIMENT 6 The Solubility of a Salt

PRE-LABORATORY QUESTIONS

1. Using a handbook of chemical data, look up the solubilities of the following salts, per 100 g of water at 20°C.

   NH_4Cl ______________________ Reference ______________________

   $(NH_4)_2SO_4$ ______________________ Reference ______________________

   K_2SO_4 ______________________ Reference ______________________

   KCl ______________________ Reference ______________________

   KBr ______________________ Reference ______________________

2. Find a specific definition in your textbook for the following terms:

   Saturated solution

   ____

   ____

   ____

   Solubility

   ____

   ____

   ____

3. Why does stirring affect the rate at which a salt dissolves in water, but not the solubility of the salt in water?

   ____

   ____

   ____

   ____

Name ______________________________ Section ______________

Lab Instructor ______________________________ Date ______________

EXPERIMENT 6 The Solubility of a Salt

RESULTS/OBSERVATIONS

mL of water used	Saturation temperature	Solubility (g/100 g H_2O)

Identity of the salt ______________________________

Literature solubility (20°C) ______________ Reference ______________

Percentage error in solubility at 20°C ______________ %

QUESTIONS

1. The solubility of many salts increases as the temperature increases. How do the solubilities of gases vary with temperature?

2. Why is it better to determine the saturation temperature while the temperature is dropping, rather than while it is rising?

3. When adding water to the salt initially, you attempted to find the minimum amount of water the salt would dissolve in at 100°C. Why was it necessary that the solution to be used be almost saturated?

____

____

____

____

____

4. The procedure indicated that the amount of time the test tube was kept in the boiling water bath should be minimized. Why was this necessary?

____

____

____

____

____

____

EXPERIMENT

7 Identification of a Substance

Objective

Selected physical properties of an unknown substance will be measured experimentally and compared with the tabulated properties of known substances.

Introduction

Modern instrumental methods permit the routine analysis and identification of unknown substances. Because of the high cost of precision instruments, and due to the cost and time required for maintenance and calibration of such instruments, however, instrumental methods of analysis are primarily used for *repetitive* determinations of *similar* samples, in which case the instrumental method is relatively fast and the cost *per analysis* moderate. For a *single* sample, however, it is found that classical "wet" laboratory methods of analysis are often preferred.

Most commonly, a thorough determination of the *physical properties* of a sample will suffice for an identification. The properties determined for the *unknown* sample are compared to the properties of *known* substances as tabulated in the chemical literature. If the properties match, an identification is assumed. The physical properties most commonly used in identifications are *density* (as discussed in Experiment 3), the *boiling and melting points* (as discussed in Experiments 4 and 5), and the *solubility* of the substance in water or in some other appropriate solvent (as studied in Experiment 6).

In general terms, one substance is likely to be soluble in another substance if the two substances have similar structural features (for example, a similar group of atoms) or have comparable electronic properties (which lead to similar interparticle forces). Generally solutes are divided into two major classes, depending on whether they dissolve in very *polar* solvents like water, or in very *nonpolar* solvents such as any of the hydrocarbons (the hydrocarbon cyclohexane is used in this experiment). Generally ionic and very polar covalent substances will dissolve in water, whereas very nonpolar substances will dissolve in cyclohexane. Some solutes that are intermediate in polarity may dissolve well in a solvent of intermediate polarity (such as ethyl alcohol).

The physical properties discussed above are almost always reported in the literature when new substances are prepared, and are tabulated for previously known substances in the various handbooks of chemical data found in most laboratories and science libraries. Other gross properties that may also be helpful in identifications include unusual *color* (e.g., for transition metal ions of characteristic color), *odor* (e.g., some types of chemical substances have very characteristic odors), *hardness* (e.g., crystals of ionic substances are very hard compared to most covalently bonded substances), and so forth.

In this experiment, you will be provided with an unknown sample chosen from those substances listed in Table 7-1. By careful determination of the physical properties of your sample, you should be able to match those properties to one of the substances listed in the table. The unknown samples have been purified and contain only a single substance. In real practice, an unknown sample may contain more than one major component and is likely also to contain minor impurities. The identification of unknown samples is a vital, important, and interesting application of modern chemistry.

Table 7-1. Physical Properties of Selected Substances

Liquid Substances (under normal room conditions)

Substance	*Solubility* *W*	*A*	*C*	*Density* g/mL	*Boiling Point* °C (760 mm)
acetone	s	s	s	0.791	56
1-butanol	s	s	s	0.810	117
1-butanol, 2-methyl	s	s	s	0.816	128
2-butanol	s	s	s	0.808	100
2-butanol, 2-methyl	s	s	s	0.806	102
cyclohexane	-	s	s	0.778	81
ethyl acetate	s	s	s	0.900	78
hexane	-	s	s	0.660	68
methanol	s	s	s	0.793	65
1-propanol	s	s	s	0.780	97
2-propanol	s	s	s	0.786	83
2-propanol, 2-methyl	s	s	s	0.789	82

Solid Substances (under normal room conditions)

Substance	*Solubility* *W*	*A*	*C*	*Density* g/mL	*Melting Point* °C
acetamide	s	s	-	1.16	82
acetanilide	-	s	s	1.22	113
benzoic acid	-	s	s	1.27	123
benzophenone	-	s	s	1.15	48
biphenyl	-	s	s	0.867	71
camphor	-	s	s	0.990	176
1,4-dichlorobenzene	-	s	s	1.25	53
diphenylamine	-	s	s	1.16	53
lauric acid	-	s	s	0.867	44
naphthalene	-	s	s	1.03	80
phenyl benzoate	-	s	s	1.24	71
sodium acetate·$3H_2O$	s	s	-	1.45	58
stearic acid	-	s	s	0.941	72
thymol	-	s	s	0.925	52

Solubility legend: *W* = water, *A* = ethyl alcohol, *C* = cyclohexane
s = soluble - = insoluble (or very low solubility)

Safety Precautions

- Wear safety glasses at all times while in the laboratory.
- Many of the unknown substances in this experiment are *flammable*; keep the unknown substances away from open flames.
- Cyclohexane, ethyl alcohol, acetone, hexane, methanol, ethyl acetate, and propanol are all highly flammable; keep these solvents in the exhaust hood.
- Many of the unknown substances in this experiment are *toxic* if inhaled or absorbed through the skin.
- Oil is used as the fluid in the Thiele tube used for boiling/melting point determinations. The oil may spatter or ignite if heated too quickly or to a temperature above 200°C. Examine the oil for contamination before use.

Apparatus/Reagents Required

Unknown sample, distilled water, ethyl alcohol, cyclohexane, sodium chloride, copper(II) sulfate, pentane, oleic acid, naphthalene, melting/boiling point apparatus (oil-filled Thiele tube), melting point capillaries, rubber bands, semimicro test tubes, thermometer

Procedure

Record all data and observations directly in your notebook in ink.

Obtain an unknown substance for identification, and record its code number in your notebook and on the report page.

Record your gross observations of the color, odor, physical state, volatility, viscosity, clarity, and so forth for the unknown substance.

A. Solubility

Before determining the solubility of your unknown sample, you will perform some preliminary solubility studies with known solutes and several solvents.

Set up three clean, dry semimicro test tubes in a test tube rack. Place 10 drops of distilled water in one test tube, 10 drops of ethyl alcohol in a second test tube, and 10 drops of cyclohexane in the third test tube.

Obtain a sample of sodium chloride, and transfer approximately equal small amounts to each of the test tubes containing the three solvents. Stopper the test tubes and shake for at least 30 seconds to attempt to dissolve the solid. Record your observations on the solubility of NaCl in the three solvents.

Repeat the solubility tests with samples of naphthalene, copper(II) sulfate, pentane, and oleic acid, determining the solubility of each solute in each of the three solvents (water, ethyl alcohol, cyclohexane). Some of the test solutes are liquids. Add 5 drops of the liquid solute to each of the test tubes containing the solvents. If a liquid does not dissolve in a given solvent, it will form a separate layer in the test tube. Record your observations of solubility.

Repeat the solubility test, using small samples of your *unknown* material in each of the three solvents. Record your observations.

B. Density

Techniques for the measurement of the densities of solids and liquids were discussed in detail in Experiment 3. Review the discussion given there.

If the unknown is a liquid, determine its density by measuring the mass of a specific volume of the liquid. For example, a weighed empty graduated cylinder could be filled with the liquid to a particular volume and then reweighed. Alternatively, a specific volume of liquid can be taken with a transfer pipet, and the liquid weighed in a clean, dry beaker.

If the unknown is a solid, the volume of a weighed sample of solid may be determined by displacement of a liquid in which the solid is *not soluble* (see Part A of this experiment). If this is performed in a graduated cylinder, the change in liquid levels reflects the volume of the solid directly.

C. Melting and Boiling Point Determinations

Semimicro methods for the determination of boiling and melting points were discussed in detail in Experiments 4 and 5 of this manual. Review these methods.

If your unknown sample is a solid, determine its melting point as you did in Experiment 5. Fill two or three melting point capillaries with finely powdered solid to a height of about 1 cm. Attach one of the capillaries to a thermometer with a rubber band, and lower the sample and thermometer into the oil bath of a Thiele tube. Heat the oil bath in such a manner that the temperature rises only one or two degrees per minute, and watch carefully for the crystals to melt. Record the melting *range* if the crystals do not melt sharply at a single temperature. Repeat the determination until you get two values that check to within one degree.

If your unknown is a liquid, determine its boiling point. Fill a micro test tube to a height of 3–4 cm with the liquid, and insert a short length of melting point capillary as described in Experiment 4. Attach the test tube to a thermometer with a rubber band, and lower into the oil bath of a Thiele tube. Heat the oil bath until bubbles come in a steady stream from the capillary, remove the heat, and record the exact temperature at which bubbles stop coming from the capillary. Add liquid to preserve the 3–4-cm height, add a fresh capillary tube, and repeat the determination as a check. If your results do not agree within one degree, perform a third determination.

Compare the properties determined for your unknown with the properties of those substances listed in Table 7-1. Make a tentative identification of your unknown substance.

Name ______________________________ Section ____________________

Lab Instructor ______________________________ Date ____________________

EXPERIMENT 7 Identification of a Substance

PRE-LABORATORY QUESTIONS

1. Why are the freezing point of a liquid and the melting point of a solid the same temperature for a given substance?

____

____

____

____

____

2. Familiarity with the various handbooks of chemical and physical data is very important in the identification of unknown substances. Using such a handbook, find information about the melting point, boiling point, density, and qualitative solubility of each of the following substances. Give your reference, including the page of the reference on which the data can be found.

Salicylic acid Reference ____________________ Page ____________

Melting point ______________________________ Density ____________________

Boiling point ______________________________ Solubility ____________________

2-Pentanone Reference ____________________ Page ____________

Melting point ______________________________ Density ____________________

Boiling point ______________________________ Solubility ____________________

1,2-Dibromomethane Reference ____________________ Page ____________

Melting point ______________________________ Density ____________________

Boiling point ______________________________ Solubility ____________________

Isopropyl acetate Reference____________________ Page__________

Melting point____________________________ Density____________________

Boiling point____________________________ Solubility__________________

3. A common mistake many students make is to confuse *whether or not* a solute is soluble in a solvent with how *quickly* a solute dissolves. Use your textbook to find a specific definition of solubility. What factors affect the solubility of a solute in a solvent? What factors affect the rate at which a solute dissolves in a solvent?

Name ______________________ Section ______________________

Lab Instructor ______________________ Date ______________________

EXPERIMENT 7 Identification of a Substance

RESULTS/OBSERVATIONS

Unknown identification number ______________________

Qualitative observations (color, etc.) of unknown

____

____

____

____

____

A. Solubility Test Observations

Solubility of known solutes observations

____

____

____

____

____

Solubility of unknown observations

____

____

____

B. Density Determination

Describe the method by which the density of the unknown was determined. ______________________

____

Density of unknown____

C. Melting/Boiling Point Determination

Describe the method used for the melting or boiling point determination. ______________________

____

Melting or boiling (circle) point determined ______________________________________

Barometric pressure (for boiling point) ____

Summarize in the following table the properties determined for your unknown sample, along with the properties of the substance you believe your unknown to be. Calculate the percentage difference between the known and unknown substance's properties for each quantitative determination.

Substance unknown is believed to be ____

	Unknown	Known	% difference
Solubility	____________	____________	____________
Density	____________	____________	____________
Melting/boiling point	____________	____________	____________

QUESTIONS

1. What *error* is introduced into a melting point determination if the oil bath is heated too quickly? Will the observed melting point be too high or too low?

____

____

____

2. In this experiment you qualitatively determined whether or not a given solute would dissolve in each of three solvents. Suggest a method by which the solubility of a solute in a solvent might be measured on a *quantitative* basis.

____

____

____

EXPERIMENT

8 Resolution of Mixtures 1: Filtration and Distillation

Objective

The separation of mixtures into their constituent components defines an entire subfield of chemistry referred to as **separation science**. In this experiment, techniques for the resolution of mixtures of solids and liquids will be examined.

Introduction

Mixtures occur very commonly in chemistry. When a new chemical substance is synthesized, for example, oftentimes the new substance first must be *separated* from a mixture of various side-products, catalysts, and any excess starting reagents still present. When a substance must be isolated from a natural biological source, the substance of interest is generally found in a very complex mixture with many other substances, all of which must be removed. Chemists have developed a series of standard methods for resolution and separation of mixtures, some of which will be investigated in this experiment. Methods of separation based on the processes of chromatography are found in Experiment 9.

Mixtures of solids often may be separated on the basis of differing *solubilities* of the components. If one of the components of the mixture is very soluble in water, for example, while the other components are insoluble, the water-soluble component may be removed from the mixture by simple *filtration* through ordinary filter paper. A more general case occurs when *all* the components of a mixture are soluble, to different extents, in water or some other solvent. The solubility of substances in many cases is greatly influenced by *temperature*. By controlling the temperature at which solution occurs, or at which the filtration is performed, it may be possible to separate the components. For water-soluble solutes, commonly a sample is added to a small amount of water and heated to boiling. The hot sample is then filtered to remove completely insoluble substances. The sample is then cooled, either to room temperature or below, which causes crystallization of those substances whose solubilities are very temperature-dependent. These crystals can then be isolated by a second filtration, and the filtrate remaining can be concentrated to reclaim those substances whose solubilities are *not* so temperature-dependent.

Mixtures of liquids are most commonly separated by *distillation*. In general, **distillation** involves heating a liquid to its boiling point, then collecting, cooling, and condensing the vapor produced into a separate container. For example, salt water may be desalinated by boiling off and condensing the water. For a mixture of liquids, however, in which several of the components of the mixture are likely to be volatile, a separation is not so easy to effect.

If the components of the mixture differ reasonably in their boiling points, it may be possible to separate the mixture simply by monitoring the temperature of the vapor produced as the mixture is heated. The components of a mixture will each boil in turn as the temperature is gradually raised, with a sharp *rise* in the temperature of the vapor being distilled indicating when a new component of the mixture has begun to boil. By changing the receiving flask at this point, a separation will be accomplished. For liquids whose boiling points only differ by a few degrees, the mixture can be passed through a **fractionating column** as it is being heated. Fractionating columns generally are packed with glass beads or short lengths of glass tubing that provide a large amount of surface area to the liquid being boiled. In effect, a fractionating column permits a mixture to be redistilled repeatedly while in the column, allowing for better separation of the components of the mixture.

Safety Precautions

- Wear safety glasses at all times while in the laboratory.
- The solid mixture contains benzoic acid, which may be irritating to the skin and respiratory tract. Never ingest any chemical in the lab.
- When moving hot containers, use metal tongs or a towel to avoid burns. Beware of burns from steam while solutions are being heated.
- The liquids used for fractional distillation may be flammable. Keep the liquids away from open flames. Perform the distillation in the exhaust hood to help remove fumes. The liquids may be toxic if absorbed through the skin or inhaled.
- Portions of the distillation apparatus may be very hot as the distillation takes place.
- **Never heat a distillation flask to complete dryness.** The distillation flask may break. Distillation to dryness also poses an explosion hazard for certain unstable organic substances, or for substances which may be contaminated with organic peroxides.
- Dispose of all solids and liquids as directed by the instructor.
- Silver nitrate, $AgNO_3$, stains the skin. All nitrates are strong oxidizers, toxic, and may be carcinogenic.

Apparatus/Reagents Required

Impure benzoic acid sample (benzoic acid which has been colored with charcoal), 1% sodium chloride solution, 0.1 *M* silver nitrate, unknown mixture of two volatile liquids for fractional distillation

Procedure

Record all observations and data directly in your notebook in ink.

A. Resolution of a Solid Mixture

Obtain a sample of impure benzoic acid for recrystallization. Benzoic acid is fairly soluble in hot water, but has a much lower solubility in cold water. The benzoic acid has been contaminated with charcoal and sand, which are not soluble under either temperature condition. Transfer the benzoic acid sample to a clean 150-mL beaker.

Set up a short-stem gravity funnel in a small metal ring clamped to a ringstand. Fit the filter funnel with a piece of filter paper folded in quarters to make a cone. See Figure 8-1.

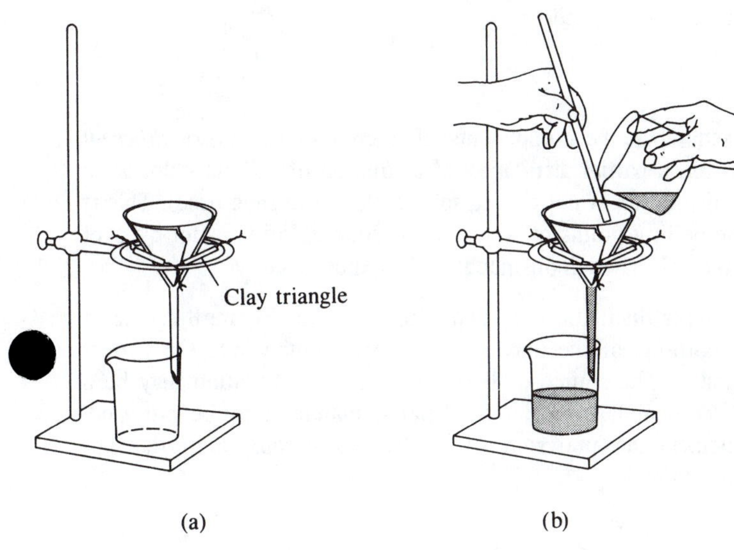

Figure 8-1. Filtration of a hot solution. Use a stirring rod as a guide for running the solution into the funnel. Do not fill the funnel more than half-full at a time, to prevent solution being lost over the rim of the paper cone.

Moisten the filter paper slightly so that it will remain in the funnel. Place a clean 250-mL beaker beneath the stem of the funnel.

Set up a 250-mL beaker about half filled with distilled water on a wire gauze over a metal ring. Heat the water to boiling.

When the water is boiling, pour about two-thirds of the water into the beaker containing the benzoic acid sample. Use a towel to protect your hands from the heat.

Pour the remainder of the boiling water through the gravity funnel to heat it. If the funnel is not preheated, the benzoic acid may crystallize in the stem of the funnel rather than passing through it. Discard the water that is used to heat the funnel.

Transfer the beaker containing the benzoic acid mixture to the burner and reheat it gently until the mixture just begins to boil again. Stir the mixture to make sure that the benzoic acid dissolves to the greatest extent possible.

Using a towel to protect your hands, pour the benzoic acid mixture through the preheated funnel. Catch the filtrate in a clean beaker.

Allow the benzoic acid filtrate to cool to room temperature.

When the benzoic acid solution has cooled to room temperature, filter the crystals to remove water. Wash the crystals with two 10-mL portions of cold water.

Transfer the liquid filtrate from which the crystals have been removed to an ice bath to see if additional crystals will form at the lower temperature. Examine, but do not isolate, this second crop of crystals.

Transfer the filter paper containing the benzoic acid crystals to a watch glass, and dry the crystals under a heat lamp or over a 400-mL beaker of boiling water. You can monitor the drying of the crystals by watching for the filter paper to dry out as it is heated. If a heat lamp is used, do not let the paper char or the crystals melt.

When the benzoic acid has been dried, determine the melting point of the recrystallized material using the method discussed in Experiment 5. Compare the melting point of your benzoic acid with that indicated in the handbook. If the melting point you obtain is significantly lower than that reported in the handbook, dry the crystals for an additional period under the heat lamp or over the hot water bath.

B. Simple Distillation

Simple distillation can be used when the components of a mixture have *very different* boiling points. In this experiment, a partial distillation of a solution of sodium chloride in water will be performed (the distillation is not carried to completion to save time). This is an extreme example, since the boiling points of water and sodium chloride differ by over 1000°C, but the technique will be clearly demonstrated by the experiment.

Your instructor has set up a simple distillation apparatus for you. (See Figure 8-2.) He or she will explain the various portions of the apparatus and will demonstrate the correct procedure for using the apparatus. The source of heat used for the distillation may be a simple burner flame, or an electrical heating device (heating mantle) may be provided. Generally electrical heating elements are preferred for distillations, because often the substances being distilled are flammable.

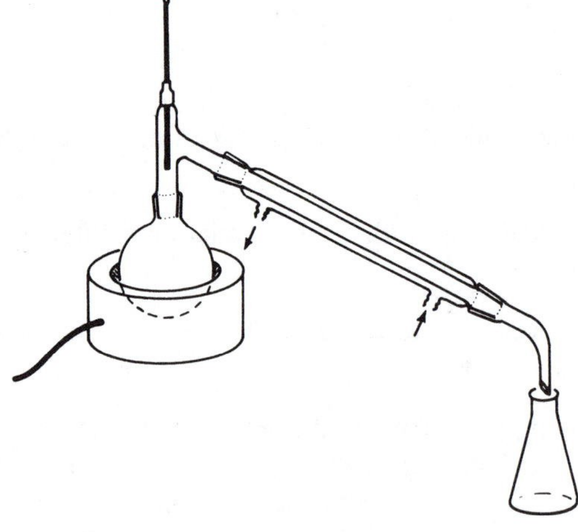

Figure 8-2. Simple distillation apparatus. Cold water entering the lower inlet of the condenser causes the vapor being distilled to liquefy.

Obtain about 50 mL of 1% sodium chloride solution. Place 1 mL of this solution in a small test tube, and transfer the remainder of the solution to the distilling flask.

Place a clean dry beaker under the mouth of the condenser of the distillation apparatus to collect the water as it distills from the salt solution.

Begin heating the sodium chloride solution as directed by the instructor, and continue distillation until approximately 20 mL of water has been collected. Transfer approximately 1 mL of the distilled water to a clean small test tube.

To demonstrate that the distilled water is now free of sodium chloride, test the sample of original 1% sodium chloride solution that was reserved before the distillation, as well as the 1-mL sample of water that has been distilled, with a few drops of 0.1 *M* silver nitrate solution (*Caution!*). Silver ion forms a *precipitate* of insoluble AgCl when added to a chloride ion solution. No precipitate should form in the water that has been distilled.

C. Fractional Distillation

Fractional distillation may be used to separate mixtures of volatile substances that differ by at least several degrees in their boiling points. The vapor of the liquid being boiled passes into a *fractionating column*, which provides a great deal of surface area and the equivalent of many separate simple distillations.

Your instructor has set up a fractional distillation apparatus in one of the exhaust hoods. (See Figure 8-3.) Compare the fractional distillation apparatus with the simple distillation apparatus used in Part B, and note the differences. Your instructor will explain the operation of the fractional distillation apparatus. The apparatus is set up in the hood, since the mixture you will distill is very volatile and may be flammable.

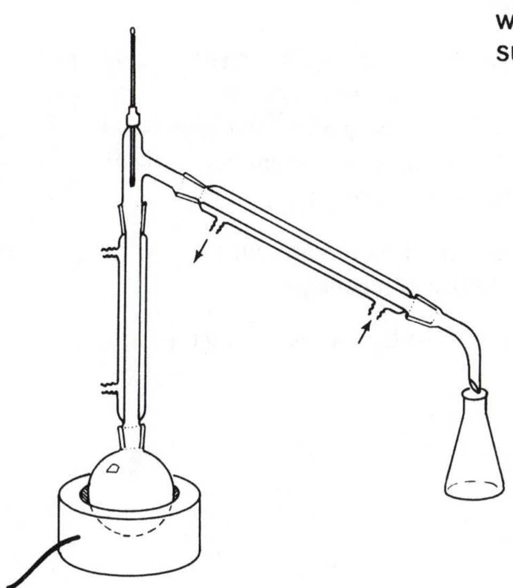

Figure 8-3. Fractional distillation. The tall vertical fractionating column is packed with bits of glass that provide a large surface area.

Obtain an unknown mixture for fractional distillation and record its identification number. Use a graduated cylinder to transfer 40 mL of the unknown mixture to the distillation flask. Record the exact volume of the mixture used. During the distillation, carefully watch the thermometer that is part of the apparatus. The temperature is used to monitor the distillation, since the temperature will *increase very suddenly* as one component finishes distilling, and another component begins to distill.

Place a clean dry flask under the mouth of the condenser to collect the first component of the mixture as it distills. Have ready a second clean dry flask for collection of the second component. Have ready corks or rubber stoppers that fit snugly in the two collection flasks.

Have your instructor approve the apparatus, and then begin heating the distillation flask with very low heat until vapor begins to rise into the fractionating column.

Allow the vapor to rise to the level of the thermometer bulb, and adjust the heat so that the thermometer will remain bathed in droplets of liquid as the mixture distills. Record the temperature indicated by the thermometer as the first component of the mixture begins to distill. Collect the distillate coming from the condenser.

Continue heating the distillation flask, using the smallest amount of heat that will maintain distillation. Monitor the temperature constantly. At the point at which the first component of the mixture has finished distilling, the temperature will rise *suddenly* and abruptly by several degrees. At this point, remove the flask used to collect the first component of the mixture, and replace it with the second flask. Stopper the flask containing the first component to prevent its evaporation.

Record the temperature indicated by the thermometer as the second component of the mixture begins to distill. Continue the distillation *until approximately 5 mL of liquid remains in the distillation flask.* Remove the source of heat, but do not remove the collection flask until distillation stops.

Do not heat the distillation flask to complete dryness, or it may break from the heat. When distillation is complete, stopper the flask containing the second component of the mixture.

With a graduated cylinder, determine the respective volumes of each of the two components of the mixture. Calculate the *approximate composition* of the original mixture, in terms of the percentages of low-boiling and high-boiling components. This percentage is only approximate, since some of the vapor being distilled may have been lost, and not all of the high-boiling component was isolated.

Report the approximate composition of your mixture to the instructor, along with the boiling temperatures of the two components.

Turn in the two flasks of distillate to the instructor for proper disposal.

Name ______________________________ Section ____________________

Lab Instructor ______________________________ Date ____________________

EXPERIMENT 8 Resolution of Mixtures 1: Filtration and Distillation

PRE-LABORATORY QUESTIONS

1. This experiment examines the techniques of *recrystallization* of solids, and also the techniques of simple and fractional *distillation* of liquids, as examples of methods by which mixtures are resolved into their components. Use your textbook to find three additional methods by which mixtures may be resolved, and describe the techniques briefly, including the sorts of mixtures to which the techniques are applied.

____

____

____

____

____

____

2. Why is a fractionating column packed with small glass beads or short pieces of glass tubing? How does this help improve a distillation?

____

____

____

____

____

Name ______________________________ Section ______________________

Lab Instructor ______________________________ Date ______________________

EXPERIMENT 8 Resolution of Mixtures 1: Filtration and Distillation

RESULTS/OBSERVATIONS

A. Resolution of a Solid Mixture

Observation of dissolving sample in hot water

____

____

Observation of material remaining on filter paper

____

____

Observation of filtrate as it cools

____

____

Appearance of crystals collected

____

____

Appearance of second crop of crystals

____

____

Melting point of dry benzoic acid ______________________________

Literature value for melting point ______________________________

Error in melting point determination ______________________________

B. Simple Distillation

Observation of distillation

____

____

Silver nitrate test:

On original sample ______________________________

On distilled sample ______________________________

C. Fractional Distillation

ID number of mixture for fractional distillation ______________

Observation of fractional distillation

____

____

Boiling point of first (low boiling) component ______________

Volume of first component collected ______________

Boiling point of second (high boiling) component ______________

Approximate volume of second component ______________

Approximate percentage composition of mixture ______________

QUESTIONS

1. The solid in Part A consisted primarily of one component (benzoic acid), with a small amount of an insoluble contaminant added. How might a solid mixture containing *two* major components be separated?

____

____

____

____

____

2. How do simple and fractional distillation *differ*? Under what circumstances is one method likely to be used in preference to the other method?

3. Fractional distillation can be used to separate liquid mixtures whose components have different boiling points in many, but not all, instances. The ordinary ethyl alcohol used in laboratories is actually only 95% ethyl alcohol, with the remainder of the mixture being water. Even very careful distillation of 95% ethyl alcohol does not permit removal of the water, even though the boiling points of water and pure ethyl alcohol differ by more than 20°C. Use a chemical encyclopedia to find the definition of an *azeotrope*, and why azeotropes such as 95% ethyl alcohol cannot be separated by distillation. Record your findings here.

4. In spite of such a major portion of the earth's surface being covered by water, surprisingly little use has been made of distillation of seawater as a source of drinking water in arid areas. Why do you suppose this is so?

EXPERIMENT

9 Resolution of Mixtures 2: Chromatography

Objective

The field of separation science is one of the most important in chemistry today. The particular branch of chemistry called **analytical chemistry** is concerned with the separation of mixtures and the analysis of the amount of each component in the mixture. In this experiment, you will perform two chromatographic separations of mixtures. Choice I studies the resolution of felt-tip pen inks by paper chromatography. Choice II deals with the resolution of a mixture of colored indicator dyes by thin-layer chromatography.

General Introduction to Chromatography

The word **chromatography** means color-writing. The name was chosen at the beginning of this century when the method was first used to separate colored components from plant leaves. Chromatography in its various forms is perhaps the most important known method of chemical analysis of mixtures.

Paper and thin-layer chromatography are simple techniques that can be used to separate mixtures into the individual *components* of the mixture. The methods are very similar in operation and principle, differing primarily in the medium used for the analysis.

Paper chromatography uses ordinary filter paper, which consists primarily of the polymeric carbohydrate *cellulose*, as the medium on which the mixture to be separated is applied. **Thin-layer chromatography** (universally abbreviated as **TLC**) uses a thin coating of aluminum oxide (alumina) or silicagel on a glass microscope slide or plastic sheet to which the mixture to be resolved is applied.

A single drop or spot of the unknown mixture to be analyzed is applied about half an inch from the end of a strip of filter paper or a TLC slide. The filter paper or TLC slide is then placed in a shallow layer of solvent or solvent mixture in a jar or beaker. Since filter paper or the coating of the TLC slide is permeable to liquids, the solvent begins rising by capillary action.

As the solvent rises to the level at which the spot of mixture was applied, various effects can occur, depending on the constituents of the spot. Those components of the spot that are completely soluble in the solvent will be swept along with the solvent front as it continues to rise. Those components that are not at all soluble in the solvent will be left behind at the original location of the spot. Most components of the unknown spot mixture will take an intermediate approach as the solvent front passes. Components in the spot that are *somewhat* soluble in the solvent will be swept along by the solvent front, but to *different extents,* reflecting their specific solubilities. By this means, the original spot of mixture is spread out into a series of spots or bands, with each spot or band representing one single component of the original mixture.

The separation of a mixture by chromatography is not solely a function of the solubility of the components in the solvent used, however. The filter paper or TLC slide coating used in chromatography is not inert, but consists of molecules that may *interact* with the molecules of the components of the mixture being separated. Each component of the mixture is likely to have a different extent of interaction with the filter paper or slide coating. This differing extent of interaction between the components of a mixture and the molecules of the support forms an equally important basis for the separation. Filter paper or the TLC slide coating adsorbs molecules on its surface to differing extents, depending on the structure and properties of the molecules involved.

To place a paper chromatography or TLC separation on a quantitative basis, a mathematical function called the **retention factor**, R_f, is defined:

$$R_f = \text{distance traveled by spot/distance traveled by solvent front}$$

The retention factor depends on what solvent is used for the separation and on the specific composition of the filter paper or slide coating used for a particular analysis. Because the retention factors for particular components of a mixture may vary if an analysis is repeated under different conditions, a *known* sample is generally analyzed at the *same time* as an *unknown* mixture on the same sheet of filter paper or slide. If the unknown mixture produces spots having the same R_f values as spots from the known sample, then an identification of the unknown components has been achieved.

Paper chromatography and TLC are only two examples of many different chromatographic methods. Mixtures of volatile liquids are commonly separated by a method called **gas chromatography**. In this method, a mixture of liquids is vaporized and passed through a long tube (column) of solid adsorbent material coated with an appropriate liquid, by the action of a carrier gas (usually helium). As with paper chromatography, the components of the mixture will have different solubilities in the liquid coating and different attractions for the solid adsorbent material. Separation of the components of the mixture thus occurs as the mixture progresses through the tube. The individual components of the mixture exit the tube one by one and are usually detected by electronic means. A final very important chromatographic technique is called high performance liquid chromatography (HPLC). In HPLC, liquid mixtures to be analyzed are blown through a column of adsorbent material under high pressure from a pump, resulting in a very quick passage through the column. HPLC is routinely used in medical and forensic laboratories to analyze biological samples. For example, blood samples can be analyzed for the presence of alcohol or illicit drugs in just a few minutes using HPLC.

Choice I. Paper Chromatography of Inks

Introduction

Although paper chromatography is a very simple technique, it is still used frequently for analyses of mixtures of colored substances (or for substances which can be made colored by treatment with an appropriate reagent). For example, biologists often use paper chromatography for quick analysis of plant pigments. Paper chromatography can also be used for simple analyses of protein extracts (amino acids).

In this Choice, you will do some very simple paper chromatographic analyses of some felt-tip pen inks. As you know, inks for such pens come in many different, bright colors—particularly in pen sets used by small children for working on their coloring books. Such brightly colored inks, however, are often *mixtures* of primary color inks. For example, a felt-tip pen having what appears to be bright purple ink may actually contain a mixture of blue and red inks. Similarly, what appears to be orange ink may be a mixture of red and yellow inks. Although this Choice is very simple, you will clearly see the basis for chromatographic analyses, and will perhaps gain some insight into the great importance of the various chromatographic methods.

Apparatus/Reagents Required

filter paper for chromatography (5 × 10 cm), latex surgical gloves, ruler, pencil, heat gun, acetone-water mixture (50% v/v), felt-tip pens (water-soluble and permanent)

Safety Precautions

- Wear safety glasses at all times while in the laboratory.
- Acetone is highly flammable. No flames are permitted in the laboratory. Acetone may be toxic if inhaled or absorbed through the skin.

Procedure

Record all observations directly in your notebook in ink.

Because the skin contains oils which can interfere with the chromatogram, latex surgical gloves should be worn from this point onward in the procedure to prevent contamination of the chromatogram. A pencil is used in the following procedure for marking the chromatogram, since ink from a ball-point pen would also undergo chromatography.

Obtain a sheet of filter paper prepared for the chromatographic analysis. Draw a light pencil line across the paper about a half inch from each end. On the lower pencil line, lightly mark three or four small circles. These circles will be where the felt-tip pen inks are to be applied to the paper.

From your instructor, obtain several felt-tip pens containing *water-soluble* inks. Apply a single small spot of a different ink to each of the pencil circles you drew on the filter paper. Allow the spots to dry completely. Record in your notebook the original color of each ink applied to the filter paper, as well as any code numbers marked on the barrel of the pen.

When the spots are completely dry, apply a second spot of each ink to its respective place on the filter paper. Applying a second spot of the same dye builds up a larger sample of the dye on the filter paper.

Clean a 400-mL beaker (or special chromatography jar) for use in developing the chromatogram. Cut a square of plastic wrap for use as a cover for the beaker.

Add distilled water to the beaker to a depth of approximately one-quarter of an inch.

Make certain that the ink spots on the filter paper are completely dry. If the spots are dry, fold the filter paper in half lengthwise.

Carefully lower the filter paper (with the spots at the bottom) into the water in the beaker. Make certain not to wet the spots as you lower the filter paper, and do not move the beaker to avoid sloshing water onto the spots. Cover the beaker with the plastic wrap.

Allow the water to rise in the filter paper until it reaches the upper pencil line. Then carefully remove the filter paper and set it on a clean paper towel. Quickly dry the filter paper using a heat-gun or hair dryer.

Determine R_f for each of the ink spots and record.

Obtain from your instructor several *permanent* ink markers (which are *not* water soluble). Repeat the chromatographic procedure using a new strip of filter paper and a 50% acetone-water mixture as the solvent (*Caution*). Determine and record the R_f values for the permanent inks.

Save your two chromatograms for submitting with your lab report.

Choice II. Thin-Layer Chromatography of Indicator Dyes

Introduction

Indicators are organic compounds that are typically used to signal a change in pH in acid/base titration analyses. Such indicators are dyes that exist in different colored forms at different pHs, and the change in color of the indicator is the signal that the titration analysis is complete. In most cases, indicator dyes are very *intensely* colored, and only a very tiny quantity of the indicator is needed.

In this experiment, you will perform a thin-layer chromatographic analysis of a mixture of the dyes bromcresol green, methyl red, and xylenol orange. These dyes have been chosen because they have significantly different retention factors, and a nearly complete separation should be possible in the appropriate solvent system. You will also investigate the effect of the solvent on TLC analyses, by attempting the separation in several different solvent systems.

In real practice, thin-layer chromatography has several uses. When a new compound is synthesized, for example, a TLC of the new compound is routinely done to make certain that the new compound is pure (a completely pure compound should only give a single TLC spot; impurities would result in additional spots). TLC is also used to separate the components of natural mixtures isolated from biological systems: for example, the various pigments in plants can be separated by TLC of an extract made by boiling the plant leaves in a solvent. Once the components of a mixture have been separated by TLC, it is even possible to isolate small quantities of each component by scraping its spot from the TLC slide and redissolving the spot in some suitable solvent.

Safety Precautions

- Wear safety glasses at all times while in the laboratory.
- The organic indicator dyes used in this experiment will stain skin if spilled; many such dies are toxic or mutagenic.
- The solvents used for the chromatographic separation are highly flammable and their vapors are toxic. No flames should be burning in the room while these solvents are in use. Use the solvents only in the exhaust hood.

Apparatus/Reagents Required

Bakerflex plastic TLC slides (1 ×4 inch),latex surgical gloves, ruler, pencil, plastic wrap or parafilm, micropipets, hot plate, heat gun, ethanolic solutions of the indicator dyes (methyl red, xylenol orange, bromocresol green), acetone, ethyl acetate, hexane, ethanol

Procedure

Clean and dry six 400-mL beakers to be used as the chambers for the chromatography. Obtain several squares of plastic wrap or parafilm to be used as covers for the beakers.

The chromatographic separation will be attempted in several solvent mixtures to investigate which gives the most complete resolution of the three dyes. All of these solvents are highly flammable: no flames should be open in the lab. A total of only 10–15 mL of each solvent mixture is necessary. Prepare mixtures of the solvents below, in the proportions indicated by volume, and transfer each to a separate 400-mL flask. Cover the flasks after adding the solvent mixture, and label the flask with the identity of the mixture it contains.

Acetone 60% / hexane 40%
Ethyl acetate 60% / hexane 40%
Acetone 50% / ethyl acetate 50%
Acetone 50% / ethanol 50%
Ethyl acetate 50% / ethanol 50%
Hexane 50% / ethanol 50%

Wearing plastic surgical gloves to avoid oils from the fingers, prepare six plastic TLC slides by marking *lightly* with pencil (not ink) a line across both the top and bottom of the slide. Do not mark the line too deeply or you will remove the coating of the slide. See Figure 9-1.

On one of the lines you have drawn on each slide, mark four small pencil dots (to represent where the spots are to be applied). *Above* the other line on each slide, mark the following letters: *R* (methyl red), *X* (xylenol orange), *G* (bromocresol green), and *M* (mixture). See Figure 9-1.

Figure 9-1. Plastic TLC slide with spots of the three dyes and the mixture applied. Keep the spots applied small.

Obtain small samples of the ethanol solutions of the three dyes (methyl red, xylenol orange, bromcresol green). Also obtain several micropipets: use a separate micropipet for each dye, and be careful not to mix up the pipets during the subsequent application of the dyes.

Apply a single small droplet of the appropriate dye to its pencil spot on each of the TLC slides you have prepared (wipe the outside of the micropipet if necessary before applying the drop to remove any excess dye solution). Keep the spots of dye as small as possible.

Apply one droplet of each dye to the spot labeled *M* (mixture) on each slide, being sure to allow each previous spot to dry before applying the next dye. Allow the spots on the TLC slides to dry before proceeding.

Gently lower one of the TLC slides, spots downward, into one of the solvent systems. Be careful not to wet the spots, or to slosh the solvent in the beaker; do not move or otherwise disturb the beaker after adding the TLC slide. Carefully cover the beaker with plastic wrap.

Allow the solvent to rise on the TLC slide until it reaches the upper pencil line (this will not take very long).

When the solvent has risen to the upper pencil mark, remove the TLC slide and quickly mark the exact solvent front before it evaporates. Mark the TLC slide with the identity of the solvent system used for development. Set the TLC slide aside to dry completely.

Repeat the process using the additional TLC slides and solvent systems. Be certain to mark each slide with the solvent system used.

Determine R_f for each dye in each solvent system and record. Which solvent system led to the most complete resolution of the dye mixture? If no mixture gave a complete resolution, your instructor may suggest other solvents for you to try, or other proportions of the solvents already used.

Save your TLC slides and staple to the lab report page for this experiment.

Name ______________________________ Section ______________________________

Lab Instructor ______________________________ Date ______________________________

EXPERIMENT 9 Resolution of Mixtures 2: Chromatography

PRE-LABORATORY QUESTIONS

CHOICE I. PAPER CHROMATOGRAPHY OF INKS

1. Use a chemical dictionary or encyclopedia to find a specific definition of *chromatography*.

____

____

____

____

2. Use a chemical encyclopedia to report the sorts of samples which might be analyzed by paper chromatography (as opposed to some other method).

____

____

____

____

3. On a paper chromatogram, a spot produced by a red ink traveled 5.2 cm, whereas the front of solvent used for the chromatographic separation traveled 8.9 cm during the experiment. Calculate R_f for the red ink.

____

____

____

____

____

____

CHOICE II. THIN-LAYER CHROMATOGRAPHY OF INDICATOR DYES

1. In preparing a TLC slide or filter paper for chromatography, a baseline is drawn for positioning the spots in pencil. Why is ink never used for drawing the baseline?

____

____

____

____

____

____

2. The indicator dyes used in this experiment are also used in acid/base titration analyses because they change color at particular values of pH. Use a handbook of chemical data to find the colors of each of these dyes under low and high pH ranges.

Methyl red ______________________________

Xylenol orange ______________________________

Bromocresol green ______________________________

3. TLC slides are most commonly coated with alumina or, less commonly, silicagel. Use an encyclopedia of chemistry or a handbook to find out the composition of each of these materials.

____

____

____

____

____

____

Name ______________________ Section ______________

Lab Instructor ______________________ Date ______________

EXPERIMENT 9 Resolution of Mixtures 2: Chromatography

CHOICE I. PAPER CHROMATOGRAPHY OF INKS

RESULTS/OBSERVATIONS

Water-soluble inks

Spot	Pen color	Identification number
a.	______________	______________
b.	______________	______________
c.	______________	______________
d.	______________	______________

Distance traveled by solvent front ______________________

Spot	Distance traveled by spot	R_f
a.	______________	________
b.	______________	________
c.	______________	________
d.	______________	________

Acetone-soluble inks

Spot	Pen color	Identification number
a.	______________	______________
b.	______________	______________
c.	______________	______________
d.	______________	______________

Distance traveled by solvent front ____________________

Spot	Distance traveled by spot	R_f
a.	____________	______
b.	____________	______
c.	____________	______
d.	____________	______

QUESTIONS

1. Why is a *mixture* of acetone and water used to separate the second set of inks (non water-soluble)?

2. Why was it necessary for you to wear plastic surgical gloves while preparing and developing your chromatogram?

3. List those samples which are single-component inks.

4. List those samples which are multicomponent inks, identifying the colors of the components used to give the overall color of the ink.

Name ______________________ Section ______________________

Lab Instructor ______________________ Date ______________________

EXPERIMENT 9 Resolution of Mixtures 2: Chromatography

CHOICE II. THIN-LAYER CHROMATOGRAPHY OF INDICATOR DYES

RESULTS/OBSERVATIONS

For each of the solvent mixtures studied, calculate R_f for each of the spots:

Acetone/hexane — Distance traveled by solvent front ______________________

	Distance traveled by spot	Calculated R_f
Xylenol orange	______________	______________
Bromcresol green	______________	______________
Methyl red	______________	______________

Ethyl acetate/hexane — Distance traveled by solvent front ______________________

	Distance traveled by spot	Calculated R_f
Xylenol orange	______________	______________
Bromcresol green	______________	______________
Methyl red	______________	______________

Acetone/ethyl acetate — Distance traveled by solvent front ______________________

	Distance traveled by spot	Calculated R_f
Xylenol orange	______________	______________
Bromcresol green	______________	______________
Methyl red	______________	______________

Acetone/ethanol — Distance traveled by solvent front ____________________

	Distance traveled by spot	Calculated R_f
Xylenol orange	____________	____________
Bromcresol green	____________	____________
Methyl red	____________	____________

Ethyl acetate/ethanol — Distance traveled by solvent front ____________________

	Distance traveled by spot	Calculated R_f
Xylenol orange	____________	____________
Bromcresol green	____________	____________
Methyl red	____________	____________

Hexane/ethanol — Distance traveled by solvent front ____________________

	Distance traveled by spot	Calculated R_f
Xylenol orange	____________	____________
Bromcresol green	____________	____________
Methyl red	____________	____________

Other mixture — Distance traveled by solvent front ____________________

	Distance traveled by spot	Calculated R_f
Xylenol orange	____________	____________
Bromcresol green	____________	____________
Methyl red	____________	____________

Which solvent mixture gave the most complete resolution of the three dyes? Which solvent mixture gave the poorest resolution?

____

____

QUESTIONS

1. Why is it important to keep the spots applied to filter paper for chromatography as small as possible?

2. Why is it necessary to keep the beaker used for chromatography tightly covered with plastic wrap while the solvent is rising through the TLC slide ?

3. Of the solvents used, some were very polar (e.g., acetone, ethanol) while others were very nonpolar (e.g., hexane). Did the polarity of the various solvent mixtures seem to affect the completeness of the separation of dyes? Why might this be so?

EXPERIMENT

10 Counting by Weighing

Objective

Chemists determine how many individual atoms or molecules are in a sample of matter from the *mass* of the sample. This activity will give you some insight into how this can be done.

Introduction

The concept of "counting by weighing" introduced in your textbook may be something new to you. Such an approach is a standard business practice, however, particularly when a business needs to take an inventory of small materials on hand. For example, if a company that makes carpentry nails had to count each individual item in their inventory the situation would be untenable. Rather than having to count each individual nail separately, the mass of a single nail is determined, and then the number of nails present in a bulk sample can be calculated easily. Similarly, in a high-volume "super" drugstore, a prescription for a large number of tablets or capsules of medication may be dispensed by this method.

In this experiment, you will demonstrate the process of "counting by weighing" for yourself using pennies as the item to be counted. This should give you some insight into why we can use average atomic masses in chemical reaction calculations. You will also investigate the effect that the presence of several isotopes has on the average atomic mass of an element.

Safety Precautions	• Wear safety glasses at all times while in the laboratory.

Apparatus/Reagents Required

25 pennies, balance, felt-tip pen

Procedure

A. Counting by Weighing

Obtain 25 pennies. In 1982, the composition of the United States penny coin was changed from one of nearly pure copper to a mostly zinc "sandwich" (the center of the coin is zinc, but the surfaces are layered in copper). Separate the pennies into two piles: those that have dates through 1981, and those that have dates 1983 and following. Since pennies minted in 1982 may be of either type, set aside any such pennies and do not use them in the rest of the experiment.

With a felt-tip pen, number each of the remaining pennies so that you can identify them.

1. Pre-1982 Pennies

Weigh each of the pre-1982 pennies (to the nearest 0.01 g) and record the masses on the data page.

Calculate the average mass of your pre-1982 pennies.

To get a truly representative sample, it would be useful if all the students in the laboratory combined their data on the average mass of the pre-1982 pennies. On the chalkboard of the laboratory, write your name, the average mass you determined for the pre-1982 pennies, and the number of pre-1982 pennies you used.

When all the students in the class have recorded their average masses on the chalkboard, your instructor will demonstrate how to calculate the average mass of the pre-1982 penny using everyone's data. Record this average mass on the data page.

Based on the average mass of the pre-1982 penny as determined from all the students' data, what would be the mass of 55 pennies? How many pre-1982 pennies are contained in a pile of pennies that has a total mass of 310. g?

2. Post-1982 Pennies

You separated the pennies into two groups above, based on their date of minting, and then determined the average mass of the pre-1982 pennies only. Although there may have been minor variations in the masses of the pre-1982 pennies due to different degrees of wear and tear, you should have found that most of the pennies had virtually the same mass. Post-1982 pennies, however, have masses that are considerably less.

Weigh each of your post-1982 pennies and record the masses on the data page.

Calculate the average mass of your post-1982 pennies.

On the chalkboard, write your name, the average mass of your post-1982 pennies, and the number of post-1982 pennies you used. After all the students in the lab have contributed their data on the post-1982 pennies, calculate the overall average mass of a post-1982 penny using all students' data.

Based on the average mass of the post-1982 penny, what would be the mass of 75 such pennies? How many such pennies would be contained in a pile of post-1982 pennies having a total mass of 250. g?

B. Effect of Isotopes on Average Atomic Masses

You have learned that most elements have several isotopic forms. The various isotopes of an element all have the same number of protons and electrons (so they are chemically the same), but differ in the number of neutrons present in the nucleus (which may result in slightly different physical properties).

You have also discussed how the average atomic mass listed for a particular element on the periodic table represents a weighted average of the masses of all the isotopes of the element. By "weighted average", we mean that the abundance of the element is reflected in the average atomic mass.

You can see what we mean by "weighted average" using the data you have collected for the pennies in Parts A.1. and A.2. above.

Using the individual masses of all the pennies (both pre- and post-1982), calculate the average mass of a penny (without regard to its minting date). Record.

You can arrive at the same average mass by another method of calculation, using the average mass you calculated for each type of penny, rather than the individual masses of all the pennies. Consider the following example:

A student has 5 pennies of average mass 3.11 g and 19 pennies of average mass 2.49 g. The weighted average mass of these 24 pennies is given by

$$[5(3.11 \text{ g}) + 19(2.49 \text{ g})]/24 = [15.55 \text{ g} + 47.31 \text{ g}]/24 = 2.62 \text{ g}$$

The weighted average mass (2.62 g) is closer to 2.49 g than it is to 3.11 g because there were more pennies of the lower mass present in the sample: the weighted average has included the relative abundance of the two types of pennies.

Using the method outlined in the example above, calculate the weighted average mass of a penny. Record this average mass both on the data page and on the chalkboard.

How do the average masses reported by the students in your class compare? Are there significant differences in the average masses reported, or are they all comparable?

Name ______________________________ Section ______________

Lab Instructor ______________________________ Date ______________

EXPERIMENT 10 Counting by Weighing

PRE-LABORATORY QUESTIONS

1. The concept of "counting by weighing" as described in your textbook is often a difficult one for students to appreciate. Describe in your own terms what is meant by this idea.

2. Give a mathematical example of how "counting by weighing" might be used in everyday life.

3. Give an example of how "counting by weighing" is used by chemists in studying chemical reactions.

Name ______________________________ Section ______________________

Lab Instructor ______________________________ Date ______________________

EXPERIMENT 10 Counting by Weighing

RESULTS/OBSERVATIONS

A. Counting by Weighing

1. Pre-1982 Pennies

Number	*Mass*	*Number*	*Mass*	*Number*	*Mass*
______	______	______	______	______	______
______	______	______	______	______	______
______	______	______	______	______	______
______	______	______	______	______	______
______	______	______	______	______	______
______	______	______	______	______	______

Average mass of your pre-1982 pennies ______________________________

Class average mass of pre-1982 pennies ______________________________

Mass of 55 pre-1982 pennies ______________________________

Number of pre-1982 pennies present in total mass of 310. g ______________________________

2. Post-1982 pennies

Number	*Mass*	*Number*	*Mass*	*Number*	*Mass*
______	______	______	______	______	______
______	______	______	______	______	______
______	______	______	______	______	______
______	______	______	______	______	______
______	______	______	______	______	______
______	______	______	______	______	______

Average mass of your post-1982 pennies ______________________________

Class average mass of post-1982 pennies ______________________________

Mass of 75 post-1982 pennies ______________________________

Number of post-1982 pennies present in total mass of 250. g ______________________________

B. Effect of Isotopes on Average Atomic Masses

Average mass of a penny based on average of individual masses ______________________________

Average mass of a penny based on "weighted average" method ______________________________

QUESTIONS

1. How do the average masses of a penny reported by your classmates compare to your own average mass? Are there significant differences in the average masses reported, or are they all comparable? What factor(s) might account for *differences* in the average masses reported by your classmates?

____

____

____

____

____

____

____

2. The element boron consists primarily of two isotopes. Use a handbook to look up the exact masses and relative abundances of these two isotopes, and show how the average atomic mass of boron would be calculated from this data.

____

____

____

____

____

EXPERIMENT

11 Stoichiometric Determinations

Objective

Stoichiometric measurements are among the most important in chemistry, indicating the proportions by mass in which various substances react. In this experiment, three examples of stoichiometric determinations will be investigated. The reactions to be studied are between two common acids (HCl, H_2SO_4) and sodium hydroxide (Choice I), sodium hydrogen carbonate and hydrochloric acid (Choice II), and magnesium and molecular oxygen, O_2 (Choice III).

Choice I. Stoichiometry and Limiting Reactant

Introduction

The concept of a limiting reactant is very important in the study of the stoichiometry of chemical reactions. The **limiting reactant** is the reactant that controls the amount of product possible for a process because once the limiting reactant has been consumed, no further reaction can occur.

Consider the following balanced chemical equation:

$$A + B \rightarrow C$$

Suppose we set up a reaction in which we combine 1 mol of substance A and 1 mol of substance B. From the stoichiometry of the reaction, and from the amounts of A and B used, we would predict that exactly 1 mol of product C should form.

Suppose we set up a second experiment involving the same reaction, in which we again use 1 mol of A, but this time we only use 0.5 mol of B. Clearly, 1 mol of product C will *not* form. According to the balanced chemical equation, substances A and B react in a 1:1 ratio, and with only 0.5 mol of substance B, there is not enough substance B present to react with the entire 1 mol of substance A. Substance B would be the *limiting reactant* in this experiment. There would be 0.5 mol of unreacted substance A remaining once the reaction had been completed, and only 0.5 mol of product C would be formed.

Suppose we set up a third experiment, in which we combine 1 mol of substance A with 2 mols of substance B. In this case, substance A would be the limiting reactant (which would control how much product C is formed), and there would be an excess of substance B present once the reaction was complete.

One way of determining the extent of reaction for a chemical process is to monitor the *temperature* of the reaction system. Chemical reactions nearly always either absorb or liberate heat energy as they occur, and the amount of heat energy transferred will be *directly proportional* to how much product has formed. In this experiment, you will monitor temperature changes as an indication of the extent of reaction using a simple thermometer.

In this experiment, you will prepare solutions of hydrochloric acid and sulfuric acid, and will determine the stoichiometric ratio in which these acids react with the base sodium hydroxide. You will perform several trials of these acid/base reactions, in which the amount of each reagent used is systematically varied between the trials. By monitoring the temperature changes that take place as the reaction occurs, you will have an index of the extent of reaction. The maximum extent of reaction will occur when the reactants have been mixed together in the correct stoichiometric ratio for reaction. If the reactants for a particular trial are *not* in the correct stoichiometric ratio, then one of these reactants will *limit* the extent of reaction, and will also limit the temperature increase observed during the experiment.

Safety Precautions

- Wear safety glasses at all times while in the laboratory.
- The acid and base solutions used in the experiment may be damaging to skin, especially if concentrated through evaporation of the water. Wash after handling. If spilled on the bench, consult with the instructor about clean-up.

Reagents/Apparatus Required

3.0 *M* solutions of HCl, H_2SO_4, and NaOH; plastic foam cup; thermometer

Procedure

Record all data and observations directly on the report pages in ink.

Check your pre-laboratory calculations with the instructor before preparing the solutions needed for this experiment.

A. Reaction Between HCl and NaOH

When preparing dilute acid/base solutions, slowly add the more concentrated stock solution to the appropriate amount of water with stirring.

Using 600-mL beakers for storage, prepare 500. mL each of 1.0 *M* NaOH and 1.0 *M* HCl, using the stock 3.0 *M* solutions available. Be sure to measure the amounts of concentrated acid or base and water required with a graduated cylinder.

Stir the solutions vigorously with a stirring rod for one minute to mix. Keep the solutions covered with a watch glass when not in use.

Frequently, diluting a concentrated solution with water will result in a temperature increase. *Allow the solutions to stand for 5–10 minutes until they have come to the same temperature* (within ±0.2°C). Be sure to rinse and wipe the thermometer before switching between solutions.

Obtain a plastic foam cup for use as a reaction vessel: an insulated cup is used so that the heat liberated by the reactions will not be lost to the room during the time frame of the experiment.

Using a graduated cylinder, measure 45 mL of 1.0 *M* HCl. Pour it into the plastic cup and determine the temperature of the HCl (to the nearest 0.2°C). Record this temperature on your report sheet.

Measure 5.0 mL of 1.0 *M* NaOH in a 10-mL graduated cylinder.

Add the NaOH to the HCl in the plastic cup all at once and *carefully* stir the mixture with the thermometer. Determine and record the *highest temperature reached* as the reaction occurs.

Rinse and dry the plastic cup.

Perform the additional reaction trials indicated in Table I on the report sheet. In each case, measure the amounts of each solution carefully with a graduated cylinder. For each trial, record the initial temperature of the solution in the plastic cup, as well as the highest temperature reached during the reaction.

B. Reaction Between H_2SO_4 and NaOH

When preparing dilute acid/base solutions, slowly add the more concentrated stock solution to the appropriate amount of water with stirring.

Using 600-mL beakers for storage, prepare 500. mL each of 1.0 *M* NaOH and 1.0 *M* H_2SO_4, using the stock 3.0 *M* solutions available. Be sure to measure the amounts of concentrated acid or base and water required with a graduated cylinder.

Stir the solutions vigorously with a stirring rod for one minute to mix. Keep the solutions covered with a watch glass when not in use.

Allow the solutions to stand for 5–10 minutes until they have come to the same temperature (within ±0.2°C). Be sure to rinse and wipe the thermometer before switching between solutions.

Using the procedure described earlier, perform the reaction trials indicated in Table II on the report sheet for H_2SO_4 and NaOH. In each case, measure the amounts of each solution carefully with a graduated cylinder. For each trial, record the initial temperature of the solution in the plastic cup, as well as the highest temperature reached during the reaction.

Interpretation of Results

Based on the volume and concentration of the solutions used in each experiment, calculate the number of moles of each reactant used in each trial. For example, if you used 25.0 mL of 3.0 *M* HCl solution, the number of moles of HCl present would be

$$25.0 \text{ mL} \times \frac{1 \text{ L}}{1000 \text{ mL}} \times 3.0\,M = 0.075 \text{ mol}$$

Record these values in Tables I and II on the report sheet as appropriate.

To most clearly demonstrate how the extent of reaction in each of the experiments is determined by the limiting reactant, prepare two **graphs** of your experimental data. One graph should be for the reaction between HCl and NaOH; the second graph should be for the reaction between H_2SO_4 and NaOH.

Set up your graphs so that the vertical axis represents the *temperature change* measured for a given trial. Set up the horizontal axis to represent the *number of moles of NaOH* used in the trial. Plot each of the data points with a sharp pencil.

You should notice that each of your graphs consists of a set of *ascending* points and a set of *descending* points. For each of these sets of points, use a ruler to draw the best straight line through the points.

The **intersection** of the lines through the ascending points and the descending points in each graph represents the maximum extent of reaction for the particular experiment. From the maximum point on your graph, draw a straight line down to the horizontal axis. Read off the horizontal axis the number of moles of NaOH (and acid) that have reacted at this point. According to your graph, in what stoichiometric ratio do HCl and NaOH react? According to your graph, in what stoichiometric ratio do H_2SO_4 and NaOH react?

For each of your graphs, indicate which points represent the acid as limiting reactant, and which points represent NaOH as limiting reactant. Staple your two graphs to the report pages.

Choice II. Stoichiometry of a Gas Evolution/Neutralization Reaction

Introduction

Several common anions, when acidified, evolve *gases*. The net ionic reactions of the carbonate, hydrogen carbonate, and sulfite ions are examples of this:

$$CO_3^{2-} + 2H^+ \rightarrow H_2CO_3 \rightarrow H_2O + CO_2(g)$$
$$HCO_3^- + H^+ \rightarrow H_2CO_3 \rightarrow H_2O + CO_2(g)$$
$$SO_3^{2-} + 2H^+ \rightarrow H_2SO_3 \rightarrow H_2O + SO_2(g)$$

Now consider the reactions

$$Na_2CO_3 + 2HCl \rightarrow 2NaCl + H_2O + CO_2(g)$$
$$NaHCO_3 + HCl \rightarrow NaCl + H_2O + CO_2(g)$$

These reactions are basically those given in the first two of the preceding net ionic reactions. Since one of the products of the reaction is a gas (CO_2), and a second of the products can be conveniently vaporized by heating (H_2O), the stoichiometry of the reaction can be studied by collecting and weighing the third product of the reaction (NaCl). A similar study could be made of the reaction of the sulfite ion, but since the product gas SO_2 is toxic and noxious, this will not be done in this experiment.

In this experiment, you will treat weighed samples of sodium carbonate and sodium hydrogen carbonate dropwise with dilute hydrochloric acid until the reactions are complete. You will then evaporate the water from the samples. The quantity of sodium chloride produced in each reaction will be determined and compared to the theoretical amount of NaCl that should have been produced from the stoichiometric ratios of the balanced chemical equations for the reactions.

The carbonate ion and the hydrogen carbonate ion are both bases in aqueous solution, and the reactions outlined earlier are typical acid/base reactions, in which one of the products of the reaction happens to be a *gas* under ordinary laboratory conditions. As in many acid/base reactions, an indicator—a dye that is sensitive to changes in pH—will be used to signal when the reactions are complete.

Safety Precautions

- Wear safety glasses at all times while in the laboratory.
- Hydrochloric and sulfuric acids are damaging to the skin and vapors of HCl are toxic. Keep theHCl in the exhaust hood.
- Considerable frothing will result as hydrochloric acid is added to the samples. Keep the casserole covered with a watch glass during addition of the acid. Beware of spattering.
- Evaporation of water from the samples must be done only over a boiling water bath or under a heating lamp. Dangerous spattering of hot salts will result if the evaporation is done over an open flame.

Apparatus/Reagents Required

Porcelain or glass evaporating dish, sodium carbonate, sodium bicarbonate, 3.0 *M* hydrochloric acid, Pasteur pipet/bulb, methyl red indicator

Procedure

Clean a 50–60-mm-diameter casserole dish with soap and water. If any solid material in the casserole cannot be removed with simple washing, consult with the instructor about other methods for cleaning. Rinse the casserole with distilled water and wipe dry with a towel.

On a wire gauze on a ringstand, heat the casserole on a low flame for 5 minutes to dry it. Move the flame occasionally during the heating so that all portions of the casserole are heated. Allow the casserole to cool completely to room temperature.

Weigh the casserole to at least the nearest 0.01 g and record. Reheat the casserole on the wire gauze for an additional 5 minutes and reweigh after cooling completely. If the second mass determined for the casserole differs from the first mass by more than 0.02 g, reheat and reweigh until constant mass is achieved (within 0.02 g).

Add about half a teaspoon of sodium carbonate, Na_2CO_3, to the casserole and reweigh (to the nearest 0.01 g). Record. Calculate the mass of sodium carbonate taken.

Moisten the sodium carbonate with 4–5 mL of distilled water and add 2 drops of methyl red indicator (the mixture will be yellow). Cover the casserole with a watch glass to catch any material that may spatter. Obtain about 25 mL of 3.0 *M* hydrochloric acid in a clean beaker.

Based on the mass of sodium carbonate taken, calculate the volume of 3.0 *M* hydrochloric acid that should be required to react with the sodium carbonate.

Transfer the calculated volume of 3.0 *M* hydrochloric acid from the beaker to a graduated cylinder.

When adding HCl to the sample in the casserole, use a medecine dropper, and add the HCl down the pouring spout of the casserole without removing the watch glass. The sodium carbonate will froth and fizz as carbon dioxide is generated, and the watch glass will prevent loss of solid. Begin adding 3.0 *M* hydrochloric acid dropwise to the casserole from the portion measured in the graduated cylinder.

Continue adding HCl with the dropper from the graduated cylinder until there is approximately 1 mL remaining in the graduate. During the initial addition of the HCl, the indicator may change to red. This may *not* signal completion of the reaction, however, because some carbon dioxide may remain in solution at this point, thereby affecting the pH of the mixture.

Transfer the casserole to the wire gauze/ringstand and heat with a low flame until the mixture *just begins to boil*. This heating is only to drive off carbon dioxide: do *not* attempt to boil off the water from the mixture at this point. As carbon dioxide is evolved on heating, the mixture should turn yellow again.

Add additional HCl dropwise from the graduated cylinder until the mixture in the casserole turns a permanent pale red.

Use a stream of distilled water from a plastic wash bottle to rinse any solids that may have collected on the bottom of the watch glass into the casserole.

Set up a 400- or 600-mL beaker that can accommodate the casserole on the wire gauze/ringstand, add about 300 mL of tap water, and bring the water to boiling to provide a steam bath.

Place the casserole in the mouth of the beaker of boiling water and begin heating to evaporate water. The evaporation of water will take a considerable amount of time, so go on to any other work that may have been assigned. Replace the water in the beaker as needed to maintain the steam bath.

Do not attempt to hurry the evaporation of water by using a direct flame on the casserole. As the mixture in the casserole becomes more concentrated by evaporation, it will tend to "bump" and spatter badly if directly heated, leading to loss of product. Spattering of hot solids is also dangerous.

After the *solid* (NaCl) in the casserole is almost completely dry, remove the casserole from the steam bath. Begin heating the casserole directly with a *very small* flame.

If the casserole contents begins to spatter, not enough water has been evaporated and the casserole should be returned to the steam bath for additional slow heating. If no spattering occurs, continue heating the casserole directly with a small flame for 5 minutes.

Increase the size of the flame somewhat, and continue heating the casserole for an additional 5-minute period to remove all moisture. Let the casserole cool completely to room temperature.

When the casserole has cooled completely, determine the mass of the casserole and contents (to the nearest 0.01 g). Calculate the mass of sodium chloride in the casserole.

Based on the mass of sodium carbonate taken originally, calculate the mass of sodium chloride that should in theory be produced by the reaction. Based on the actual yield of NaCl and the calculated theoretical yield, calculate the percent yield for your experiment.

If time permits and your instructor so directs, repeat the experiment, substituting sodium hydrogen carbonate (bicarbonate) for the sodium carbonate.

Choice III. Determination of a Formula

Introduction

Magnesium metal is a moderately reactive elementary substance. At room temperature, magnesium reacts only very slowly with oxygen and can be kept for long periods of time without appreciable oxide buildup. At elevated temperatures, however, magnesium will ignite in an excess of oxygen gas, burning with an intensely white flame and producing magnesium oxide. Because of the brightness of its flame, magnesium is used in flares and in photographic flashbulbs.

In this experiment, however, you will be heating magnesium in a closed container called a **crucible**, exposing it only gradually to the oxygen of the air. Under these conditions, the magnesium will undergo a more controlled oxidation, gradually turning from shiny metal to grayish-white powdered oxide. Because the air also contains a great deal of nitrogen gas, a portion of the magnesium being heated may be converted to magnesium nitride, Mg_3N_2, rather than magnesium oxide. Magnesium nitride will react with water and, on careful heating, is converted into magnesium oxide

$$Mg_3N_2 + 3H_2O \rightarrow 3MgO + 2NH_3$$

The ammonia produced by this reaction can be detected by its odor, which is released on heating the mixture.

Magnesium is a Group IIA metal, and its oxide should have the formula MgO. Based on this formula, magnesium oxide should consist of approximately 60% magnesium by weight. By comparing the weight of magnesium reacted, and the weight of magnesium oxide that results from the reaction, this will be confirmed.

Safety Precautions

- Wear safety glasses at all times while in the laboratory.
- Magnesium produces an intensely white flame if ignited, which may be damaging to the eyes. If the magnesium used in this experiment ignites in the crucible by accident, *immediately* cover the crucible and stop heating. Do not look directly at magnesium while it is burning.
- When water is added to the crucible to convert magnesium nitride to magnesium oxide, the contents of the crucible may spatter when heated. Use only gentle heating in evaporating the water. Do not heat the crucible strongly until nearly all the water has been removed.
- Use crucible tongs to handle the hot crucible and cover.
- Hydrochloric acid is damaging to skin and clothing. If it is spilled, wash immediately and inform the instructor.

Apparatus/Reagents Required

Porcelain crucible and cover, crucible tongs, clay triangle, magnesium turnings (or ribbon), pH paper, 6 *M* HCl

Procedure

Record all data and observations directly in your notebook in ink.

Obtain a crucible and cover and examine. The crucible and cover are extremely fragile and expensive. Use caution in handling them.

If there is any loose dirt in the crucible, moisten and rub it gently with a paper towel to remove the dirt. If dirt remains in the crucible, bring it to the hood, add 5–10 mL of 6 *M* HCl and allow the crucible to stand for 5 minutes. Discard the HCl and rinse the crucible with water. If the crucible is not clean at this point, consult with the instructor about other cleaning techniques, or replace the crucible. After the crucible has been cleaned, use tongs to handle the crucible and cover.

Set up a clay triangle on a ringstand. Transfer the crucible and cover to the triangle. The crucible should sit firmly in the triangle (the triangle's arms can be bent slightly if necessary).

Begin heating the crucible and cover with a small flame to dry them. When the crucible and cover show no visible droplets of moisture, increase the flame to full intensity, and heat the crucible and cover for 5 minutes.

Remove the flame, and allow the crucible and cover to cool to room temperature.

When the crucible and cover are completely cool, use tongs to move them to a clean dry watch glass or flat glass plate. Do not place the crucible directly on the lab bench. Weigh the crucible and cover to the nearest milligram (0.001 g).

Return the crucible and cover to the clay triangle. Reheat in the full heat of the burner flame for 5 minutes. Allow the crucible/cover to cool completely to room temperature.

Reweigh the crucible after it has cooled. If the weight this time differs from the earlier weight by more than 5 mg (0.005 g), reheat the crucible for an additional 5 minutes and reweigh when cool. Continue the heating/weighing until the weight of the crucible and cover is constant to within 5 mg.

Add approximately 1 teaspoon of magnesium turnings (or about 8 inches of magnesium ribbon coiled into a spiral) to the crucible.

Using tongs, transfer the crucible/cover and magnesium to the balance and weigh to the nearest mg (0.001 g).

Set up the crucible on the clay triangle with the cover very slightly ajar. (See Figure 11-1.) With a very small flame, begin heating the crucible gently.

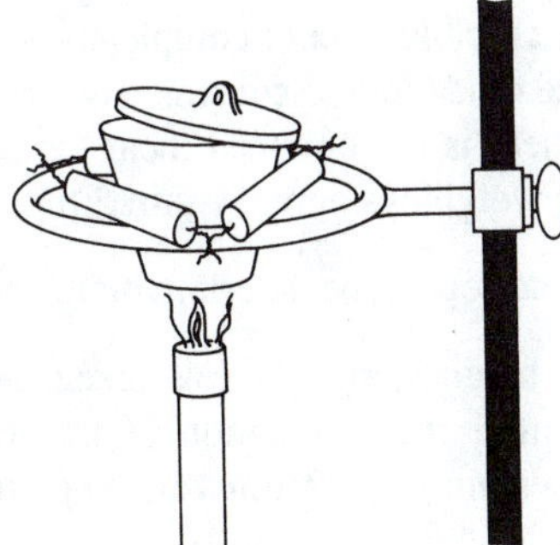

Figure 11-1. Set-up for oxidation of magnesium

If the crucible begins to smoke when heated, *immediately* cover the magnesium completely and remove the heat for 2–3 minutes. The smoke consists of the magnesium oxide product and must not be lost from the crucible.

Continue to heat gently for 5–10 minutes with the cover of the crucible slightly ajar. Remove the heat and allow the crucible to cool for 1–2 minutes.

Remove the cover and examine the contents of the crucible. If portions of the magnesium still demonstrate the shiny appearance of the free metal, return the cover and heat with a small flame for an additional 5 minutes; then reexamine the metal. Continue heating with a small flame until no shiny metallic pieces are visible.

When the shiny magnesium metal appears to have been converted fully to the dull gray oxide, return the cover to its slightly ajar position, and heat the crucible with the full heat of the burner flame for 5 minutes. Then slide the cover to about the half-open position and heat the crucible in the full heat of the burner flame for an additional 5 minutes.

Remove the heat and allow the crucible and contents to cool completely to room temperature. Remove the crucible from the clay triangle and set it on a sheet of clean paper on the lab bench.

With a stirring rod, gently break up any large chunks of solid in the crucible. Rinse any material that adheres to the stirring rod into the crucible with a few drops of distilled water. With a dropper, add about 10 drops of distilled water to the crucible, spreading the water evenly throughout the solid.

Return the crucible to the clay triangle, and set the cover in the slightly ajar position. With a very small flame, begin heating the crucible to drive off the water that has been added. Beware of spattering during the heating. Remove the flame and close the cover of the crucible if spattering occurs.

As the water is driven off, hold a piece of moistened pH paper (with forceps) in the stream of steam being expelled from the crucible. Any nitrogen that had reacted with the magnesium is driven off as ammonia during the heating and should give a basic response with pH paper (you may also note the odor of ammonia).

When it is certain that all the water has been driven off, slide the cover so that it is in approximately the half-open position, and increase the size of the flame. Heat the crucible and contents in the full heat of the burner for 5 minutes.

Allow the crucible and contents to cool completely to room temperature. When completely cool, weigh the crucible and contents to the nearest milligram (0.001 g).

Return the crucible to the triangle and heat for another 5 minutes in the full heat of the burner flame. Allow the crucible to cool completely to room temperature and reweigh. The two measurements of the crucible and contents should give weights that agree within 5 mg (0.005 g). If this agreement is not obtained, heat the crucible for additional 5-minute periods until two successive weighings agree within 5 mg.

Clean out the crucible, and repeat the determination.

Calculate the weight of magnesium that was taken, as well as the weight of magnesium oxide that was present after the completion of the reaction. Calculate the percentage of magnesium in the magnesium oxide from your experimental data. Calculate the mean for your two determinations.

Calculate the theoretical percentage of magnesium (by mass) in magnesium oxide, and compare this to the mean experimental value. Calculate the percent error in your determination.

Name ______________________ Section ______________________

Lab Instructor ______________________ Date ______________________

EXPERIMENT 11 Stoichiometric Determinations

PRE-LABORATORY QUESTIONS

CHOICE I. STOICHIOMETRY AND LIMITING REACTANT

1. Magnesium metal reacts with chlorine gas to produce magnesium chloride, $MgCl_2$.

a. Write the balanced chemical equation for the reaction.

____

b. If 5.00 g magnesium is combined with 10.0 g of chlorine, show by calculation which substance is the limiting reactant, and calculate the theoretical yield of magnesium chloride for the reaction.

____

____

____

____

____

____

____

2. How many mL of 3.0 *M* HCl are required to prepare 500. mL of 1.0 *M* HCl solution? Show the method of calculation clearly.

____

____

____

____

____

____

3. Consider the following data table, which is similar to the sort of data you will collect in this experiment. Complete the entries in the table. On graph paper from the end of this manual, construct a graph of the data, plotting the *temperature change* measured for each run versus the *number of moles* of substance A used for the run. Use your graph to determine the stoichiometric ratio in which substances A and B react. Explain your reasoning.

mL 1.0 *M* A	Moles of A	mL 1.0 *M* B	Moles of B	Temperature change, °C
5.0	________	45.0	________	1.7
10.0	________	40.0	________	3.4
15.0	________	35.0	________	5.2
20.0	________	30.0	________	6.8
25.0	________	25.0	________	8.5
30.0	________	20.0	________	6.7
35.0	________	15.0	________	5.1
40.0	________	10.0	________	3.3
45.0	________	5.0	________	1.8

CHOICE II. STOICHIOMETRY OF A GAS EVOLUTION/NEUTRALIZATION REACTION

1. If 5.00 g of pure Na_2CO_3 is treated with HCl solution, how many grams of NaCl will be produced if the water is subsequently evaporated from the salt?

____

____

____

____

2. If 5.00 g of pure $NaHCO_3$ is treated similarly, how many g of NaCl will result?

____

____

____

____

____

____

CHOICE III. DETERMINATION OF A FORMULA

1. Calculate the percentage by mass of magnesium and oxygen in magnesium oxide, MgO.

2. Suppose 2.033 g of magnesium is heated in air. What is the theoretical amount of magnesium oxide that should be produced?

Name ______________________ Section ______________________

Lab Instructor ______________________ Date ______________________

EXPERIMENT 11 Stoichiometric Determinations

CHOICE I. STOICHIOMETRY AND LIMITING REACTANT

RESULTS/OBSERVATIONS

A. Reaction Between Hydrochloric Acid and Sodium Hydroxide

Table I

mL 1.0 *M* HCl	Moles HCl	mL 1.0 *M* NaOH	Moles NaOH	Temperature change, °C
45.0	________	5.0	________	________
40.0	________	10.0	________	________
35.0	________	15.0	________	________
30.0	________	20.0	________	________
25.0	________	25.0	________	________
20.0	________	30.0	________	________
15.0	________	35.0	________	________
10.0	________	40.0	________	________
5.0	________	45.0	________	________

Intersection of ascending and descending portions of graph, mol NaOH ______________________

Stoichiometric ratio, mol HCl/mol NaOH at intersection ______________________

Balanced chemical equation for the reaction between HCl and NaOH

____

B. Reaction Between Sulfuric Acid and Sodium Hydroxide

Table II

mL 1.0 M H_2SO_4	Moles H_2SO_4	mL 1.0 M NaOH	Moles NaOH	Temperature change, °C
45.0	________	5.0	________	________
40.0	________	10.0	________	________
35.0	________	15.0	________	________
30.0	________	20.0	________	________
25.0	________	25.0	________	________
20.0	________	30.0	________	________
15.0	________	35.0	________	________
10.0	________	40.0	________	________
5.0	________	45.0	________	________

Intersection of ascending and descending portions of graph, mol NaOH ________________

Stoichiometric ratio, mol H_2SO_4/mol NaOH at intersection ________________

Balanced chemical equation for the reaction between H_2SO_4 and NaOH

____

QUESTIONS

1. Why was it important to wait until the solutions to be mixed had come to the same temperature?

____

____

____

2. Suppose a similar experiment had been done using 1.0 M H_3PO_4 solution as the acid. What would be the stoichiometric ratio, mol H_3PO_4/mol NaOH, at the intersection of the ascending and descending portions of the graph for such an experiment? Explain.

____

____

____

Name ______________________________ Section ______________________

Lab Instructor ______________________________ Date ______________________

EXPERIMENT 11 Stoichiometric Determinations

CHOICE II. STOICHIOMETRY OF A GAS EVOLUTION/NEUTRALIZATION REACTION

RESULTS/OBSERVATIONS

Reaction of Sodium Carbonate

Mass of casserole after first heating, g ______________________

Mass of casserole after second heating, g ______________________

Mass of casserole plus sodium carbonate, g ______________________

Mass of sodium carbonate taken, g ______________________

Approximate volume of 3 *M* HCl required, Ml ______________________

Mass of casserole and sodium chloride, g ______________________

Mass of sodium chloride formed, g ______________________

Theoretical yield of sodium chloride, g ______________________

Percentage yield of sodium chloride ______________________

Reaction of Sodium Hydrogen Carbonate

Mass of casserole after first heating, g ______________________

Mass of casserole after second heating, g ______________________

Mass of casserole plus sodium carbonate, g ______________________

Mass of sodium carbonate taken, g ______________________

Approximate volume of 3 *M* HCl required, mL ______________________

Mass of casserole and sodium chloride, g ______________________

Mass of sodium chloride formed, g ______________________

Theoretical yield of sodium chloride, g ______________________

Percentage yield of sodium chloride ______________________

QUESTIONS

1. What volume of 3 *M* HCl would be required to react with 5.00 g of a mixture that is 50.0% by mass sodium carbonate and 50.0% sodium hydrogen carbonate?

2. If small portions of NaCl were lost to spattering during the evaporation of the solvent in this experiment, would the percentage yield be higher or lower? Explain.

Name ______________________ Section ______________

Lab Instructor ______________________ Date ______________

EXPERIMENT 11 Stoichiometric Determinations

CHOICE III. DETERMINATION OF A FORMULA

RESULTS/OBSERVATIONS

	Trial 1	Trial 2
Weight of empty crucible (after first heating)	________	________
Weight of empty crucible (after second heating)	________	________
Weight of crucible with Mg	________	________
Weight of Mg taken	________	________
Weight of crucible/MgO (after first heating)	________	________
Weight of crucible/MgO (after second heating)	________	________
Weight of MgO produced	________	________
Weight of oxygen gained	________	________
% magnesium in the oxide	________	________

Mean % magnesium ______________________

Error ______________________%

QUESTIONS

1. If water had not been added to your initial product, what error in the percentage magnesium determined would have resulted (that is, if part of the product had been magnesium nitride)? Explain.

2. If the magnesium oxide smoke had been lost in too great an amount during the heating of the crucible, would this have made the calculated % Mg in the product too high or too low? Explain.

EXPERIMENT

12 Hydrates

Objective

Chemical compounds that contain discrete water molecules as part of their crystalline structure are called **hydrates**. Hydrates occur quite commonly among chemical substances, especially among ionic substances. More often than not, such compounds are either prepared in, or are recrystallized from, aqueous solutions. Hydrates exist for ionic compounds most commonly, but hydrates of polar and nonpolar covalent molecules are also known. In this experiment, you will study some of the properties and characteristics of several ionic hydrates.

Introduction

Hydrates are most commonly encountered in the study of metal salts, especially those salts of the *transition metals*. Water is bound in most hydrates in definite, *stoichiometric* proportions, and the number of water molecules bound per metal ion is often *characteristic* of a particular metal ion.

A very common hydrate often encountered in the general chemistry laboratory is copper(II) sulfate pentahydrate, $CuSO_4 \cdot 5H_2O$. The word "*penta*hydrate" in the name of this substance indicates that *five* water molecules are bound in this substance per copper sulfate formula unit. Hydrated water molecules are generally indicated in formulas as shown above for the case of the copper sulfate, using a *dot* to separate the water molecules from the formula of the salt itself.

Many hydrated salts can be transformed to the **anhydrous** (without water) compound by heat. For example, if a sample of copper sulfate pentahydrate is heated, the bright blue crystals of the hydrate are converted to the white, powdery, anhydrous salt.

$$\underset{\text{blue}}{CuSO_4 \cdot 5H_2O(s)} \rightarrow \underset{\text{white}}{CuSO_4(s)} + 5H_2O(g)$$

During the heating of copper sulfate pentahydrate, the water of crystallization is clearly seen escaping as steam from the crystals.

It is also possible to *reconstitute* the hydrate of copper sulfate: if water is added to the white anhydrous salt, the solid will reassume the blue color of the hydrated salt. Anhydrous salts are sometimes used as chemical drying agents, or **desiccants**, because of their ability to combine with and remove water from their surroundings. For example, most electronic equipment (and even at least one brand of ground coffee) comes packed with small envelopes of a desiccant to protect the equipment from moisture. Substances that absorb moisture and are able to be used as desiccants are said to be **hygroscopic.**

Not all hydrated salts are converted simply into the anhydrous compound when heated, however. Some hydrated metal salts will *decompose* upon losing the water of crystallization; subsequently they are usually converted to the metal *oxide* if the heating is carried out in air. Most covalent hydrates decompose rather than simply lose water when heated.

The water molecules contained within the crystals of a hydrate may be bound by several different means. For the case in which water molecules are bound with a metal salt, generally a nonbonding pair of electrons on the oxygen atom of the water molecule forms a coordinate covalent bond with empty, relatively low energy *d*-orbitals of the metal ion. In the case of copper sulfate pentahydrate, for example, four of the five water molecules form such coordinate bonds with the copper(II) ion. In other situations, the water molecules of the hydrate may be hydrogen-bonded to one or more species of the salt. This is especially common for covalently bonded hydrates.

Safety Precautions

- Wear safety glasses at all times while in the laboratory.
- Copper, cobalt, nickel, chromium, and barium compounds are all highly toxic. Wash hands after use.
- When you are heating the hydrated metal salts, they may *spatter* if heated too strongly. To avoid this, heat the solids with as *small* a flame as possible at first, and do not heat strongly with the full heat of the burner until most of the water has been driven from the hydrate. Make certain that the mouth of the test tube used to heat the hydrate is not pointed at yourself or anyone else.
- Dispose of the metal salts as directed by the instructor. Do not wash the salts down the drain, and do not place them in the wastebasket.

Apparatus/Reagents Required

Nickel(II) chloride hexahydrate, cobalt(II) chloride hexahydrate, copper(II) sulfate pentahydrate, chromium(III) chloride hexahydrate, anhydrous calcium chloride

Procedure

Record all data and observations directly in your notebook in ink.

Determine the mass of a clean, dry casserole or small evaporating dish to the nearest milligram (0.001 g). Add to the casserole or evaporating dish a spatula tipful of copper(II)

sulfate pentahydrate, $CuSO_4 \cdot 5H_2O$, and reweigh. Calculate the mass of $CuSO_4 \cdot 5H_2O$ taken. Record the appearance of the crystals.

Based on the mass of $CuSO_4 \cdot 5H_2O$ taken and its formula, calculate the *theoretical mass* of water that should be lost from the crystals if all the water were driven off.

Set up a wire gauze on a metal ring, and prepare to heat the casserole/evaporating dish in the burner flame. Begin the heating with a very small flame. If there is any evidence that the material is about to spatter, remove the heat immediately. Record any changes in appearance/color as the hydrate is heated.

When it is apparent that most of the water has been driven from the sample, increase the size of the flame. Stir the salt with a clean stirring rod until the sample is uniform in texture and appearance.

Remove the heat and allow the casserole to cool completely to room temperature. When the casserole has cooled completely, reweigh and calculate the mass of water driven off from the crystals. Using the theoretical mass loss calculated above, along with the experimentally determined mass loss, determine the percent error in your experiment.

After all mass determinations for the $CuSO_4 \cdot 5H_2O$ sample have been completed, add water dropwise to the sample. Record any changes in appearance/color.

For the hydrates listed below, first record the appearance of the crystals. Then transfer tiny amounts of the hydrates each to separate, clean borosilicate test tubes.

Using a test tube clamp to protect your hands, and making sure that the mouth of the test tube is not pointed at yourself or anyone else, heat each hydrate sample in turn and record any color changes or other changes in appearance that take place on heating. Allow the test tubes to cool completely to room temperature and then add a few drops of water to each test tube. Record any changes that take place on adding water. The hydrate samples to be used are: $NiCl_2 \cdot 6H_2O$; $CoCl_2 \cdot 6H_2O$; $CrCl_3 \cdot 6H_2O$.

As a vivid demonstration of the ability of anhydrous salts to absorb moisture, do the following: weigh an empty clean watch glass (to the nearest 0.01 g), then add about a teaspoon of *anhydrous* calcium chloride to the watch glass and reweigh. Examine the salt from time to time during the remainder of the lab period, and reweigh the watch glass and contents before leaving lab. Calcium chloride is an excellent desiccant and is able to absorb so much moisture from the air that it usually forms a solution of itself. A salt that absorbs such a great deal of water is said to be **deliquescent**. Calculate the mass of water absorbed by the anhydrous calcium chloride. Calculate what percentage of its own weight the $CaCl_2$ sample was able to absorb in moisture during the lab period.

Name ______________________ Section ______________________

Lab Instructor ______________________ Date ______________________

EXPERIMENT 12 Hydrates

PRE-LABORATORY QUESTIONS

1. Use a handbook of chemical data to find the number of water molecules bound per formula unit in the common hydrates of the following salts:

   Strontium chloride, $SrCl_2$ ______________________

   Sodium chromate, Na_2CrO_4 ______________________

   Nickel(II) nitrate, $Ni(NO_3)_2$ ______________________

   Iron(II) ammonium sulfate, $Fe(NH_4)_2(SO_4)_2$ ______________________

2. As described in the introduction to this experiment, when copper(II) sulfate pentahydrate is heated, the deep *blue* color of the hydrate changes to the *white* color of the anhydrous salt. Use the sections of your textbook discussing the chemistry of the transition elements to determine why such a vivid change in color is common when such elements' hydrated compounds are heated.

   ____

   ____

   ____

   ____

   ____

   ____

   ____

3. Suppose 2.3754 g of copper(II) sulfate pentahydrate is heated to drive off the water of crystallization. Calculate what weight of anhydrous salt will remain.

   ____

   ____

   ____

   ____

4. In $CuSO_4 \cdot 5H_2O$, it was mentioned that four of the five water molecules held per formula unit of the salt were attached by coordinate covalent bonds to the copper ion. The fifth water molecule is attached to the sulfate ion, but by a different mechanism. Use your textbook or a chemical encyclopedia to determine how a water molecule might be bonded to a sulfate ion.

Name ______________________ Section ______________________

Lab Instructor ______________________ Date ______________________

EXPERIMENT 12 Hydrates

RESULTS/OBSERVATIONS

Copper(II) sulfate pentahydrate

Mass of empty casserole/evaporating dish, g ______________________

Mass of casserole/evaporating dish plus $CuSO_4 \cdot 5H_2O$, g ______________________

Mass of $CuSO_4 \cdot 5H_2O$ taken, g ______________________

Appearance of $CuSO_4 \cdot 5H_2O$ ______________________

Theoretical mass loss expected on heating $CuSO_4 \cdot 5H_2O$ ______________________

Mass of casserole/evaporating dish after heating, g ______________________

Mass of water lost from $CuSO_4 \cdot 5H_2O$ crystals, g ______________________

% error in mass water lost from $CuSO_4 \cdot 5H_2O$, g ______________________

Appearance of $CuSO_4$ after heating ______________________

Appearance of $CuSO_4$ on adding water ______________________

Nickel(II) chloride hexahydrate

Observation before heating ______________________

Observation after heating ______________________

Observation on adding water ______________________

Cobalt(II) chloride hexahydrate

Observation before heating ______________________

Observation after heating ______________________

Observation on adding water ______________________

Chromium(III) chloride hexahydrate

Observation before heating ______________________

Observation after heating ______________________

Observation on adding water ______________________

Barium chloride dihydrate

Observation before heating ______________________

Observation after heating ______________________

Observation on adding water ______________________

Calcium chloride (anhydrous)

Observation on absorbing moisture from air ______________________

Mass of empty watch glass, g ______________________

Mass of watch glass plus anhydrous $CaCl_2$, g ______________________

Mass of watch glass plus $CaCl_2$ on standing, g ______________________

Mass of water absorbed, g ______________________

Percent water absorbed ______________________%

QUESTIONS

1. Use a chemical dictionary or your textbook to distinguish between the terms *desiccant*, *hygroscopic*, and *deliquescent*.

____

____

____

2. Sugars and starches belong to a class of biological compounds called *carbohydrates*, indicating that the general formula for such compounds is of the sort $(CH_2O)_n$. Use your textbook to find out why such compounds are not really hydrates of carbon as the family name suggests and record your findings here.

____

____

____

____

EXPERIMENT

13 Preparation and Properties of Some Common Gases

Objective

Two of the most common gaseous elemental substances (hydrogen and oxygen gases) will be prepared by chemical reaction. A gaseous compound substance (carbon dioxide) will also be prepared. In Choice I, hydrogen gas will be prepared by the action of acid on metallic zinc. The hydrogen will be collected by displacement of water. The flammability of pure hydrogen and of hydrogen/oxygen mixtures will be investigated. In Choice II, oxygen gas will be generated by the catalytic decomposition of hydrogen peroxide. The oxygen will be collected by displacement of water. The chemical properties of the oxygen collected will be examined. In Choice III, carbon dioxide will be generated by treatment of calcium carbonate with acid. The gas will be collected by water displacement. The properties of the carbon dioxide collected will be investigated.

General Technique for the Generation of the Gases

In each of the choices of this experiment, the gas generated will be collected by the technique of **displacement of water**. The gases to be generated—hydrogen, oxygen, and carbon dioxide—are not very soluble in water, so if they are bubbled from a closed reaction system into an inverted bottle filled with water, the water will be pushed from the bottle. The gas thus collected in the bottle will be saturated with water vapor, but is otherwise free of contaminating gases from the atmosphere.

Before beginning the specific choices assigned for your lab work, construct the gas generating apparatus shown in Figure 13-1 (it will be used for each of the choices). The tall vertical tube shown inserted into the Erlenmeyer flask is called a **thistle tube**. It is a convenient means of delivering a liquid to the flask. If a thistle tube is not available, a long-stemmed gravity funnel may be used.

When inserting glass tubing or the stem of your thistle tube/funnel through the two-hole stopper, use plenty of glycerine to lubricate the stopper, and protect your hands with a towel in case the glass breaks. Make certain that the two-hole stopper fits snugly in the mouth of the Erlenmeyer flask. Rinse all glycerine from the glassware before proceeding.

Set up a trough filled with water. Fill three or four gas bottles or flasks to the rim with water. Cover the mouth of the bottles/flasks with the palm of your hand and invert into the water trough. Set aside several stoppers or glass plates; use them to cover the mouths of the bottles/flasks once they have been filled with gas. In the following procedure, be sure to keep the gas collection bottles or flasks under the surface of the water in the trough until a stopper or glass plate has been placed over the mouth of the bottle/flask to contain the gas.

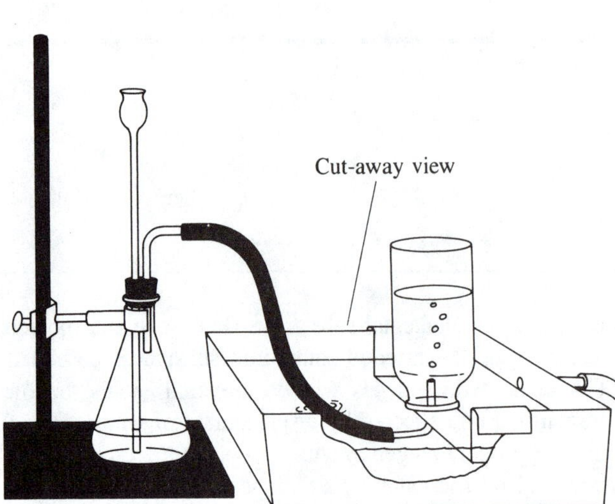

Figure 13-1. Apparatus for the generation and collection of a water-insoluble gas. A long-stemmed funnel may be substituted for the thistle tube shown. Be certain that the thistle tube or funnel extends to almost the bottom of the flask.

When the trough and gas collection equipment are ready, begin the evolution of gas by the specific method discussed for each choice. Liquids required for the reaction are added to the gas generating flask through the thistle tube or funnel. The liquid level in the Erlenmeyer flask must *cover the bottom* of the thistle tube/funnel during the generation of the gas, or gas will escape up the stem of the funnel rather than passing through the rubber tubing. Add additional liquid as needed to ensure this.

Gas should begin to bubble from the mouth of the rubber tubing as the chemical reaction begins. Allow the gas to bubble out of the rubber tubing for 2–3 minutes to sweep air out of the system. Add additional portions of liquid through the thistle tube/funnel as needed to continue the production of gas, and do not allow gas evolution to cease (or water may be sucked from the trough into the mixture in the Erlenmeyer flask).

After air has been swept from the system, collect three or four bottles full of gas by inserting the end of the rubber tubing into the bottles of water in the trough. Gas will displace the water from the bottles. Remember to keep the mouth of the bottles under the surface of the water in the trough.

When the bottles of gas have been collected, stopper them while they are still under the surface of the water; then remove them. If glass plates are used to contain the gas rather than stoppers, slide the plates under the mouths of the gas bottles while under water and remove the bottles from the water. For hydrogen only, keep the bottles in the inverted position after removing them from the water trough (hydrogen is less dense than air).

Choice I. Preparation and Properties of Elemental Hydrogen

Introduction

Elemental hydrogen is a colorless, odorless, tasteless gas. Hydrogen can be generated by many methods, most notably during the electrolysis of water or by replacement from acids by active metals:

$$2H_2O(l) \rightarrow 2H_2(g) + O_2(g)$$

$$M(s) + nH^+(aq) \rightarrow M^+(aq) + (n/2)\ H_2(g) \quad \text{(M is an active metal)}$$

Pure elemental hydrogen burns quietly itself, but forms *explosive mixtures* when mixed with air prior to ignition. Hydrogen gas is considerably less dense than air, and for this reason was used as the buoyant medium for airships in the earlier part of this century. Because hydrogen forms explosive mixtures with air, however, it was replaced by helium for this purpose after the explosion of the German airship *Hindenberg*: helium is considerably more expensive than hydrogen, but is not at all flammable.

Elemental hydrogen is used extensively in the commercial production of the substance ammonia, which is a vital component in the production of fertilizers, explosives, and other important chemical substances.

$$N_2 + 3H_2 \rightarrow 2NH_3$$

Typically, ammonia manufacturing plants are located near sources of inexpensive electricity (such as hydroelectric stations) so that elemental hydrogen can be produced for the synthesis of ammonia by the electrolysis of water.

Elemental hydrogen has been used as the reducing agent in the production of pure metallic substances from their oxide ores, but generally cheaper and easier to handle reducing agents are preferred.

$$MO(s) + H_2(g) \rightarrow M(s) + H_2O(g)$$

Elemental hydrogen will be produced in this experiment by treating metallic zinc with an excess quantity of hydrochloric acid. Since hydrogen gas is relatively insoluble in water, the hydrogen gas will be collected by displacement of water from an inverted bottle.

Safety Precautions

- Wear safety glasses at all times while in the laboratory.
- Elemental hydrogen is flammable and forms highly explosive mixtures with air. Although the quantity of hydrogen generated in this experiment is quite small, exercise caution when dealing with the gas.
- Hydrochloric acid is damaging to skin. Wash immediately if it is spilled and inform the instructor.
- Examine the bottles/flasks to be used to collect the hydrogen gas. Replace any bottles/flasks which are cracked, chipped, or otherwise imperfect.

Apparatus/Reagents Required

250-mL Erlenmeyer flask with tightly fitting two-hole stopper, glass tubing, thistle tube or long-stemmed funnel, rubber tubing, water trough, gas collection bottles (or flasks), stoppers to fit the gas collection bottles (or glass plates to cover the mouth of the bottles), wood splints, mossy zinc, 3 *M* hydrochloric acid, glycerine

Procedure

Record all data and observations directly in your notebook in ink.

A. Generation and Collection of Hydrogen

Construct the gas generator and water trough as indicated in Figure 13-1 for use in collecting hydrogen.

Add approximately a half-inch layer of mossy zinc chunks to the Erlenmeyer flask. Replace the two-hole stopper (with thistle tube/funnel and glass delivery tube) in the Erlenmeyer flask, and make certain that the stopper is set tightly in the mouth of the flask. Make certain that the stem of the funnel/thistle tube extends almost to the bottom of the flask.

When the trough and gas collection equipment are ready, begin the evolution of hydrogen by adding approximately 30 mL of 3 *M* hydrochloric acid to the zinc through the thistle tube or funnel. The liquid level must cover the bottom of the thistle tube/funnel or hydrogen will escape up the stem of the funnel rather than passing through the rubber tubing. Add additional HCl as needed to ensure this.

After sweeping air from the system for several minutes, collect three or four bottles of hydrogen. When the hydrogen has been collected, stopper the gas bottles under the surface of the water; then remove the bottles. Since hydrogen is lighter than air, keep the bottles inverted on your lab bench to prevent loss of the gas.

B. Tests on the Hydrogen

Bring two bottles of hydrogen to your instructor, who will peform two tests on the hydrogen for you. Be sure to wear your safety glasses at all times

To demonstrate that hydrogen forms explosive mixtures with air, the instructor will hold one bottle of hydrogen, mouth downward, and will remove the stopper or glass plate. The bottle of hydrogen will be tilted *momentarily* at a 45-degree angle to allow some hydrogen to escape from the bottle and a quantity of air to enter the bottle. The instructor will then hold the bottle of hydrogen/air mixture at arm's length, and will bring the mouth of the bottle near a burner flame. A loud "pop" will result as the hydrogen explodes.

The instructor will light a wooden splint. The instructor will then hold a second bottle of hydrogen in the inverted position, will remove the stopper, and will quickly insert the burning splint deeply into the bottle of hydrogen. Observe whether or not the splint continues to burn when deep in the hydrogen. The instructor will then slowly withdraw the splint to ignite the hydrogen as it escapes from the mouth of the bottle. If this is done carefully, the hydrogen should burn *quietly* with a pale blue flame (rather than exploding as in the first test).

Open the third bottle of hydrogen and turn the bottle mouth upward for 30 seconds. After the 30-second period, invert the bottle and attempt to ignite the gas in the bottle. There will be no "pop." Because hydrogen is considerably lighter than air, the hydrogen will have already escaped from the bottle.

Choice II. Preparation and Properties of Elemental Oxygen

Introduction

Elemental oxygen is essential to virtually all known forms of living creatures. Oxygen is used on the cellular level in the oxidation of carbohydrates. For example, the sugar glucose reacts with oxygen according to

$$C_6H_{12}O_6 + 6O_2 \rightarrow 6CO_2 + 6H_2O$$

This reaction is the major source of energy in the cell (the reaction is exergonic). The chief source of oxygen gas in the earth's atmosphere is the activity of green plants. Plants contain the substance chlorophyll, which permits the reverse of the preceding reaction to take place. Plants convert carbon dioxide and water vapor into glucose and oxygen gas. This reverse reaction is endergonic and requires the input of energy from sunlight (photosynthesis).

Elemental oxygen is a colorless, odorless, and practically tasteless gas. While elemental oxygen itself does not burn, it supports the combustion of other substances. Generally a substance will burn much more vigorously in pure oxygen than in air (which is only 20% oxygen by volume). Hospitals, for example, ban smoking in patient rooms in which oxygen is in use.

Elemental oxygen is the classic oxidizing agent after which the process of oxidation was named. For example, metallic substances are oxidized by oxygen, resulting in the production of the metal oxide. Iron rusts in the presence of oxygen, resulting in the production of the rust-colored iron oxides.

$$2Fe + O_2 \rightarrow 2FeO$$

$$4Fe + 3O_2 \rightarrow 2Fe_2O_3$$

Safety Precautions

- Wear safety glasses at all times while in the laboratory.
- Hydrogen peroxide and manganese(IV) oxide may be irritating to the skin. Wash after use.
- Oxygen gas supports vigorous combustion.
- Sulfur dioxide gas is toxic and irritating to the respiratory tract. Generate and use this gas only in the exhaust hood.

Apparatus/Reagents Required

250-mL Erlenmeyer flask with tightly fitting two-hole stopper, glass tubing, thistle tube or long-stemmed funnel, rubber tubing, water trough, gas collection bottles (or flasks), stoppers to fit the gas collection bottles (or glass plates to cover the mouth of the bottles), wood splints, deflagrating spoon, pH paper, 6% hydrogen peroxide solution, manganese(IV) oxide, sulfur, clean iron/steel nails, 1 *M* hydrochloric acid

Procedure

Record all data and observations directly in your notebook in ink.

A. Generation and Collection of Oxygen Gas

Construct the gas generator and set up the water trough as indicated in Figure 13-1 for use in collecting oxygen.

Add approximately 1 g of manganese(IV) oxide to the Erlenmeyer flask. Add also about 15 mL of water, and shake to wet the manganese(IV) oxide (it will *not* dissolve). Replace the two-hole stopper (with thistle tube/funnel and glass delivery tube) in the Erlenmeyer flask, and make certain that the stopper is set tightly in the mouth of the flask. Make certain that the stem of the funnel/thistle tube extends almost to the bottom of the flask.

When the trough and gas collection equipment are ready, begin the evolution of oxygen by adding approximately 30 mL of 6% hydrogen peroxide to the manganese(IV) oxide (through the thistle tube or funnel). Make certain that the liquid level in the flask covers the bottom of the stem of the thistle tube/funnel or the oxygen gas will escape from the system. Add more hydrogen peroxide as needed to ensure this.

The hydrogen peroxide should immediately begin bubbling in the flask, and oxygen gas should begin to bubble from the mouth of the rubber tubing. After sweeping air from the system for several minutes, collect three or four bottles of oxygen. Add 6% hydrogen peroxide as needed to maintain the flow of gas. When the oxygen has been collected, stopper the gas bottles under the surface of the water; then remove the bottles.

B. Tests on the Oxygen Gas

Ignite a wooden splint. Open the first bottle of oxygen, and slowly bring the splint near the mouth of the oxygen bottle in an attempt to ignite the oxygen gas as it diffuses from the bottle. Is oxygen itself flammable?

Ignite a wooden splint in the burner flame. Blow out the flame, but make sure that the splint still shows some glowing embers. Insert the glowing splint deeply into a second bottle of oxygen gas. Explain what happens to the splint.

Take the third stoppered bottle of oxygen to the exhaust hood. Place a small amount of sulfur in the bowl of a deflagrating spoon, and ignite the sulfur in a burner flame in the hood. Remove the stopper from the bottle of oxygen, and insert the spoon of burning sulfur.

Allow the sulfur to burn in the oxygen bottle for at least 30 seconds. Then add 15–20 mL of distilled water to the bottle, and stopper. Shake the bottle to dissolve the gases produced by the oxidation of the sulfur. Test the water in the bottle with pH paper. Write the equation for the reaction of sulfur with oxygen, and explain the pH of the solution measured.

Clean an iron (not steel) nail free of rust by dipping it briefly in 1 *M* hydrochloric acid; rinse with distilled water. If the surface of the nail still shows a coating of rust, repeat the cleaning process until the surface of the nail is clean.

Using tongs, heat the nail in the burner flame until it glows red. Open the fourth bottle of oxygen and drop the nail into the bottle. After the nail has cooled to room temperature, examine the nail. What is the substance produced on the surface of the nail?

Choice III. Preparation and Properties of Carbon Dioxide

Introduction

The amount of carbon dioxide in the earth's atmosphere is still relatively small but is increasing at an alarming rate. Carbon dioxide is a gaseous product of the combustion of petroleum-based fuels, and with the continued dependency of civilization on these fuels, the amount of carbon dioxide discharged into the atmosphere each year continues to increase. Although carbon dioxide is odorless, tasteless, colorless, and nontoxic, when present in the atmosphere, it enables radiant energy from the sun to be trapped on the earth as heat. In the past few decades, the mean temperature of the earth has increased by several degrees. If this increase in temperature continues, such consequences as partial melting of the polar ice caps could follow, which in turn would lead to serious flooding of low-lying coastal areas. The increase in temperature of the earth due to increasing levels of CO_2 is called the **greenhouse effect**.

Carbon dioxide is also produced by living creatures as an end product of the metabolism of carbohydrates. For example, when the sugar glucose is metabolized in cells, the reaction is

$$C_6H_{12}O_6 + 6O_2 \rightarrow 6H_2O + 6CO_2$$

Nature provided for a balance of carbon dioxide in the atmosphere. Whereas animals use oxygen from the atmosphere and produce carbon dioxide as a waste product, green plants are able to use carbon dioxide (through photosynthesis) and release oxygen gas. However, the quantity of carbon dioxide released from combustion of hydrocarbon fuels is too large for green plants to use fully. Also, the fraction of the earth's surface covered by green plants decreases each year as civilization spreads, as does the quantity of plankton in the earth's oceans.

In the laboratory, carbon dioxide can be generated chemically by treatment of a carbonate or hydrogen carbonate (bicarbonate) salt with acid. For example, if sodium hydrogen carbonate or sodium carbonate is treated with hydrochloric acid, the reactions are

$$NaHCO_3 + HCl \rightarrow NaCl + H_2O + CO_2$$

$$Na_2CO_3 + 2HCl \rightarrow 2NaCl + H_2O + CO_2$$

Carbon dioxide finds use in many situations. The most common type of fire extinguisher contains pressurized carbon dioxide (CO_2 does not burn and does not support combustion). The solid form of carbon dioxide (called dry ice) is used in the food industry to keep perishables frozen during shipment. The beverage industry uses carbon dioxide to carbonate sodas.

Safety Precautions

- Wear safety glasses at all times while in the laboratory.
- Hydrochloric acid is damaging to skin. If it is spilled, wash immediately and inform the instructor.
- Calcium hydroxide (limewater) is caustic and damaging to skin. If the solution is spilled, wash immediately and inform the instructor.
- Excess levels of carbon dioxide in the atmosphere of the lab may cause lightheadedness and cardiac distress. Make certain that the room is well ventilated, or perform the generation of CO_2 in the exhaust hood.
- Dry ice will cause frostbite if handled with the fingers. Use tongs or forceps to pick up the dry ice.

Apparatus/Reagents Required

250-mL Erlenmeyer flask with tightly fitting two-hole stopper, glass tubing, thistle tube or long-stemmed funnel, rubber tubing, water trough, gas collection bottles (or flasks), stoppers to fit the gas collection bottles (or glass plates to cover the mouth of the bottles), wood splints, pH paper, calcium carbonate, 3 *M* hydrochloric acid, glycerine, saturated calcium hydroxide solution (limewater), drinking straw

Procedure

Record all data and observations directly in your notebook in ink.

A. Generation and Collection of Carbon Dioxide

Construct the gas generator and set up the water trough as indicated in Figure 13-1 for use in collecting carbon dioxide.

Add approximately 5 g of calcium carbonate (chalk) to the Erlenmeyer flask. Add about 10 mL of distilled water and shake to make a slurry (the salt will *not* dissolve). Replace the two-hole stopper (with thistle tube/funnel and glass delivery tube) in the Erlenmeyer flask, and make certain that the stopper is set tightly in the mouth of the flask.

When the trough and gas collection equipment are ready, begin the evolution of carbon dioxide by adding approximately 30 mL of 3 *M* hydrochloric acid to the calcium carbonate through the thistle tube or funnel. Make certain that the liquid level covers the bottom of the stem of the thistle/tube funnel, or carbon dioxide will escape from the system. Add additional HCl to ensure this.

The calcium carbonate should start bubbling immediately in the flask, and carbon dioxide gas should begin to bubble from the mouth of the rubber tubing. After sweeping air from the system for several minutes, collect three or four bottles of carbon dioxide. Add 3 *M* hydrochloric acid as needed to maintain the flow of gas. When the carbon dioxide has been collected, stopper the gas bottles under the surface of the water; then remove the bottles.

B. Tests on the Carbon Dioxide

Ignite a wooden splint. Open a bottle of carbon dioxide and slowly bring the burning splint near the mouth of the bottle to try to ignite the carbon dioxide as it mixes with air. Does carbon dioxide itself burn?

Ignite a wooden splint. Open a second bottle of carbon dioxide and insert the burning splint deeply into the bottle. Does carbon dioxide support combustion?

Open a third bottle of carbon dioxide; *quickly* add 25 mL of distilled water to the bottle and restopper. Shake the bottle for 5 minutes to allow carbon dioxide to dissolve in the water. Open the bottle and test the carbon dioxide solution with pH paper. Why is the solution acidic?

Open the fourth bottle of carbon dioxide and add 25 mL of clear limewater solution (saturated calcium hydroxide). Stopper the bottle *quickly* and shake for several minutes. What is the identity of the solid that forms in the solution?

Place about 25 mL of limewater in a beaker. Obtain a clean drinking straw, take a deep breath, and exhale through the straw into the limewater solution. Why does a precipitate form in the limewater?

Place about 25 mL of limewater in a beaker. Pick up a small piece of dry ice with tongs or forceps; add it to the limewater. Let the beaker stand until the dry ice has completely disappeared. Why does a precipitate form in the limewater?

Name ______________________ Section ______________________

Lab Instructor ______________________ Date ______________________

EXPERIMENT 13 Preparation and Properties of Some Common Gases

PRE-LABORATORY QUESTIONS

1. In this experiment, various laboratory techniques are used to generate small quantities of some common gases. Use your textbook or a chemical encyclopedia to look up the major industrial methods of preparing the three gases (hydrogen, oxygen, and carbon dioxide) when very large quantities of the gases are needed. Summarize your findings here.

____

____

____

____

____

2. When collecting hydrogen gas in this experiment, you are cautioned to keep the bottles or flasks in the *inverted* position because hydrogen is lighter than air. Use a handbook of chemical data or the ideal gas law to determine the densities of hydrogen, oxygen, nitrogen, and carbon dioxide gases at 25°C and 1 atm pressure. Determine also the density of air (approximately 80% nitrogen, 20% oxygen by volume) under these conditions, and compare the density of hydrogen to the densities of each of the other gases.

____

____

____

____

____

____

____

____

____

____

Name ______________________________ Section ______________________

Lab Instructor ______________________________ Date ______________________

EXPERIMENT 13 Preparation and Properties of Some Common Gases

RESULTS/OBSERVATIONS

Choice I. Hydrogen

Observation on adding acid to zinc

____

Test on igniting hydrogen/air mixture

____

Test on igniting pure hydrogen

____

Test on igniting hydrogen after inverting bottle

____

Choice II. Oxygen

Observation on adding hydrogen peroxide to MnO_2

____

Test on attempting to ignite oxygen

____

Test on inserting glowing splint into oxygen

____

Test on adding burning sulfur to oxygen

____

pH of sulfur/oxygen product solution

Equation for sulfur/oxygen reaction

Test on adding glowing iron nail to oxygen

Equation for iron/oxygen reaction

Choice III. Carbon Dioxide

Observation on adding acid to calcium carbonate

Test on trying to ignite carbon dioxide

Test on inserting burning splint into carbon dioxide

---

pH of carbon dioxide/water solution

Why is the CO_2/water solution acidic?

Test on shaking limewater with CO_2

Test on exhaling into limewater

Test on adding dry ice to limewater

____

What is the precipitate that forms with CO_2/limewater mixtures?

____

Write the balanced equation for the reaction.

____

QUESTIONS

1. Hydrochloric acid with metallic zinc was used to generate hydrogen gas. Was this reaction *specific* with HCl, or could *any* acid have been used?

____

____

____

2. What was the purpose of the manganese(IV) oxide added to the hydrogen peroxide in the generation of oxygen? Could any other substances have been used for this purpose?

____

____

____

3. Why do gases become saturated with water vapor when collected by the method used in this experiment? Suggest a method by which such water vapor could be removed from the gases.

____

____

____

EXPERIMENT

14 The Gas Laws

Objective

The gas laws describe the behavior of ideal gas samples under different pressure, volume, and temperature conditions. Boyle's law (pressure/volume) and Charles's law (volume/temperature) will be investigated. Graham's law, which describes the rates at which ideal gases will diffuse/effuse, will also be studied.

According to Boyle's law, the volume of a sample of ideal gas varies *inversely* with the pressure on the gas sample (at constant temperature). This means, in simple words, if you squeeze harder on a sample of gas, the size of the sample will decrease. In Choice I, a sample of air will be trapped under a column of mercury in a constant-bore glass tube. By adjusting the height of the column of mercury, the pressure on the trapped air sample can be varied. Also, the effect of the pressure on the volume of the gas can be determined by measuring the change in height of the gas sample in the tube.

According to Charles's law, the volume of a sample of ideal gas varies *directly* with the absolute temperature of the gas sample (at constant pressure). This simply means that if you heat a sample of gas, the sample will increase in size. In Choice II, a sample of air will be trapped in a capillary tube beneath a small droplet of mercury. The gas sample will be heated to the temperature of a boiling water bath and then will be allowed to cool spontaneously. The volume of the gas sample will be determined at regular temperature intervals, and the relationship between volume and temperature will be confirmed graphically. The graph will be extrapolated to determine its intercept with the volume axis (absolute zero).

Graham's law, which describes the relative rates at which gases diffuse/effuse, will be investigated in Choice III for the gases ammonia and hydrogen chloride. These gases will be diffused simultaneously into opposite ends of a glass tube, and the appearance of a ring of ammonium chloride will be used to indicate the distances traveled by the gases.

Choice I. Boyle's Law

Introduction

By careful measurement of the volumes of gas samples under different pressure conditions, Sir Robert Boyle determined that the volume of a sample of ideal gas varied inversely with the pressure of the gas sample (at constant temperature). If the pressure on a sample of gas is *increased*, for example, the volume of the sample of gas *decreases*. If the pressure on a sample of gas is decreased, the volume of the sample increases in proportion to the change in the pressure. In particular, if the pressure on a sample of gas is doubled, the volume of the sample will become half of what it had been before the pressure was increased.

Various mathematical statements can be written to describe a gas that follows Boyle's law. Some of these are:

$$P \times V = \text{constant (at constant } T)$$

$$P = \text{constant}/V$$

$$P_1 \times V_1 = P_2 \times V_2$$

Each of these statements corresponds to the type of mathematical function known as an **hyperbola**: the algebraic form of the general equation for an hyperbola is $xy = c$.

In this experiment, a sample of gas will be trapped in a sealed constant-diameter glass tube (a buret) beneath a column of mercury. By increasing the height of the mercury column, the pressure on the sample of gas can be increased. Since pressures of gas samples are commonly measured with reference to units of millimeters of mercury (mm Hg, or torr) as derived from the common mercury barometer, the height of the mercury column above the gas in millimeters, combined with the pressure of the atmosphere above the column of mercury, is a direct measure of the pressure of the gas sample. Thus, measuring the height of the column of mercury will provide information about the pressure and volume of the gas sample in the sealed glass tube. Since the glass tube containing the gas sample is of constant diameter, it is geometrically a regular cylinder. The *volume* of a regular cylinder is a constant times the *height* of the cylinder.

Because the apparatus for the measurement of gas volume is open to the atmosphere above the column of mercury, the pressure of the atmosphere is also being applied to the sample of gas and will have to be measured. The most common instrument for the measurement of atmospheric pressure is the **mercury barometer**. The barometer (see Figure 14-1) consists of a glass tube, sealed at one end, that has been filled with mercury and then inverted into a reservoir of mercury. The mercury does not fall out of the glass tube completely because the reservoir into which the tube has been inverted is open to the atmosphere. The pressure of the atmosphere is sufficient to hold most of the mercury in the inverted tube. However, as the pressure of the atmosphere changes from day to day with the weather, the exact height to which the mercury level is held in the tube varies. The height of the mercury in the tube is taken as a direct measurement of the atmospheric pressure at any time and is quoted in units of millimeters of mercury. The average pressure of the atmosphere can support a column of mercury to a level of approximately 760 mm. During periods of clear weather (high pressure), the mercury level in the barometer will be above 760 mm; during periods of stormy weather (low pressure), the mercury level will be below 760 mm. Radio and television stations usually report the barometric pressure in inches of mercury; 760 mm is approximately equivalent to 30 inches.

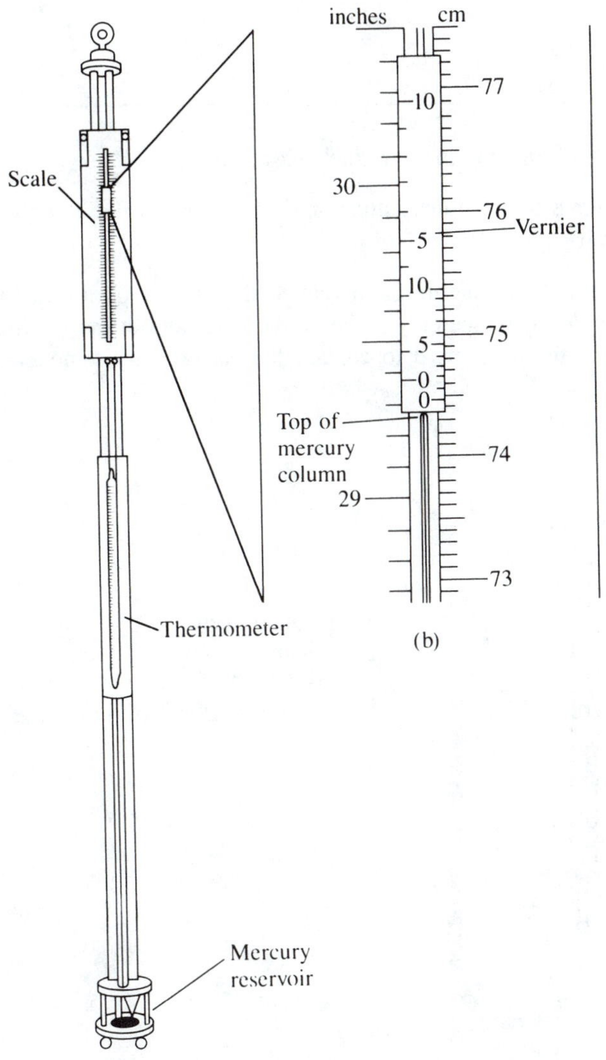

Figure 14-1. The mercury barometer. One standard atmosphere supports a column of mercury to a height of 760 mm. The vernier scale in (b) reads the pressure to a fraction of a millimeter.

Safety Precautions

- Wear safety glasses at all times while in the laboratory.
- Mercury is easily spilled, and its vapor is extremely toxic. Mercury can also be absorbed through the skin. For these reasons, Choice I will be performed with the apparatus in a trough in the exhaust hood. The instructor will manipulate the mercury reservoir for you. If any mercury is spilled, the mercury must be cleaned up by suction or a "mercury sponge."
- Use glycerine liberally when inserting glass tubing through rubber stoppers. Protect your hands with a towel. Wash the glycerine off before using the glass.

Apparatus/Reagents Required

Buret, rubber tubing, leveling bulb/mercury reservoir, barometer, meter stick, one-hole stopper to fit mouth of the buret, rubber bands

Procedure

Record all data and observations directly in your notebook in ink.

Your *instructor* will manipulate the pressure/volume apparatus for you. Do *not* attempt to operate the apparatus yourself.

The measurements that need to be taken are the *height* of the gas sample in the buret as well as the *difference in height* of the mercury levels in the buret and in the leveling bulb (see Figure 14-2). Therefore, the buret used to contain the gas sample is mounted on a meter stick.

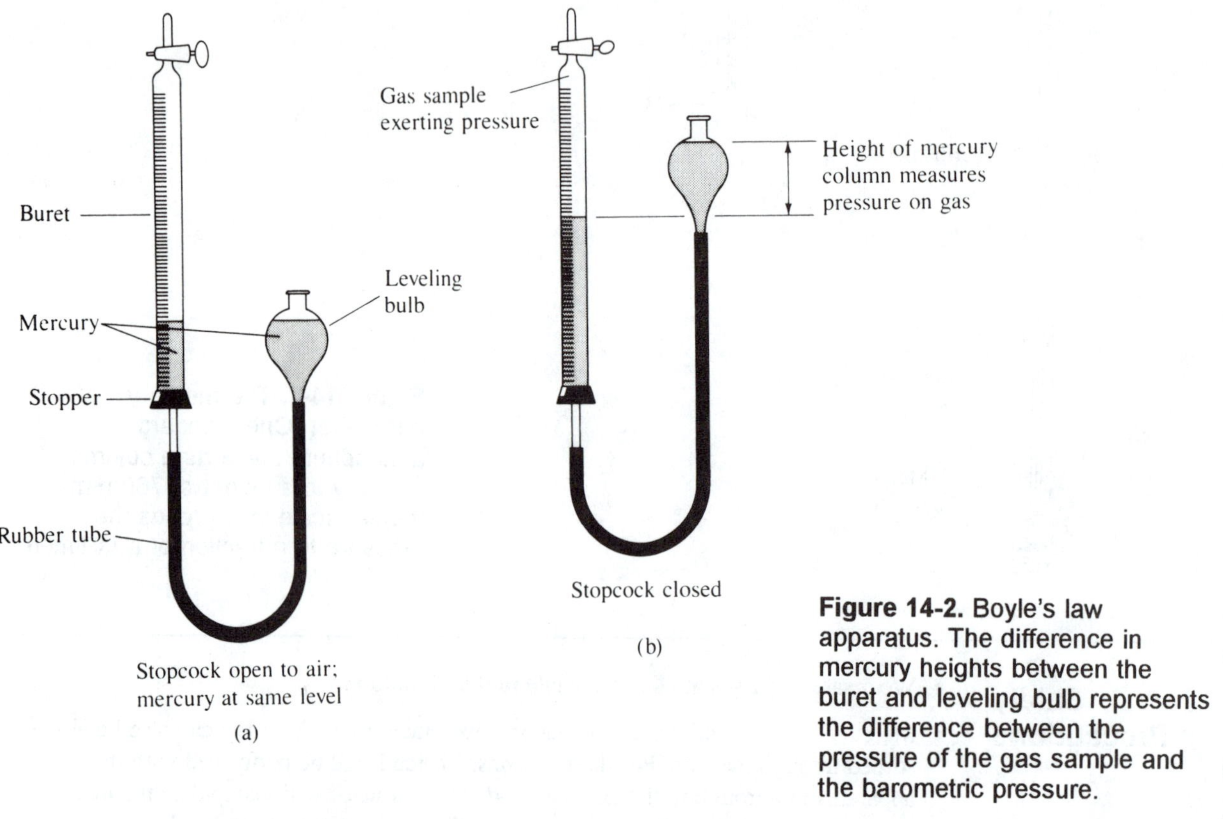

Figure 14-2. Boyle's law apparatus. The difference in mercury heights between the buret and leveling bulb represents the difference between the pressure of the gas sample and the barometric pressure.

The height of the gas sample, which is directly proportional to the volume of the sample, is measured from the *top of the mercury meniscus* at the bottom of the gas sample to the *top of the inner chamber of the buret* (directly beneath the stopcock when the buret is in the inverted position).

Your first measurement will be of the *initial* volume of the gas when the system is *open to the atmosphere.* Open the stopcock at the top of the buret. The instructor will then move the leveling bulb of mercury so that the level of mercury in the leveling bulb matches the level of mercury in the buret. While the two levels of mercury match, determine the height of the gas sample in the buret to the nearest mm by reference to the meter stick. When the height reading has been taken, *close the stopcock* on the buret.

The instructor will now move the leveling bulb to a higher position. When the bulb is raised, the gas sample is subjected to a higher pressure because of the difference in heights of the mercury columns in the leveling bulb and in the buret. The volume of the gas sample will decrease because of the additional pressure. After the instructor has moved the leveling bulb, and the system has stood for several minutes, determine the height of the gas sample in the buret to the nearest mm. Also determine the *difference* in height of the mercury columns.

The instructor will then move the leveling bulb to a higher position three more times. Each time the bulb is moved, record the height of the gas sample and the difference in the mercury levels. The instructor may have to add mercury to the leveling bulb to keep the mercury surface within the limits of the bulb itself.

Read the barometric pressure in the laboratory.

From your data, calculate the total pressure exerted on the gas sample at each position of the mercury leveling bulb. The total pressure is equal to the barometric pressure in the laboratory plus the additional pressure exerted by the column of mercury (i.e., the difference in the height of the mercury columns between the leveling bulb and the buret).

Plot a graph of volume (height) versus pressure for your data.

Calculate the *reciprocal* of the pressures ($1/P$) for each of your experimental determinations. Plot a graph of volume versus the reciprocal pressure for your data. Which of the graphs is a straight line?

Choice II. Charles's Law

Introduction

The volume of a sample of an ideal gas is *directly proportional* to the absolute temperature of the gas sample (at constant pressure). The first study of the relationship between gas volumes and temperatures was made in the late eighteenth century by Jacques Charles. Charles's original statement describing his observations was that the volume of a gas sample decreased by the same factor for each degree Celsius (centigrade) the temperature of the sample was lowered. Specifically, Charles found that the volume of a gas sample decreased by 1/273 of its volume for each degree the temperature of the sample was lowered.

The fact that the volume of a gas sample decreases in a regular way when the temperature drops led scientists to wonder what would happen if a gas sample were cooled *indefinitely*. If the gas sample's volume continued to decrease each time the temperature was lowered, then eventually the volume of the gas sample should be so small that lowering its temperature any further would cause the sample of gas to disappear. The temperature at which the volume of an ideal gas would be predicted to approach zero as a limit is called the **absolute zero** of temperature. Absolute zero is the lowest possible temperature. Absolute zero formed the basis for a scale of temperature (Kelvin or absolute), which has zero as its lowest point with all temperatures positive relative to this. The size of the degree on the Kelvin temperature scale was chosen to correspond to that of the Celsius scale. Absolute zero corresponds to –273.15°C; it is a theoretical temperature. Real gases would never reach zero volume but rather would liquefy before reaching this temperature.

Various mathematical statements of Charles's gas law have been made. Some of these are

$$V = \text{constant} \times T \quad (\text{at constant } P)$$

$$V/T = \text{constant}$$

$$V_1/T_1 = V_2/T_2$$

Mathematically, a graph of volume versus temperature should be a *straight line*. The intercept of this line with the volume axis (i.e., $V = 0$) represents absolute zero. In this experiment, you will measure the volume of a sample of gas at several temperatures that are easily achieved in the laboratory and will plot the data obtained. By *extrapolation*, the intercept of this graph with the volume axis will be calculated.

The gas sample to be determined will be contained in a small glass **capillary tube** beneath a droplet of mercury. Because mercury is a liquid, the droplet will be free to move up and down in the capillary tube as the gas sample is heated or cooled. Because the capillary is of constant diameter, the linear height of the gas sample beneath the mercury drop can be used as a direct index of the gas's volume.

Safety Precautions

- Wear safety glasses at all times while in the laboratory.
- Mercury and its vapor are very toxic. Mercury is absorbed through the skin and should not be handled. Work with mercury in the exhaust hood.
- The capillary containing the gas sample is extremely fragile, and if broken, the mercury contained may be released. If the tube is broken, inform the instructor *immediately* so that the mercury can be cleaned up at once.
- Use tongs or a towel when handling hot beakers.

Apparatus/Reagents Required

Gas sample (air) trapped beneath a drop of mercury contained in a 2-mm-diameter capillary tube, wooden ruler

Procedure

Record all data and observations directly in your notebook in ink.

Obtain a capillary tube containing the gas sample trapped beneath a droplet of mercury (see Figure 14-3). The capillary tube is extremely fragile and is broken easily. The capillary tube must be handled gently to avoid breaking up the mercury droplet. There must be a *single small unbroken droplet* of mercury in the capillary tube.

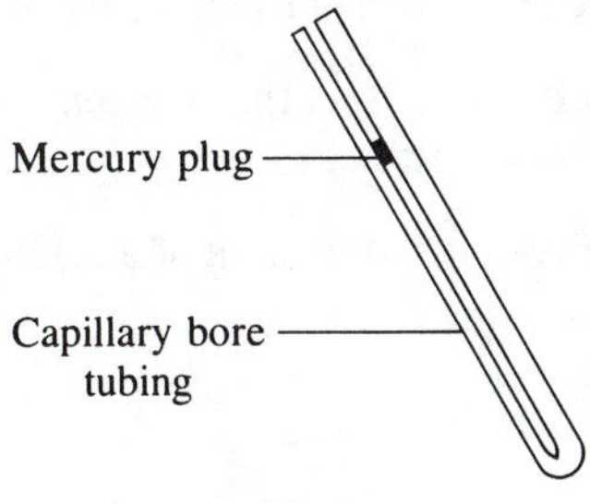

Figure 14-3. Charles's law apparatus. The mercury droplet moves as the gas is heated or cooled, giving an index as to the volume of the gas sample.

Set up a 600-mL beaker about two-thirds full of water on a ringstand in the exhaust hood and start heating the water to boiling.

Using glycerine as a lubricant, and protecting your hands with a towel, insert the *top* of your thermometer through a rubber stopper so that the thermometer can still be read from 0–100°C.

Attach the capillary tube containing the gas sample to your thermometer with at least two rubber bands (to prevent the capillary from moving during the experiment). Align the *bottom* of the capillary tube with the *mercury reservoir* of the thermometer.

Attach the capillary/thermometer assembly to a millimeter scale *wooden* (not plastic) ruler with several rubber bands. The scale of the ruler will be used for determining the position of the mercury droplet as it moves in the capillary tube.

When the water in the beaker is boiling, clamp the thermometer/capillary/ruler assembly to the ringstand and lower it into the boiling water. Allow the apparatus to heat in the boiling water for 3–4 minutes. Record the temperature of the boiling water bath to the nearest 0.2°C. Notice that the droplet of mercury moves up in the capillary tube as the gas beneath it is heated and expands in volume. While the gas is heating, record the position of the *inside bottom* of the glass capillary tube on the scale of the ruler. This measurement represents the lower end of the cylinder of gas being heated in the capillary.

When the gas sample has been heated in the boiling water for several minutes, record the position of the *bottom* of the droplet of mercury as indicated on the ruler scale. Calculate the height of the cylinder of gas in the capillary tube by subtracting the position of the bottom of the gas capillary from the position of the mercury droplet's lower end. Remove the heat from the beaker and allow the water bath to cool spontaneously while still surrounding the capillary/ruler/thermometer apparatus.

As the water bath cools, the gas sample in the capillary tube will contract in volume, and the mercury droplet will move downward. Determine the position of the lower end of the mercury droplet at approximately 10°C intervals as the gas cools. Record the actual temperature at which each measurement is made (to the nearest 0.2 degree). Continue taking readings in this manner until the temperature has dropped to 30°C.

After the temperature has reached 30°C, begin adding ice to the beaker of water in small portions, with vigorous stirring, and take readings at approximately 20°, 10°, and finally, 0°C. Do not add too much ice at any one time, or the temperature will drop too rapidly.

Construct a graph of your data, plotting the *height of the gas sample* (in mm) versus the *Celsius temperature*. The graph should be a *straight line*. If the graph is not a straight line, you delayed too long before reading the height of the gas sample, thereby allowing the volume of the gas sample to change. If this happens, *repeat* the measurement procedure.

Calculate the slope of the line, as well as the intercept of the line with the volume axis (where $V = 0$), as directed by your instructor.

Calculate the percent error in your determination of absolute zero.

Choice III. Graham's Law

Introduction

Effusion, strictly speaking, describes the passage of the molecules of a gas through a small hole (a "pinhole") into an evacuated chamber. This term is often confused with a similar word—diffusion. **Diffusion** is the spreading out of gas molecules through space when a container of gas is opened, allowing the gas to mix freely with any other gases present. Suppose a balloon were filled with hydrogen sulfide gas (which has the obnoxious odor of rotten eggs): if there were a tiny hole in the balloon, the hydrogen sulfide would *effuse* slowly through the hole and would then *diffuse* into the air of the room.

According to the kinetic-molecular theory of gases, the average velocity (root mean square velocity) of the particles in a sample of gas is inversely related to the square root of the molar mass of the gas

$$u_{rms} = (3RT/M)^{1/2}$$

While it is not possible experimentally to determine easily and directly the average speed of the molecules in a sample of gas, the average speed at which gases diffuse or effuse can be measured readily. The speed of diffusion in centimeters per second can be determined by measuring how long it takes a gas to pass through a tube of known length.

Thomas Graham, a Scottish chemist, determined experimentally in the nineteenth century that the relative rate of diffusion of two different gases at the same temperature was given by the relationship

$$r_1/r_2 = (M_2/M_1)^{1/2}$$

in which r represents the rate of diffusion of a gas and M its molar mass. This equation, called Graham's law, is consistent with the postulate of the kinetic-molecular theory describing the average speed of molecules in a gas sample.

In this experiment, you will determine the relative rates of diffusion of the gases hydrogen chloride and ammonia, by measuring the *distances* traveled by the two gases in the same time period. For a given period of time, a lighter weight gas should be able to diffuse *farther* than a heavier gas (distance traveled in a given time period is directly related to speed). Cotton balls dipped in concentrated hydrochloric acid (HCl gas in water) and concentrated aqueous ammonia (NH_3 gas in water) will be placed in opposite ends of a glass tube. The two gases will diffuse through the tube toward each other. Hydrogen chloride and ammonia gases *react* with each other, forming the salt ammonium chloride

$$HCl(g) + NH_3(g) \rightarrow NH_4Cl(s)$$

As the gases meet and react, a white *ring* of $NH_4Cl(s)$ will appear in the tube. The position of this white ring along the length of the tube can be used to determine which of the two gases has diffused farther.

Apparatus/Reagents Required

3–4 feet of 1-cm-diameter glass tubing, concentrated HCl, concentrated NH_3, cotton balls, forceps, latex surgical gloves, rubber stoppers that fit the ends of the glass tube, plastic wrap, meter stick, china marker

Safety Precautions

- Wear safety glasses at all times while you are in the laboratory.
- Concentrated hydrochloric acid and concentrated ammonia solutions are each damaging to skin; wear rubber gloves while handling them.
- The fumes of both HCl and NH_3 are extremely irritating and are dangerous to the respiratory tract. *Use these substances only in the exhaust hood.*
- When cutting glass, protect your hands with a towel and fire-polish all rough edges.

Procedure

Obtain a length of 1-cm-diameter glass tubing and make certain that the ends have been fire-polished. Obtain two rubber stoppers that will fit snugly in the ends of the tubing. Set up the tubing in the exhaust hood, using two adjustable clamps to hold the tubing in a steady horizontal position (see Figure 14-4).

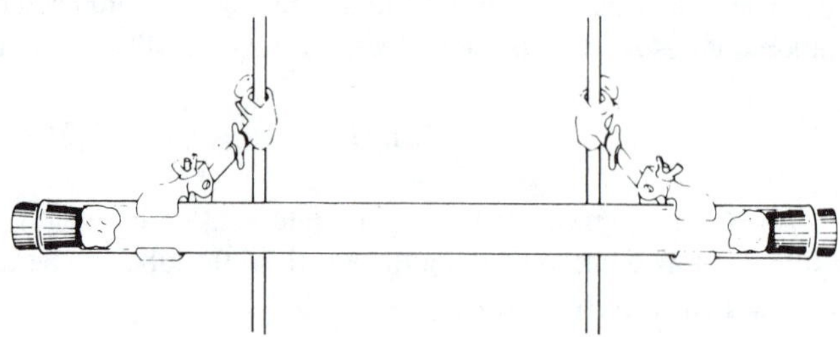

Figure 14-4. Graham's law apparatus. A white ring of ammonium chloride will appear in the tube where the two gases have reacted with each other.

Wear rubber gloves during the following procedure and work in the exhaust hood.

In separate small beakers, obtain 3–4 mL each of concentrated HCl and concentrated NH_3 (*Caution!*). Keep the beakers covered loosely with plastic wrap or a watchglass when not in use. Place a cotton ball in each beaker.

Using forceps, transfer the cotton balls from the beakers to the opposite ends of the glass tubing apparatus. Do this as *quickly* as possible so that one gas will not get too much of a head start over the other (perhaps ask a friend to help you to insert the cotton balls simultaneously). Stopper the ends of the tubing and do *not* disturb or move the glass tube.

Allow the gases to diffuse toward each other until a white ring of ammonium chloride is evident in the tube. Mark the *first* appearance of the ring of ammonium chloride with a china marker before the gases diffuse too much and blur the location of the ring.

Remove the rubber stoppers. Using forceps, remove the cotton balls, and transfer them to a beaker of tap water to dilute the reagents (dispose of the cotton balls in the wastebasket after soaking in water).

Measure the distance diffused by each of the gases to the nearest millimeter, measuring from the respective ends of the glass tube to the center of the ammonium chloride ring.

Rinse out the glass tube with distilled water, allow it to dry, and repeat the determination twice.

Calculate the mean distance diffused by HCl and by NH_3 from your three sets of data.

Realizing that the distances diffused by two gases in the same time period should be directly related to the rates of diffusion of the gases, determine the percent error in your mean experimental data compared to Graham's law as expressed in the introduction to this choice.

Name ______________________________ Section ______________________________

Lab Instructor ______________________________ Date ______________________________

EXPERIMENT 14 The Gas Laws

PRE-LABORATORY QUESTIONS

CHOICE I. BOYLE'S LAW

1. Given the following pressure/volume data for a sample of an ideal gas, construct the following two graphs on graph paper from the back of the lab manual:

   a. Plot the volume of the gas versus the pressure.

   b. Plot the volume of the gas versus the reciprocal of the pressure (1/pressure).

Volume, L	Pressure, atm
14.0	1.57
13.0	1.69
12.0	1.83
11.0	2.00
10.0	2.20
9.0	2.44
8.0	2.75
7.0	3.14
6.0	3.66
5.0	4.40

   Attach your graphs to this page.

2. The gas laws are defined only for *ideal* gases. Use your textbook to determine why real gases do not always follow the gas laws exactly. Under what conditions do real gases most closely approximate ideal gases?

____

____

____

____

____

____

____

CHOICE II. CHARLES'S LAW

1. The volume of a sample of ideal gas at 25°C is 372 mL. What will the volume of the gas be if it is heated at constant pressure to 50°C? What will the volume of the gas be if it is cooled to -272°C?

2. For the following volume/temperature data:

   a. Plot the data on graph paper from the back of this manual.

   b. Determine the slope of the line.

   c. Determine the intercept of the line with the volume axis (i.e., $V = 0$)

Volume, mL	Temperature, °C
83.5	30.0
86.2	40.0
88.9	50.0
91.7	60.0
94.4	70.0
97.2	80.0
100.0	90.0

Attach your graphs to this page.

CHOICE III. GRAHAM'S LAW

1. How many times faster will helium gas effuse through a pinhole than will nitrogen gas under the same conditions?

2. The fact that gases of different molecular weights effuse and diffuse at different speeds was used during World War II as a means of separating the isotopes of uranium for the first atomic bomb. Use your textbook or an encyclopedia of chemistry to determine in what form the uranium isotopes were separated and how the process was accomplished.

Name ______________________ Section ______________

Lab Instructor ______________________ Date ______________

EXPERIMENT 14 The Gas Laws

CHOICE I. BOYLE'S LAW

RESULTS/OBSERVATIONS

Initial height of gas at atmospheric pressure ______________________

Barometric pressure ______________________

Gas height after first movement of leveling bulb ______________________

Difference in mercury levels ______________________

Calculated pressure of gas ______________________

Gas height after second movement of leveling bulb ______________________

Difference in mercury levels ______________________

Calculated pressure of gas ______________________

Gas height after third movement of leveling bulb ______________________

Difference in mercury levels ______________________

Calculated pressure of gas ______________________

Gas height after fourth movement of leveling bulb ______________________

Difference in mercury levels ______________________

Calculated pressure of gas ______________________

Product of pressure × height of gas for your five sets of data:

Atmospheric pressure ______________________

First movement ______________________

Second movement ______________________

Third movement ______________________

Fourth movement ______________________

Mean value of pressure × height of gas ______________________________

Standard deviation ______________________________

Attach your two graphs to this report page.

QUESTIONS

1. What effect would have been observed in the pressure/volume of the gas sample if the *temperature* had increased markedly during the course of your measurements?

____

____

____

____

____

2. Why is it that the *height* of the gas sample can be used in the determination of Boyle's law in this experiment, rather than the actual volume of the gas in milliliters or liters?

____

____

____

____

____

____

Name ______________________________ Section ______________

Lab Instructor ______________________________ Date ______________

EXPERIMENT 14 The Gas Laws

CHOICE II. CHARLES'S LAW

RESULTS/OBSERVATIONS

Height of gas sample	*Temperature*
____________	____________
____________	____________
____________	____________
____________	____________
____________	____________
____________	____________
____________	____________
____________	____________
____________	____________
____________	____________
____________	____________

Attach your graph to this report form.

Slope of volume versus temperature graph ______________________________

Intercept of graph with volume axis ______________________________

Percent error in absolute zero determination ______________________________

QUESTIONS

1. Occasionally, when the glass tubes containing the gas sample are prepared, water vapor is inadvertently trapped beneath the mercury plug. What error in the volumes measured would the presence of water vapor cause?

2. Why were you able to use the height of the gas sample in the glass tube, rather than the actual volume of the gas, in the plotting of the Charles's law graph?

3. In this experiment, you *extrapolated* data over several hundred degrees to calculate a value for absolute zero. What errors might be introduced in an experiment by such a large extrapolation?

Name ______________________________ Section ______________________

Lab Instructor ______________________________ Date ______________________

EXPERIMENT 14 The Gas Laws

CHOICE III. GRAHAM'S LAW

RESULTS/OBSERVATIONS

Observation of NH_4Cl appearance ______________________________

____

Distance of NH_4Cl ring from ends of tube

	NH_3 end	HCl end
Trial 1	______________	______________
Trial 2	______________	______________
Trial 3	______________	______________
	Mean ______________	Mean ______________

Ratio predicted from Graham's Law for rates of diffusion for NH_3/HCl ______________

Experimental ratio for diffusion of NH_3/HCl based on mean distances ______________

Percent error ______________________________

QUESTIONS

1. The two gases used diffused into air, rather than into a vacuum. Is any error likely to be introduced into the experiment by this?

____

____

____

____

____

2. You were told to try to introduce the cotton balls containing HCl and NH_3 into the glass tube simultaneously. What error would be introduced if one substance had been introduced to the glass tube substantially before the other substance?

3. What effect on the distances measured would have occurred if the experiment had been done at a higher temperature?

EXPERIMENT

15 Molar Mass of a Volatile Liquid

Objective

The molar mass of a volatile liquid will be determined by measuring what mass of vapor of the liquid is needed to fill a flask of known volume at a particular temperature and pressure.

Introduction

The most common instrument for the determination of molar masses in modern chemical research is the **mass spectrometer**. Such an instrument permits very precise determination of molar mass and also gives a great deal of structural information about the molecule being analyzed; this is of great help in the identification of new or unknown compounds.

Mass spectrometers, however, are extremely expensive and take a great deal of time and effort to calibrate and maintain. For this reason, many of the classical methods of molar mass determination are still widely applied. In this experiment, a common modification of the ideal gas law will be used in the determination of the molar mass of a liquid that is easily evaporated.

The ideal gas law ($PV = nRT$) indicates that the observed properties of a gas sample [pressure(P), volume(V), and temperature(T)] are directly related to the quantity of gas in the sample (n, moles). For a given container of fixed volume at a particular temperature and pressure, only one possible quantity of gas can be present in the container:

$$n = \frac{PV}{RT}$$

By careful measurement of the mass of the gas sample under study in the container, the molar mass of the gas sample can be calculated, since the molar mass, M, merely represents the number of grams, g, of the volatile substance per mole:

$$n = \frac{g}{M}$$

In this experiment, a small amount of easily volatilized liquid will be placed in a flask of known volume. The flask will be heated in a boiling water bath and will be equilibrated with atmospheric pressure. From the volume of the flask used, the temperature of the boiling water bath, and the atmospheric pressure, the number of moles of gas contained in the flask may be calculated. From the mass of liquid required to fill the flask with vapor when it is in the boiling water bath, the molar mass of the liquid may be calculated.

A major assumption is made in this experiment that may affect your results. We assume that the vapor of the liquid behaves as an *ideal* gas. Actually, a vapor behaves *least* like an ideal gas under conditions near which the vapor would liquefy. The unknown liquids provided in this experiment have been chosen, however, so that the vapor will approach ideal gas behavior.

Safety Precautions

- Wear safety glasses at all times while in the laboratory.
- Assume that the vapors of your liquid unknown are toxic. Work in an exhaust hood or other well-ventilated area.
- The unknown may also be flammable. All heating is to be performed using a hotplate.
- The liquid unknowns may be harmful to skin. Avoid contact, and wash immediately if the liquid is spilled.
- A boiling water bath is used to heat the liquid, and there may be a tendency for the water to splash when the flask containing the unknown liquid is inserted. Exercise caution.
- Use tongs or a towel to protect your hands from hot glassware.

Apparatus/Reagents Required

500-mL Erlenmeyer flask and 1000-mL beaker, aluminum freezer foil, needle or pin, oven (110°C), unknown liquid sample, hotplate

Procedure

Record all data and observations directly in your notebook in ink.

Prepare a 500-mL Erlenmeyer flask by cleaning the flask and then drying it *completely*. The flask must be *completely dry*, since any water present will vaporize under the conditions of the experiment and will adversely affect the results. An oven may be available for heating the flask to dryness, or your instructor may describe another technique.

Cut a square of thick (freezer) aluminum foil to serve as a cover for the flask. Trim the edges of the foil so that it neatly covers the mouth of the flask but does not extend far down the neck.

Prepare a large beaker for use as a heating bath for the flask. The beaker must be large enough for most of the flask to be *covered* by boiling water when in the beaker. Add the required quantity of water to the beaker. Set up the beaker on a hot plate in the exhaust hood, but do not begin to heat the water bath yet.

Weigh the dry, empty flask with its foil cover to the nearest mg (0.001 g).

Obtain an unknown liquid and record its identification number.

Add 3–4 mL of liquid to the dry Erlenmeyer flask. Cover the flask with the foil cover, making sure that the foil cover is tightly crimped around the rim of the flask. Punch a single small hole in the foil cover with a needle or pin.

Heat the water in the beaker to boiling. When the water in the beaker begins to boil, adjust the temperature of the hotplate so that the water remains boiling but does not splash from the beaker.

Immerse the flask containing the unknown liquid in the boiling water so that most of the flask is covered with the water of the heating bath (see Figure 15-1). Clamp the neck of the flask to maintain the flask in the boiling water.

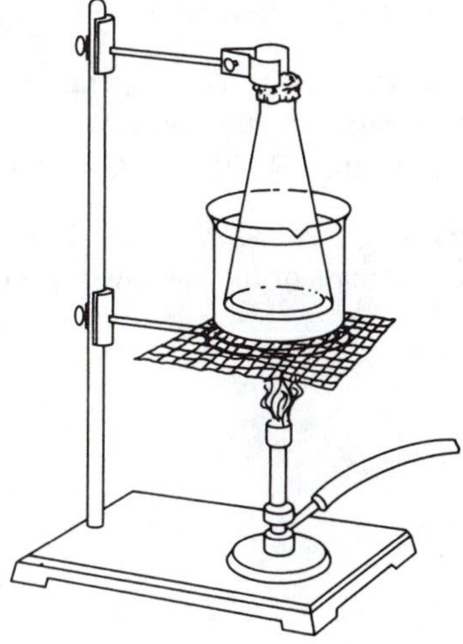

Figure 15-1. Apparatus for determination of the molar mass of a volatile liquid. Most of the flask containing the unknown liquid must be beneath the surface of the boiling water bath.

Watch the unknown liquid carefully. The liquid will begin to evaporate rapidly, and its volume will decrease. The amount of liquid placed in the flask is much *more* than will be necessary to fill the flask with vapor at the boiling water temperature. Excess vapor will be observed escaping through the pinhole made in the foil cover of the flask.

When it appears that all the unknown liquid has vaporized, and the flask is filled with vapor, continue to heat for 1–2 more minutes. Then remove the flask from the boiling water bath; use the clamp on the neck of the flask to protect your hands from the heat.

Set the flask on the lab bench, remove the clamp, and allow the flask to cool to room temperature. Liquid will *reappear* in the flask as the vapor in the flask cools. While the flask is cooling, measure and record the exact temperature of the boiling water in the beaker, as well as the barometric pressure in the laboratory.

When the flask has cooled *completely* to room temperature, carefully dry the outside of the flask to remove any droplets of water. Then weigh the flask, foil cover, and condensed vapor to the nearest mg (0.001 g).

Repeat the determination by adding another 3–4-mL sample of unknown liquid. Reheat the flask until it is filled with vapor; cool, and reweigh the flask. The weight of the flask after the second sample of unknown liquid is vaporized should agree with the first determination within 0.05 g. If it does not, do a third determination.

When two acceptable determinations of the weight of vapor needed to fill the flask have been obtained, remove the foil cover from the flask and clean it out.

Fill the flask to the very rim with tap water, cover with the foil cover, and weigh the flask, cover, and water to the nearest 0.1 g. Determine the temperature of the tap water in the flask. Using the density of water at the temperature of the water in the flask and the weight of water the flask contains, calculate the exact volume of the flask.

If no balance with the capacity to weigh the flask when filled with water is available, the volume of the flask may be approximated by pouring the water in the flask into a 1-L graduated cylinder and reading the water level in the cylinder.

Using the volume of the flask (in liters), the temperature of the boiling water bath (in kelvins), and the barometric pressure (in atmospheres), calculate the number of moles of vapor the flask is capable of containing. $R = 0.0821$ L atm/mol K.

Using the weight of unknown vapor contained in the flask, and the number of moles of vapor present, calculate the molar mass of the unknown liquid.

Name ______________________ Section ______________

Lab Instructor ______________________ Date ______________

EXPERIMENT 15 Molar Mass of a Volatile Liquid

PRE-LABORATORY QUESTIONS

1. The method used in this experiment is sometimes called the **vapor density method**. Beginning with the ideal gas equation, show how the *density* of a vapor may be determined by this method.

2. If 2.31 g of the vapor of a volatile liquid is able to fill a 498-mL flask at 100°C and 775 mm Hg, what is the molecular weight of the liquid? What is the density of the vapor under these conditions?

3. Why is a vapor unlikely to behave as an ideal gas near the temperature at which the vapor would liquefy?

Name ______________________________ Section ______________________

Lab Instructor ______________________________ Date ______________________

EXPERIMENT 15 Molar Mass of a Volatile Liquid

RESULTS/OBSERVATIONS

Identification number of unknown liquid ______________________________

Mass of empty flask and cover ______________________________

	Sample 1	Sample 2
Mass of flask/cover/vapor	______________	______________
Temperature of vapor, °C	______________	______________
Temperature of vapor, K	______________	______________
Pressure of vapor, mm Hg	______________	______________
Pressure of vapor, atm	______________	______________
Mass of flask/cover with water	______________	______________
Mass of water in flask	______________	______________
Temperature of water in flask	______________	______________
Density of water	______________	______________
Volume of flask,	mL ______________	L ______________
Moles of vapor in flask	______________	______________
Molar mass of vapor	______________	______________

Mean value of molar mass of vapor ______________________________

Density of vapor in flask	______________	______________

Mean value of density of vapor ______________________________

QUESTIONS

1. Two methods were described for determining the volume of the flask used for the molar mass determination. Which method will give a more precise determination of the volume? Why?

2. It was important that the flask be completely dry before the unknown liquid was added so that water present would not vaporize when the flask was heated. A typical single drop of liquid water has a volume of approximately 0.05 mL. Assuming the density of liquid water is 1.0 g/mL, how many moles of water are in one drop of liquid, and what volume would this amount of water occupy when vaporized at 100°C and 1 atm?

EXPERIMENT

16 Vapor Pressure

Objective

In this experiment, the vapor pressure of several volatile liquids will be demonstrated.

Introduction

When a volatile liquid evaporates into a vacuum in a closed container, evaporation continues only until the space above the liquid becomes *saturated* with the vapor of the liquid. Beyond this point, an *equilibrium* exists between the vapor and any remaining liquid. The maximum number of molecules of the vapor of a volatile liquid that can exist above the liquid in a container of a fixed volume is a *constant number* at a given temperature. The pressure of vapor that exists in a closed container above such a liquid is called the **saturation vapor pressure** of the liquid, and is also constant at a given temperature.

To a macroscopic observer of this situation, it would appear that the liquid had stopped evaporating once the space above the liquid had been saturated with the vapor of the liquid. On a microscopic level, however, evaporation is still taking place. Evaporation *appears* to have stopped because the vapor of the liquid is *condensing* at the same rate at which evaporation is occurring. Evaporation and condensation are opposing processes, and in a system in which a liquid and its vapor have come to equilibrium, every time a molecule of liquid enters the vapor phase, somewhere else in the system a molecule of vapor enters the liquid phase. Studies using radioactive tracers have shown unequivocally that there is a constant interchange between liquid and vapor molecules in such a system.

For the situation in which a liquid evaporates into a container that already contains some other gas (air, for example), the total pressure in the container will be the sum of the partial pressures of all the components, that is

$$P_{\text{total}} = P_1 + P_2 + \ldots + P_n \quad \text{where } n \text{ is the number of components.}$$

If the container of vapor is equilibrated with the atmosphere (see Figure 16-1, in which the bottom of the tubes containing vapor are open to the atmosphere), then the total pressure in the container is equal to the barometric pressure of the laboratory.

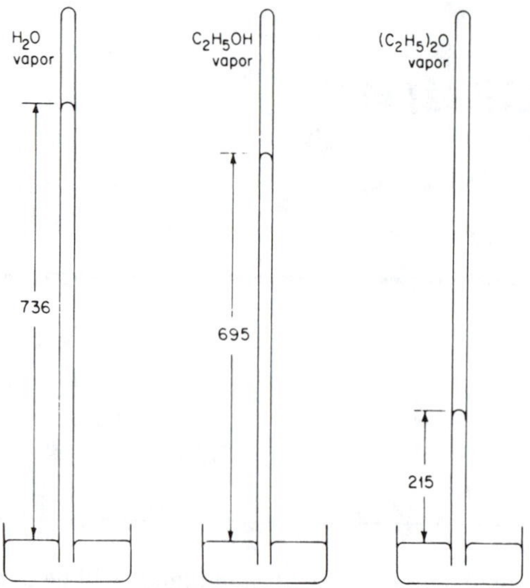

Figure 16-1. The vapor pressure of a liquid can be easily demonstrated using an inverted buret. The three liquids shown—water, ethanol, and diethyl ether—have quite different vapor pressures. Ether is by far the most volatile of the three, causing the largest depression of the mercury level.

Safety Precautions

- Wear safety glasses at all times while in the laboratory.
- The liquids used for vapor pressure determination are highly volatile and extremely flammable. For this reason, the apparatus should be set up in the *exhaust hood* to minimize the build-up of fumes. *Absolutely no flames are permitted in the laboratory under any circumstances.*
- The liquids used in the vapor pressure determination should be assumed to be toxic. Do not spill them on the skin, and confine the transfer of the liquids to the exhaust hood.
- Dispose of the liquid samples as directed by the instructor.

Apparatus/Reagents Required

Buret, funnel, rubber tubing, large beaker or basin, samples of volatile liquids, medicine dropper or Pasteur pipet

Procedure

Record all data and observations directly in your notebook in ink.

Obtain a buret and clean it with soap and water. Rinse thoroughly with tap water to remove all traces of soap. Make a final rinse of the buret with distilled water.

Pour water to a depth of 6–7 inches into a large basin. Place the basin at the base of a ringstand. Determine and record the temperature of the water. Place a glass funnel, mouth downward, into the basin of water. See Figure 16-2.

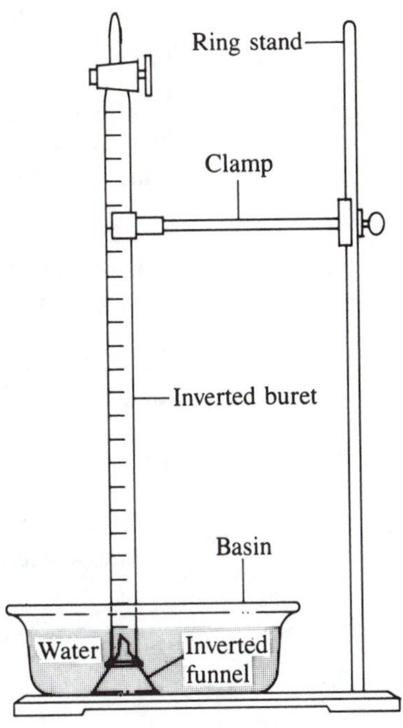

Figure 16-2. Apparatus for vapor pressure determination

Fill the buret to the rim with distilled water, place your finger over the rim of the buret, and invert the buret into the basin of water, with the mouth of the buret placed *over the stem of the funnel.* Clamp the buret securely to the ring stand. If you have performed this maneuver correctly, very little (if any) water should run out of the buret. See Figure 16-2.

Carefully, and very slowly, open the stopcock of the buret until the water level drops to approximately 6 inches below the stopcock (in the inverted position). If you open the stopcock too far, or too quickly, and the water level falls by more than 6 inches, you should refill the buret completely before continuing.

Record the position of the water level in the buret (remember that the normal scale of the buret has been inverted).

Obtain a sample of one of the volatile liquids and record its identity. The liquid samples have been chosen to be highly volatile and of low density, but are relatively insoluble in water. When a sample of one of the liquids is introduced into the inverted buret, it will rise to the surface of the water in the buret and will then evaporate.

A medicine dropper or pipet will be used to transfer 1–2 mL of the volatile liquid to the buret. Lift the funnel/buret assembly about an inch from the bottom of the basin, insert the tip of the dropper/pipet containing the liquid beneath the rim of the funnel, and squeeze the bulb of the dropper/pipet.

If the transfer of the liquid has been successful, the bulk of the water in the buret will *not* fall from the buret.

As the volatile liquid rises to the surface of the water in the buret, and begins to evaporate, it will cause the water level in the buret to *change*. Allow the system to stand until the water level no longer changes, and record the position of the the water level on the scale of the buret. At this point, the air in the buret is saturated with the vapor of the volatile liquid. Record the change in the water level as an index of the vapor pressure of the volatile liquid.

Add another 1–2-mL portion of the volatile liquid to the buret through the funnel. Since the air in the buret was already saturated with the vapor of the liquid, the addition of more liquid should *not* cause any further change in the water level.

Slightly increase the temperature of the system by grasping the buret with your hand in the region of the volatile liquid for approximately 5 minutes. Does an increase in temperature result in a change in the vapor pressure of the volatile liquid?

Clean out the buret, and repeat the procedure using each of the other available volatile liquid samples. While this procedure does not easily permit a very precise determination of the vapor pressure of the liquids tested, you should be able to rank order the liquids in order of increasing vapor pressure.

Using a handbook of chemical data, compare the results of your ordering of the liquids with the actual vapor pressures of those liquids at the temperature of the water in the basin.

Name ______________________________ Section ______________________

Lab Instructor ______________________________ Date ______________________

EXPERIMENT 16 Vapor Pressure

PRE-LABORATORY QUESTIONS

1. Find the specific definitions from your textbook for the following terms:

   Vapor pressure

   ____

   ____

   Evaporation

   ____

   ____

   Condensation

   ____

   ____

2. How does the vapor pressure of a pure liquid vary with *temperature*?

   ____

   ____

3. How does the presence of a solute influence the vapor pressure of a solvent, that is, how does the vapor pressure of a solution differ from that of the pure solvent used in making the solution?

   ____

   ____

4. What does it mean to say that the development of a vapor pressure above a liquid in a closed container represents a *dynamic equilibrium*?

   ____

   ____

Name ______________________________ Section ______________________________

Lab Instructor ______________________________ Date ______________________________

EXPERIMENT 16 Vapor Pressure

RESULTS/OBSERVATIONS

Liquid ID	Initial water level, mL	Final water level, mL	Change in water level

Arrange the liquids determined in order of increasing vapor pressure:

____

Literature value of vapor pressures:

Liquid	Vapor pressure	Temperature	Reference

QUESTIONS

1. What effect might be introduced into the qualitative determination of vapor pressure (as in this experiment) by the saturation of the air in the buret with water vapor?

2. The liquids used in this experiment are *insoluble* in water. What effect would be seen if the liquids tested by this method were very soluble in water? For example, ethyl alcohol is fairly volatile but is also completely miscible with water in all proportions.

3. What effect on the water level in the buret would have been seen if you had cooled the samples of liquid?

4. Some solids pass directly into the vapor state (without first liquefying) and also have appreciable vapor pressures at room temperature. For example, dry ice is said to be dry because it turns directly into gaseous carbon dioxide without melting. What is the direct phase transition from solid to vapor called?

EXPERIMENT

17 Calorimetry

Objective

The calorimeter constant for a simple coffee-cup calorimeter will be determined, and then the calorimeter will be used to measure the quantity of heat that flows in several physical and chemical processes.

Choice I. Determination of a Calorimeter Constant

Introduction

Chemical and physical changes are always accompanied by a change in *energy*. Most commonly, this energy change is observed as a flow of heat energy either into or out of the system under study. Heat flows are measured in an instrument called a **calorimeter**. There are specific types of calorimeters for specific reactions, but all calorimeters contain the same basic components. They are *insulated* to prevent loss or gain of heat energy between the calorimeter and its surroundings. For example, the simple calorimeter you will use in this experiment is made of a heat-insulating plastic foam material. Calorimeters contain a *heat sink* that can absorb or provide the energy for the process under study. The most common material used as a heat sink for calorimeters is water, because of its simple availability and large heat capacity. Calorimeters also must contain some device for the measurement of *temperature*, because it is from the temperature change of the calorimeter and its contents that the magnitude of the heat flow is calculated. Your simple calorimeter will use an ordinary thermometer for this purpose.

To determine the heat flow for a process, the calorimeter typically is filled with a weighed amount of water. The process that releases or absorbs heat is then performed within the calorimeter, and the temperature of the water in the calorimeter is monitored. From the *mass of water* in the calorimeter, and from the *temperature change* of the water, the quantity of heat transferred by the process can be determined.

The calorimeter for this experiment is pictured in Figure 17-1. The calorimeter consists of two nested plastic foam coffee cups and a cover, with thermometer and stirring wire inserted through holes punched in the cover. As you know, plastic foam does not conduct heat well and will not allow heat generated by a chemical reaction in the cup to be lost to the room. (Coffee will not cool off as quickly in such a cup compared to a china or paper cup.) Other sorts of calorimeters might be available in your laboratory; your instructor will demonstrate such calorimeters if necessary. The simple coffee-cup calorimeter generally gives quite acceptable results, however.

Although the plastic foam material from which your calorimeter is constructed does not conduct heat well, it does still absorb some heat. In addition, a small quantity of heat may be transferred to or from the metal wire used for stirring the calorimeter's contents or to the glass of the thermometer used to measure temperature changes. Some heat energy may also be lost through the openings for these devices. Therefore, the calorimeter will be *calibrated* using a known system before it is used in the determination of the heat flows in unknown systems.

As mentioned earlier, there are several mechanisms by which a calorimeter can absorb or transmit heat energy. Rather than determining the influence of each of these separately, a function called the **calorimeter constant** can be determined for a given calorimeter. The calorimeter constant represents what portion of the heat flow from a chemical or physical process conducted in the calorimeter goes to the apparatus itself, rather than to affecting the temperature of the heat sink (water). Once the calorimeter constant has been determined for a given apparatus, the value determined can be applied whenever that calorimeter is employed in subsequent experiments.

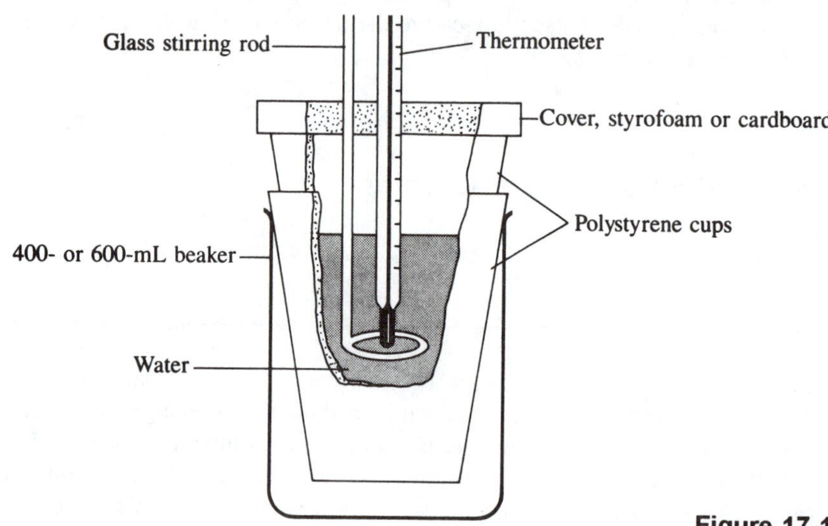

Figure 17-1. A simple calorimeter made from two nested plastic foam coffee cups. Make certain the stirring wire can be agitated easily.

As discussed, the temperature changes undergone by the heat sink are used to calculate the quantity of heat energy that flows during a chemical or physical process conducted in the calorimeter. When a sample of *any* substance changes in temperature, the quantity of heat, Q, involved in the temperature change is given by

$$Q = mC\Delta T \tag{17-1}$$

where m is the mass of the substance, ΔT is the temperature change, and C is a quantity called the **specific heat** of the substance. The specific heat represents the quantity of heat required to raise the temperature of one gram of the substance by one degree Celsius. (Specific heats for many substances are tabulated in handbooks of chemical data.) Although the specific heat is not constant over all temperatures, it remains constant for many substances over fairly broad ranges of temperatures (such as in this experiment). Specific heats are quoted in units of kilojoules per gram per degree, kJ/g°C (or in molar terms, in units of kJ/mol°C).

To determine the calorimeter constant for the simple coffee-cup apparatus to be used in the later choices of this experiment, we will make use of the conservation of energy principle: Energy cannot be created or destroyed during a process, but can only be transformed from one form to another or transferred from one part of the universe to another. A measured quantity of cold water is placed in the calorimeter to be calibrated and is allowed to come to thermal equilibrium with the calorimeter. Then a measured quantity of warm water is added to the cold water in the calorimeter. Since the energy contained in the hot water is conserved, we can make the following accounting of energy:

$$Q_{\text{warm water}} = -[Q_{\text{cold water}} + Q_{\text{calorimeter}}] \qquad (17\text{-}2)$$

The minus sign in this statement is necessary because the warm water is *losing* energy, whereas the cold water and calorimeter are *gaining* energy (these processes have the opposite sense from one another). Since the calorimeter is considered a complete single unit, the amount of heat absorbed by the calorimeter, $Q_{\text{calorimeter}}$, can be written as

$$Q_{\text{calorimeter}} = C_{\text{calorimeter}}\,\Delta T \qquad (17\text{-}3)$$

in which ΔT is the temperature change undergone by the calorimeter, and $C_{\text{calorimeter}}$ is the calorimeter constant, which represents the number of kilojoules of heat required to warm the calorimeter by 1°C.

Applying Equations 17-1 and 17-3 to the accounting of the energy transferred in the system as given in Equation 17-2, we can say the following:

$$(mC\Delta T)_{\text{warm water}} = -[(mC\Delta T)_{\text{cold water}} + (C_{\text{calorimeter}}\Delta T)] \qquad (17\text{-}4)$$

Since the specific heat of water is effectively constant over the range of temperatures in this experiment (C_{water} = 4.18 J/g°C), determination of the calorimeter constant amounts simply to making two measurements of mass and two measurements of changes in temperature.

Safety Precautions

- Wear safety glasses at all times while in the laboratory.
- Use tongs or a towel to protect your hands when handling hot glassware.

Apparatus/Reagents Required

Plastic foam coffee cups and covers, thermometer, wire for use as a stirrer, one-hole paper punch

Procedure

Record all data and observations directly in your notebook in ink.

Nest two similar-sized plastic foam coffee cups for use as the calorimeter chamber. If the cups have been rinsed with water, dry them out.

Obtain a plastic lid that tightly fits the coffee cups. Using the paper punch, make two small holes in the lid. Make one hole near the center of the lid (for the thermometer), and one hole to the side (for the stirring wire). Assemble the stirring wire and thermometer as indicated in Figure 17-1.

Since the density of water over the range of temperatures in this experiment is very nearly 1.00 g/mL, the amount of water to be placed in the calorimeter can be more conveniently measured by volume.

With a graduated cylinder, place 75.0 ± 0.1 mL of *cold* water into the calorimeter. Cover the calorimeter with the thermometer/stirrer apparatus..

Measure 75 ± 0.1 mL of water into a clean, dry beaker, and heat the water to 70–80°C. Stir the water with a glass rod occasionally during the heating to ensure that the temperature is as uniform as possible.

While the water is heating, monitor the temperature of the cold water in the calorimeter for 2–3 minutes to make certain that it has become constant. Record the temperature of the cold water in the calorimeter to the nearest 0.2°C.

When the water being heated has reached 70–80°C, use tongs or a towel to remove the beaker from the heat. Allow the beaker to stand on the lab bench for 2–3 minutes, stirring the water occasionally during this time period. After the standing period, record the temperature of the hot water to the nearest 0.2°C.

Quickly remove the lid from the calorimeter, and pour the hot water into the cold water in the calorimeter. Immediately replace the lid of the calorimeter, stir the water with the stirring wire for 30 seconds to mix, and begin monitoring the temperature of the water in the calorimeter. Record the *highest* temperature reached by the water in the calorimeter, to the nearest 0.2°C.

From the masses (volumes) of cold and hot water used, and from the two temperature changes, calculate the calorimeter constant for your calorimeter.

Repeat the experiment twice to obtain additional values for the calorimeter constant. Use the mean value of the three determinations of the calorimeter constant for the other choices of this experiment.

Choice II. Specific Heats of Metals and Glass

Introduction

The specific heat, C, of a substance represents the quantity of heat energy (in joules) required to warm one gram of the substance by one Celsius degree. Although the specific heats of many substances are relatively constant over broad ranges of temperatures (such as those likely to be encountered in the general chemistry laboratory), the specific heat is dependent on the temperature. Generally the temperature range over which a particular value of the specific heat applies is quoted in the literature.

Metallic substances generally have numerically small specific heats. Metals are good conductors of heat energy and require very little input of heat energy to cause an increase in their temperature. Insulating substances, on the other hand, are very poor conductors of heat energy and have much larger specific heats. For example, the plastic foam used in the construction of the coffee-cup calorimeter in this experiment is an insulator.

When any sample of substance undergoes a temperature change, the amount of heat energy (Q) involved in causing the temperature change is given by

$$Q = mC\Delta T$$

where m is the mass of the sample of substance, C is the specific heat of the substance, and ΔT is the temperature change undergone by the sample.

In this choice, you will determine the specific heats of an unknown metallic substance and of ordinary glass. The method used is essentially the same as in Choice I, with the sample of metal or glass replacing the hot water. A measured sample of cold water is placed in the calorimeter while a weighed metal/glass sample is being heated to 100°C (boiling water bath). The hot metal or glass is then poured into the cold water in the calorimeter, and the maximum temperature reached by the cold water as it absorbs heat from the metal/glass is determined. From the masses of cold water and metal/glass used, and from the temperature changes undergone, the specific heat of the metal/glass may be calculated:

$$Q_{\text{substance}} = -[Q_{\text{water}} + Q_{\text{calorimeter}}] \qquad (17\text{-}5)$$

$$(mC\Delta T)_{\text{substance}} = -[(mC\Delta T)_{\text{water}} + (C\Delta T)_{\text{calorimeter}}] \qquad (17\text{-}6)$$

Safety Precautions

- Wear safety glasses at all times while in the laboratory.
- Use tongs or a towel to protect your hands when handling hot glassware.
- Metal pellets are very expensive and will be collected by the instructor. Do not spill the metal pellets on the floor of the laboratory. (Clean up any accidents immediately to prevent any possible injury.)
- Use caution when handling glass fragments. The edges of the glass may be very sharp. Be careful not to spill the glass, and clean up any spills immediately.

Apparatus/Reagents Required

Coffee-cup calorimeter (as designed in Choice I), metal pellets, glass beads or rings

Procedure

Record all data and observations directly in your notebook in ink.

Choice I, in which the calorimeter constant is determined for the apparatus, must be performed before this portion of the experiment. Use the same calorimeter assembly here.

Since the density of water over the range of temperatures in this experiment is nearly 1.00 g/mL, the amount of water to be placed in the calorimeter can be more conveniently measured by volume.

With a graduated cylinder, place 75.0 ± 0.1 mL of cold water into the calorimeter. Cover the calorimeter with the thermometer/stirrer apparatus.

Obtain an unknown metal sample and record its identification code number in your notebook and on the report page. Weigh out approximately 50 g of the metal sample, and record the precise weight taken to the nearest 0.1 g.

Set up a 600-mL beaker half-filled with water on a ringstand and heat the water to boiling. When the water is boiling, record its temperature to the nearest 0.1°C.

Transfer the unknown metal sample to a clean, *dry* test tube, and heat the test tube in the boiling water bath for at least 10 minutes to allow the metal to reach the temperature of the boiling water.

While the metal sample is heating in the boiling water, monitor the temperature of the cold water in the calorimeter for 2–3 minutes to make certain that the temperature is constant. Record the temperature of the cold water to the nearest 0.2°C.

Remove the cover from the calorimeter, and quickly transfer the metal pellets to the calorimeter. Cover the calorimeter, stir the water for 60 seconds, and record the highest temperature reached to the nearest 0.2°C.

From the mass of metal sample used and the mass (volume) of cold water in the calorimeter, and from the temperature changes and calorimeter constant (see Choice I), calculate the specific heat of the unknown metal.

Repeat the determination twice more, and calculate a mean value from your three determinations for the specific heat of the unknown metal. The same metal sample may be used for these repeat determinations, but the pellets must be dried completely on a paper towel before reheating in the boiling water bath. If no pellets are lost during any point of the procedure, it is not necessary to reweigh the metal sample.

Make three determinations of the specific heat of glass beads or rings by the same method used for the metal pellets. Be careful of any sharp edges on the glass rings, and clean up any spills of glass immediately.

Choice III. Heat of Acid/Base Reactions

Introduction

Many chemical reactions are performed routinely with the reactant species dissolved in water. The use of such solutions has many advantages over the use of "dry chemical" methods. The presence of a solvent matrix permits easier and more intimate mixing of the reactant species, solutions may be measured by volume rather than mass, and the presence of water may act as a moderating agent in the reaction.

Heats of reaction between species dissolved in water are especially easy to measure because the measurement may be performed in a simple calorimeter as described in the earlier choices of this experiment. A measured volume of solution of one of the reagents is placed in the calorimeter cup and its temperature is determined. The second reagent solution is prepared in a separate container and is allowed to come to the same temperature as the solution in the calorimeter. When the two solutions have come to the same temperature, they are combined in the calorimeter, and the temperature of the calorimeter contents is monitored. If the chemical reaction that takes place is **exergonic**, the temperature of the water in the calorimeter will *increase* as energy is transferred to it from the reagents. If the chemical reaction is **endergonic**, the temperature of the water in the calorimeter will *decrease* as thermal energy is drawn from the water into the reactant substances.

For the purpose of tabulating heats of reaction in the chemical literature, such heats are usually converted to the basis of the number of *kilojoules of heat energy* that flows in the reaction *per mole* of reactant (or product). A typical experimental determination may only use a small fraction of a mole of reactant, and so only a few joules of heat energy will be involved in the experiment, but the results are converted to the basis of one mole. When such a determination of heat flow is conducted in a calorimeter that is equilibrated with the constant pressure of the atmosphere, the heat flow in kJ/mole is given the symbol ΔH, and is referred to as the **enthalpy change** for the reaction. Handbooks of chemical data list the enthalpy changes for many reactions.

You will measure the heat of reaction for four "different" reactions:

$$HCl + NaOH \rightarrow$$
$$HCl + KOH \rightarrow$$
$$HNO_3 + NaOH \rightarrow$$
$$HNO_3 + KOH \rightarrow$$

Each of these reactions represents the neutralization of an acid by a base, and although the reactions formally appear to involve different substances, the net reaction occurring in each case is the same:

$$H^+ \textit{ (from the acid)} + OH^- \textit{ (from the base)} \rightarrow H_2O$$

The actual reaction that occurs in each situation is the combination of a proton with a hydroxide ion, producing a water molecule. For this reason, the heat flows for each of these reactions should be the same.

Safety Precautions

- Wear safety glasses at all times while in the laboratory.
- Although the acids and bases used in this experiment are in relatively dilute solution, the acids/bases will concentrate if spilled on the skin and the water is allowed to evaporate. Wash immediately if these substances are spilled.

Apparatus/Reagents Required

2 *M* hydrochloric acid, 2 *M* nitric acid, 2 *M* sodium hydroxide, 2 *M* potassium hydroxide, calorimeter/thermometer/stirrer apparatus as used in the previous choices

Procedure

Record all data and observations directly in your notebook in ink.

Choice I, in which the calorimeter constant is determined for the apparatus, must be performed before this portion of the experiment. Use the same calorimeter assembly here.

The procedure that follows is written in terms of the reaction between hydrochloric acid and sodium hydroxide. Perform this determination first; then repeat the procedure for each of the other three acid/base combinations indicated in the equations in the Introduction.

Obtain 75 mL of 2 *M* NaOH and place it in the calorimeter. Obtain 75 mL of 2 *M* HCl in a clean dry beaker. Allow the two solutions to stand until their temperatures are the same (within ± 0.5 degree). Be sure to rinse off and dry the thermometer when transferring between the solutions to prevent mixing of the reagents prematurely. Record the temperature(s) of the solutions to the nearest 0.2°C.

Add the HCl from the beaker *all at once* to the calorimeter, cover the calorimeter quickly, stir the mixture for 30 seconds, and record the highest temperature reached by the mixture (to the nearest 0.2°C).

From the change in temperature undergone by the mixture upon reaction, the total mass (volume) of the combined solutions, and the calorimeter constant for the apparatus (see Choice I) calculate the quantity of heat that flowed from the reactant species into the water of the solution.

Calculate the number of moles of water produced when 75 mL of 2 *M* HCl reacts with 75 mL of 2 *M* NaOH.

Calculate ΔH in terms of the number of kilojoules of heat energy transferred when 1 mol of water is formed in the neutralization of aqueous HCl with aqueous NaOH.

Repeat the determination of ΔH for the HCl/NaOH reaction twice, and calculate a mean value for ΔH for the reaction.

Repeat the procedure for the other combinations of acids and bases:

$$HCl + KOH \rightarrow$$
$$HNO_3 + NaOH \rightarrow$$
$$HNO_3 + KOH \rightarrow$$

Calculate a mean value and standard deviation for ΔH for the net reaction occurring in each of the mixtures, i.e.,

$$H^+ + OH^- \rightarrow H_2O$$

from your twelve determinations.

Choice IV. Heat of Metal/Acid Reactions

Introduction

Some metallic elemental substances are easily able to displace hydrogen from its compounds. For example, the alkali metals and some of the alkaline earth metals react with cold water, producing the metal hydroxide and releasing gaseous elemental hydrogen:

$$2Na + 2H_2O \rightarrow 2NaOH + H_2$$

$$Ca + 2H_2O \rightarrow Ca(OH)_2 + H_2$$

Other less reactive metals, although not able to displace hydrogen from water, will evolve elemental hydrogen gas from solutions of acids. For example,

$$Mg + 2HCl \rightarrow MgCl_2 + H_2 \qquad (Mg + 2H^+ \rightarrow Mg^{2+} + H_2)$$

$$Zn + H_2SO_4 \rightarrow ZnSO_4 + H_2 \qquad (Zn + 2H^+ \rightarrow Zn^{2+} + H_2)$$

In Choice III, the enthalpy changes for various strong acids reacting with strong bases were determined, and it was noted that, because the net reaction occurring was the same in each case, the enthalpy changes should be the same for each combination. For the reactions of metallic substances with acids to be studied in this choice, based on the net reactions shown above, the enthalpy change determined should depend on which metal is used, but should not be a function of which strong acid is used.

Safety Precautions

- Wear safety glasses at all times while in the laboratory.
- Although the acids used in this experiment are in relatively dilute solution, the acids may concentrate if allowed to evaporate. Wash immediately if the acids come in contact with skin, and clean up any spills in the laboratory with large amounts of water.
- The acids used in this experiment are damaging to clothing.
- Hydrogen gas is generated in this experiment. Hydrogen gas is extremely flammable. Make certain that the room is well ventilated. No flames will be permitted in the room while hydrogen is being generated.

Apparatus/Reagents Required

1 *M* hydrochloric acid, 1 *M* nitric acid, 1 *M* sulfuric acid, calorimeter/thermometer/stirrer apparatus as used in the previous choices, magnesium turnings, mossy or granulated zinc

Procedure

Record all observations and data directly in your lab notebook in ink.

Choice I, in which the calorimeter constant is determined for the apparatus, must be performed before this portion of the experiment. Use the same calorimeter assembly here.

For the dilute acid solutions to be used in this experiment, it may be assumed that the densities of the solutions are very nearly 1.0 g/mL. For this reason, a particular mass of solution may be conveniently measured out by volume.

Obtain 150 ± 1 mL of 1 *M* HCl and place it in the calorimeter. After covering the calorimeter, determine the temperature of the acid to the nearest 0.2°C. Monitor the temperature of the acid over a 5-minute period to make certain that the temperature is constant.

Weigh out a 0.5-g sample of magnesium turnings and record the exact mass to the nearest milligram (0.001 g).

Remove the cover of the calorimeter, quickly add the magnesium turnings, and replace the cover. Stir the mixture and monitor the temperature as the reaction occurs. Record the highest temperature reached.

Repeat the reaction between magnesium and hydrochloric acid twice.

Taking the specific heat of the dilute hydrochloric acid solution to be the same as that of water, calculate the number of joules of heat energy released for each of the individual Mg/HCl reactions.

Based on the mass of magnesium taken in each individual run, calculate the enthalpy change for each run, in terms of the number of kJ of heat energy transferred per mole of magnesium reacting.

Calculate the mean value for the enthalpy change for the reaction of magnesium with hydrochloric acid.

Perform similar sets of determinations for the reaction between magnesium and 1 *M* nitric acid, and for the reaction between magnesium and 1 *M* sulfuric acid.

Based on these nine determinations, calculate a mean value and standard deviation for the reaction between magnesium metal and aqueous hydrogen ion.

Repeat the processes, determining the enthalpies of reaction of zinc metal with each of the three acids available.

Choice V. Heat of Solution of a Salt

Introduction

When salts dissolve in water, the positive and negative ions of the salt *interact* with water molecules. Water molecules are highly polar, and arrange themselves in a layer around the ions of the salt so as to maximize electrostatic attractive forces. Such a layer of water molecules surrounding an ion is called a **hydration sphere.**

For example, consider dissolving the salt potassium bromide, KBr, in water. As the ions enter solution, water molecules would orient their dipoles in a particular manner. The potassium ions would become surrounded by a layer of water molecules in which the *negative* ends of the water dipoles would be oriented toward the positive potassium ions. Similarly, the bromide ions would become surrounded by a layer of water molecules in which the *positive* ends of the dipoles would be oriented toward the *negative* bromide ions. The development of such hydration spheres can have a large effect on the properties of ions in solutions, which you will study in depth if you go on to take a course in physical chemistry.

Formation of a hydration sphere around an ion necessarily involves an energy change. In this Choice, you will determine the energy change for such a process.

Safety Precautions

- Wear safety glasses at all times while in the laboratory.
- The salts used in this Choice may be toxic. Wash after handling them and dispose of them as directed by the instructor.

Apparatus/Reagents Required

Salt sample, calorimeter/thermometer/stirrer apparatus as used in the previous choices

Procedure

Record all data and observations directly in your notebook in ink.

Choice I, in which the calorimeter constant is determined for the apparatus, must be performed before this portion of the experiment is begun. Use the same calorimeter assembly here.

Place 75.0 mL of distilled water in the calorimeter. Determine and record the temperature of the water to the nearest 0.2°C. Monitor the temperature of the water for 3 minutes to make certain that its temperature does not change.

Obtain a salt sample from your instructor. If the salt's identity is known, record the name and formula of the salt. If the salt is presented as an unknown, record the code number of the unknown sample. If an unknown is presented, the instructor will also tell you the molar mass of the unknown salt (record).

Using a clean, dry beaker, weigh out a 5–6-gram sample of the salt, recording the exact mass taken to the nearest 0.01 g.

Remove the lid from the calorimeter and quickly add the weighed salt sample. Replace the lid of the calorimeter and immediately agitate the solution with the stirring wire. Monitor the temperature of the solution while continuing to stir the solution. Record the highest—or lowest—temperature reached as the salt dissolves in the water.

From the mass of salt taken and the mass (volume) of water used, and from the temperature change and calorimeter constant, calculate the quantity of heat that flowed during the dissolving of the salt. From this quantity of heat, and from the molar mass of the salt, calculate the enthalpy change for the dissolving of the salt (heat of solution).

Dispose of the salt solution as directed by the instructor.

Clean out and dry the calorimeter, and repeat the determination twice more. Calculate a mean value for the heat of solution of your salt from your three determinations.

Name ____________________ Section ____________________

Lab Instructor ____________________ Date ____________________

EXPERIMENT 17 Calorimetry

PRE-LABORATORY QUESTIONS

CHOICE I. DETERMINATION OF A CALORIMETER CONSTANT

1. What is the definition of the *joule* in terms of the basic SI units?

____

2. A calorimeter is to be calibrated: 51.203 g of water at 55.2°C is added to a calorimeter containing 49.783 g of water at 23.5°C. After stirring and waiting for the system to equilibrate, the final temperature reached is 37.6°C. Calculate the calorimeter constant.

CHOICE II. SPECIFIC HEATS OF METALS AND GLASS

1. Using a handbook of chemical data, look up the specific heat capacities of the following materials. Give your references.

Substance	Specific heat	Reference
Al(*s*)	________	____________
C(*s*)	________	____________
Fe(*s*)	________	____________
$H_2O(l)$	________	____________
Pb(*s*)	________	____________

2. What are the *units* of specific heat capacity?

____

CHOICE III. HEAT OF ACID/BASE REACTIONS

1. The acids and bases to be determined in this experiment are all classified as *strong* acids or bases. Use your textbook to find what is meant by the word *strong* in this context.

2. What is meant by *neutralization*?

CHOICE IV. HEAT OF METAL/ACID REACTIONS

1. If adding a 0.552 g sample of Metal X to 150 mL of 1.0 *M* HCl caused a temperature increase of 7.33°C, calculate the quantity of heat that has flowed.

2. If the atomic weight of Metal X is 113, calculate ΔH for the above process.

CHOICE V. HEAT OF SOLUTION OF A SALT

1. Use your textbook or a chemical encyclopedia to write a specific definition of a *salt*.

2. What does it mean to say that an ion becomes *hydrated* when a salt is dissolved in water?

Name ______________________ Section ______________________

Lab Instructor ______________________ Date ______________________

EXPERIMENT 17 Calorimetry

CHOICE I. DETERMINATION OF A CALORIMETER CONSTANT

RESULTS/OBSERVATIONS

	Trial 1	Trial 2	Trial 3
Mass (volume) of cold water			
Temperature of cold water			
Mass (volume) of hot water			
Temperature of hot water			
Final temperature reached			
Temperature change, ΔT			
Calorimeter constant			

Mean value of calorimeter constant ______________________

QUESTIONS

1. What effect on the calorimeter constant calculated would be observed if the calorimeter cup were made of a conducting material (such as metal) rather than plastic foam?

____

____

____

2. Why is water typically used as the heat-absorbing liquid in calorimeters?

____

____

3. What is the significance of the minus sign in Equation 17-2 in the Introduction of Choice I of this experiment?

____

____

____

____

4. A unit of heat energy that was formerly used frequently was the *calorie*. Look up the definition of the calorie in your textbook or a handbook and record it here.

____

____

____

____

5. Calculate the calorimeter constant for your calorimeter in cal/°C.

____

Name ______________________ Section ______________________

Lab Instructor ______________________ Date ______________________

EXPERIMENT 17 Calorimetry

CHOICE II. SPECIFIC HEATS OF METALS AND GLASS

RESULTS/OBSERVATIONS

Specific heat of a metal

Metallic substance used ______________________

	Trial 1	Trial 2	Trial 3
Mass of substance taken	________	________	________
Mass (volume) cold water used	________	________	________
Initial temperature of water	________	________	________
Final temperature of water	________	________	________
Temperature change, ΔT	________	________	________
Specific heat of substance	________	________	________

Mean value of specific heat of substance ______________________

Literature value ______________ Reference ______________

Percent error in specific heat determination ______________________

Specific heat of glass

	Trial 1	Trial 2	Trial 3
Mass of glass taken	________	________	________
Mass (volume) cold water used	________	________	________

Initial temperature of water ____________ ____________ ____________

Final temperature of water ____________ ____________ ____________

Temperature change, ΔT ____________ ____________ ____________

Specific heat of glass ____________ ____________ ____________

Mean value of specific heat of glass ____________

Literature value ____________ Reference ____________

Percent error in specific heat determination ____________

QUESTIONS

1. In the pre-lab question, you were asked to look up the specific heat of ice [i.e., $H_2O(s)$]. Why is the value for ice not the same as for liquid water?

____

____

____

2. You determined the specific heat of glass. There are many types of glass. Use a chemical handbook to find the chemical composition of three types of glass, and tell how their specific heats differ.

____

____

____

____

____

____

Name ______________________ Section ______________________

Lab Instructor ______________________ Date ______________________

EXPERIMENT 17 Calorimetry

CHOICE III. HEAT OF ACID/BASE REACTIONS

RESULTS/OBSERVATIONS

Reaction of HCl and NaOH

	Trial 1	Trial 2	Trial 3
Volume of 2 *M* NaOH used			
Initial temperature of NaOH			
Volume of 2 *M* HCl used			
Initial temperature of HCl			
Final temperature reached			
Total mass (volume) of mixture			
Temperature change, ΔT			
Heat flow, joules			
Moles of water produced			
ΔH (kJ/mol water)			

Mean value of ΔH ______________________ Literature value ______________________

Reaction of HCl and KOH

	Trial 1	Trial 2	Trial 3
Volume of 2 *M* KOH used			
Initial temperature of KOH			
Volume of 2 *M* HCl used			
Initial temperature of HCl			

Final temperature reached			
Total mass (volume) of mixture			
Temperature change, ΔT			
Heat flow, joules			
Moles of water produced			
ΔH (kJ/mol water)			

Mean value of ΔH ______________________ Literature value ______________________

Reaction of HNO_3 and NaOH

	Trial 1	Trial 2	Trial 3
Volume of 2 M NaOH used			
Initial temperature of NaOH			
Volume of 2 M HNO_3 used			
Initial temperature of HNO_3			
Final temperature reached			
Total mass (volume) of mixture			
Temperature change, ΔT			
Heat flow, joules			
Moles of water produced			
ΔH (kJ/mol water)			

Mean value of ΔH ______________________ Literature value ______________________

Reaction of HNO_3 and KOH

	Trial 1	Trial 2	Trial 3
Volume of 2 M KOH used			
Initial temperature of KOH			
Volume of 2 M HNO_3 used			
Initial temperature of HNO_3			
Final temperature reached			

Total mass (volume) of mixture ________ ________ ________

Temperature change, ΔT ________ ________ ________

Heat flow, joules ________ ________ ________

Moles of water produced ________ ________ ________

ΔH (kJ/mol water) ________ ________ ________

Mean value of ΔH ________ Literature value________

QUESTIONS

1. The heat flows measured in this experiment were actually not for the simple neutralization of a proton and hydroxide ion (as indicated in the Introduction for Choice III); rather they include contributions based on the fact that these species are hydrated in aqueous solution. What does it mean to say that a proton is hydrated, and how will the heat of hydration affect the measured heat flow for a neutralization reaction?

2. Based on your 12 measurements, what is the mean value and the standard deviation for ΔH for the reaction between H^+ and OH^- in aqueous solution?

Name ______________________ Section ______________________

Lab Instructor ______________________ Date ______________________

EXPERIMENT 17 Calorimetry

CHOICE IV. HEAT OF METAL/ACID REACTIONS

RESULTS/OBSERVATIONS

Reaction of HCl and Mg metal

	Trial 1	Trial 2	Trial 3
Mass of Mg used, g	______	______	______
Volume of 1 *M* HCl used	______	______	______
Initial temperature of HCl	______	______	______
Final temperature reached	______	______	______
Temperature change, ΔT	______	______	______
Heat flow, joules	______	______	______
Moles of Mg used	______	______	______
ΔH (kJ/mol Mg)	______	______	______

Mean value of ΔH ______________ Literature value ______________

Reaction of HNO_3 and Mg metal

	Trial 1	Trial 2	Trial 3
Mass of Mg used, g	______	______	______
Volume of 1 *M* HNO_3 used	______	______	______
Initial temperature of HNO_3	______	______	______
Final temperature reached	______	______	______
Temperature change, ΔT	______	______	______

Heat flow, joules	______	______	______
Moles of Mg used	______	______	______
ΔH (kJ/mol Mg)	______	______	______

Mean value of ΔH ______ Literature value ______

Reaction of H_2SO_4 and Mg metal

	Trial 1	**Trial 2**	**Trial 3**
Mass of Mg used, g	______	______	______
Volume of 1 M H_2SO_4 used	______	______	______
Initial temperature of H_2SO_4	______	______	______
Final temperature reached	______	______	______
Temperature change, ΔT	______	______	______
Heat flow, joules	______	______	______
Moles of Mg used	______	______	______
ΔH (kJ/mol Mg)	______	______	______

Mean value of ΔH ______ Literature value ______

Reaction of HCl and Zn metal

	Trial 1	**Trial 2**	**Trial 3**
Mass of Zn used, g	______	______	______
Volume of 1 M HCl used	______	______	______
Initial temperature of HCl	______	______	______
Final temperature reached	______	______	______
Temperature change, ΔT	______	______	______
Heat flow, joules	______	______	______
Moles of Zn used	______	______	______
ΔH (kJ/mol Zn)	______	______	______

Mean value of ΔH ______ Literature value ______

Reaction of HNO_3 and Zn metal

	Trial 1	Trial 2	Trial 3
Mass of Zn used, g			
Volume of 1 *M* HNO_3 used			
Initial temperature of HNO_3			
Final temperature reached			
Temperature change, ΔT			
Heat flow, joules			
Moles of Zn used			
ΔH (kJ/mol Zn)			

Mean value of ΔH __________ Literature value__________

Reaction of H_2SO_4 and Zn metal

	Trial 1	Trial 2	Trial 3
Mass of Zn used, g			
Volume of 1 *M* H_2SO_4 used			
Initial temperature of H_2SO_4			
Final temperature reached			
Temperature change, ΔT			
Heat flow, joules			
Moles of Zn used			
ΔH (kJ/mol Zn)			

Mean value of ΔH __________ Literature value__________

QUESTIONS

1. Why would one not expect the heat of reaction to depend on which acid is used for the reaction?

____

____

____

2. Which metal, zinc or magnesium, releases more heat energy per gram when reacted with acid? Which metal releases more heat energy per mole?

____

____

____

Name ____________________ Section ____________________

Lab Instructor ____________________ Date ____________________

EXPERIMENT 17 Calorimetry

CHOICE V. HEAT OF SOLUTION OF A SALT

RESULTS/OBSERVATIONS

	Trial 1	Trial 2	Trial 3
Volume of water used, mL	________	________	________
Initial temperature of water, °C	________	________	________
Mass of salt taken, g	________	________	________
Moles of salt taken, mol	________	________	________
Highest/Lowest temperature, °C	________	________	________
Calculated heat flow, J	________	________	________
Enthalpy change (ΔH), kJ/mol	________	________	________

Mean heat of solution, kJ/mol ____________________

QUESTIONS

1. If the identity of the salt was provided by the instructor, use a chemical handbook to find the literature value of the heat of solution for your salt. Calculate the percent error between the literature value and your mean experimental value for ΔH of solution for your salt.

   Literature value for ΔH ____________ Reference ____________ % error ____________

2. Why is the dissolving of a salt sometimes an exothermic and sometimes an endothermic process?

   ____

   ____

   ____

   ____

   ____

Experiment

18 Atomic Spectroscopy

Objective

The emission and absorption of light energy of particular wavelengths by atoms and molecules is a common phenomenon. The emissions/absorptions are characteristic for each element's atoms and arise from transitions of electrons among the various energy levels of the particular atom under study. The apparatus used to study the wavelengths of light emitted/absorbed by atoms is called a **spectroscope**. In this experiment, you will first calibrate a spectroscope (Choice I). Then you will use the calibrated spectroscope to determine the wavelengths of the emission lines of hydrogen and nitrogen (Choice II) and of a number of metal salts (Choice III).

Choice I. Calibration of the Spectroscope

Introduction

The radiant energy emitted by the sun (or other stars) contains *all* possible wavelengths of electromagnetic radiation. The portion of this radiation to which the retina of the human eye responds is called the **visible light region** of the electromagnetic spectrum. The fact that the radiation emitted by the sun contains a *mixture* of radiation wavelengths may be demonstrated by passing sunlight through a prism. A prism *bends* light; the *degree* to which light is bent is related to the wavelength of the light. When sunlight (or other "white" light) containing all possible wavelengths is passed through a prism, each component color of the white light is bent to a different extent by the prism, resulting in the beam of white light being spread out into a complete rainbow of colors. Such a rainbow pattern is called a **continuous spectrum**.

It was discovered that the use of a narrow *slit* in the spectroscope between the prism and the source of white light sharpened and improved the quality of spectra produced by a beam of white light. See Figure 18-1.

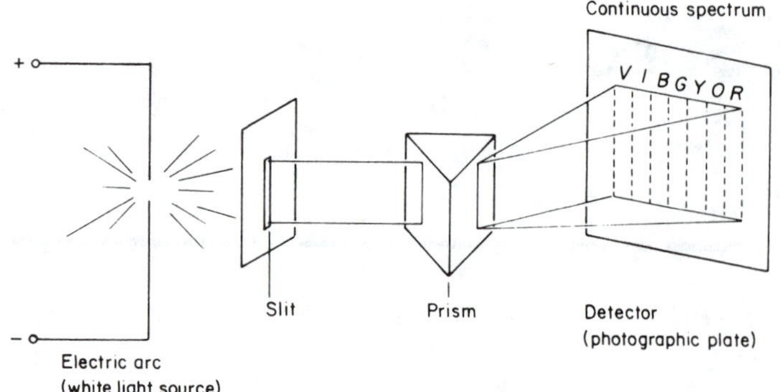

Figure 18-1. The spectrum of "white" light. When light from the sun or from a high-intensity incandescent bulb is passed through a prism, the component wavelengths are spread out into a continuous rainbow spectrum.

Most substances will emit light energy if heated to a high enough temperature. For example, a fireplace poker will glow red if left in the fireplace flame for several minutes. Similarly, neon gas will glow with a bright red color when excited with a sufficiently high electrical voltage; this is made use of in neon signs. When energy is applied to a substance, the atoms present in the substance may *absorb* some of that energy. Electrons within the atoms of the substance move from their normal positions to positions of higher potential energy, farther away from the nuclei of the atoms. Later, atoms which have been excited by the application and absorption of energy will "relax" and will *emit* the excess energy they had gained. When atoms re-emit energy, more often than not, at least a portion of this energy is visible as light.

However, atoms do *not* emit light energy *randomly*. In particular, the atoms of a given element do *not* generally emit a continuous spectrum, but rather emit visible radiation at only certain discrete, well-defined, fixed wavelengths. For example, if you have ever spilled common table salt, NaCl, in a flame, you have seen that sodium atoms emit a characteristic yellow/orange wavelength of light.

If the light being emitted by a particular element's atoms is passed through a prism and is viewed with a spectroscope, only certain sharp bright-colored *lines* are seen in the resulting spectrum. The positions of these colored lines occur in the corresponding location (wavelength region) as in the spectrum of white light. See Figure 18-2, which illustrates the bright line spectrum of hydrogen.

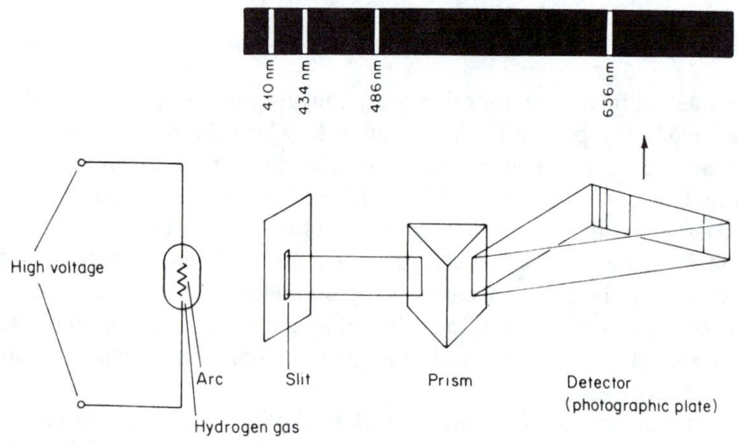

Figure 18-2. The line spectrum of hydrogen. The location of colored lines in the spectrum corresponds to the location of the same color in the spectrum of white light.

The fact that a given atom produces only *certain* fixed bright *lines* in its spectrum indicates that the atom can only undergo energy changes of certain fixed, definite amounts.

An atom cannot continuously or randomly emit radiation but can only emit energy corresponding to definite, regular changes in the energies of its component electrons. The experimental demonstration of bright line spectra implied a regular, fixed electronic microstructure for the atom and led to an enormous amount of research to discover exactly what that microstructure is.

In later choices of this experiment, you will examine the line spectra of a number of elements, using a simple spectroscope of the sort indicated in Figure 18-3. The spectroscope includes four major features: a *slit* for admitting a narrow, collimated beam of light; a *prism* or *diffraction grating* that spreads the incident light into its component wavelengths; a *telescope* for viewing the spectrum; and an illuminated *reference scale* against which the spectrum may be viewed (as an aid in locating the positions of the lines in the spectrum).

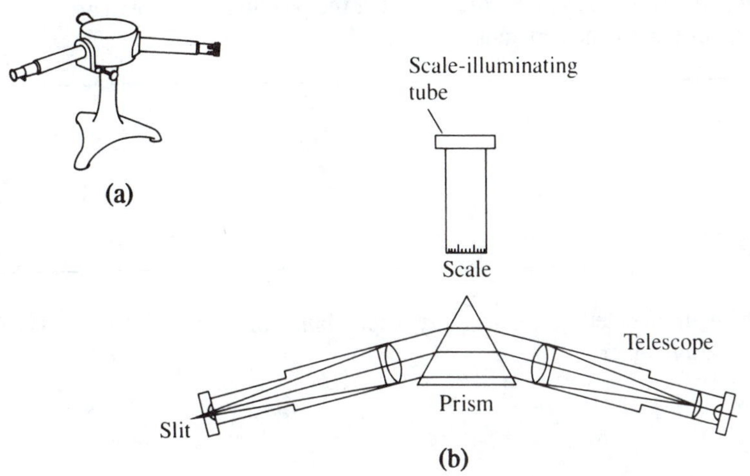

Figure 18-3. (a) a view of the spectroscope to be used. (b) A schematic representation of the spectroscope, showing its component parts. When viewed through the telescope, the spectrum will appear superimposed on the numerical scale.

The scale of the spectroscope is provided merely as a convenience, and the divisions on the scale are arbitrary. For this reason, the spectroscope must be *calibrated* before it is used to determine the spectrum of an unknown element. This is accomplished by viewing a known element that produces especially sharp lines in its spectrum and whose spectrum has been previously characterized (with the emission wavelengths being known with great precision). The positions on the spectroscope scale of the lines in the spectrum of the known element are recorded. Then a **calibration curve** that relates the wavelength of a spectral line of the known element to its position when viewed against the spectroscope scale is prepared. This calibration curve may then be used to calculate the wavelengths of emission lines in the spectra of unknown elements when viewed through the same spectroscope under the same conditions.

In this option, you will calibrate the spectroscope by viewing the spectrum of mercury. The spectrum of mercury has been intensively studied. It has several bright lines in the visible region (400–700-nanometer wavelength):

404.7 nm	violet
435.8 nm	blue
546.1 nm	green
579.0 nm	yellow

A calibration graph will then be prepared, with the apparent position of the spectral line on the scale of the spectroscope, plotted against the known wavelength of that spectral line. This calibration graph should be kept if any additional measurements are to be made with the spectroscope.

Safety Precautions

- Wear safety glasses at all times in the laboratory.
- In addition to visible light, the mercury vapor lamp emits radiation at ultraviolet wavelengths. Ultraviolet radiation is *damaging* to the eyes. Wearing your safety glasses while taking readings with the spectroscope will absorb most of this radiation; nevertheless, refrain from looking at the source of radiation for any extended period of time.
- The power supply used with the mercury lamp develops a potential of several thousand volts. *Do not touch* any portion of the power supply, wire leads, or mercury lamp unless the power supply is unplugged from the wall outlet.
- Always *unplug* the power supply before adjusting the position of the mercury lamp or any other part of the apparatus.

Apparatus/Reagents Required

Spectroscope with illuminated scale, mercury vapor lamp (discharge tube), high-voltage power pack with lamp holder

Procedure

Record all data and observations directly in your notebook in ink.

Check to make sure that the power supply pack is *unplugged.* Remember that this pack operates at high voltages and is dangerous.

Turn on the illuminated scale of the spectroscope, and look through the eyepiece to make sure that the scale is visible but not so brightly lighted as to obscure the mercury spectral lines.

Position the power supply pack containing the mercury vapor lamp so that the lamp is directly in front of the slit opening of the spectroscope.

With the instructor's permission, plug in the power supply and turn on the power supply switch to illuminate the mercury lamp.

Look through the eyepiece, and adjust the slit opening of the spectroscope so that the mercury vapor spectral lines are as bright and as sharp as possible. If necessary, adjust the illuminated scale of the spectroscope so that the numbered scale divisions are easily read but do not obscure the mercury spectral lines.

Record the *color* and *location* on the numbered scale of the spectroscope for each line in the visible spectrum of mercury.

Using the line positions you have recorded with the spectroscope and the known wavelengths of the emissions of the mercury atom given earlier, construct a calibration curve for the spectroscope using fine-scale graph paper. Plot the *observed scale reading* for each bright line versus the *wavelength* of the line.

Choice II. Spectra of Atomic Hydrogen and Nitrogen

Introduction

Hydrogen is the simplest of the atoms, consisting of a single proton and a single electron. The emission spectrum of hydrogen is of interest because this spectrum was the first to be completely explained by a theory of atomic structure, by the Danish scientist Niels Bohr.

As described in Choice I, we know that atoms absorb and emit radiation as light of fixed, characteristic wavelengths when excited. This absorption and emission of light is now known to correspond to electrons within the atom moving away from the nucleus (energy absorbed) or closer to the nucleus (energy emitted). Atoms emit and absorb energy of only certain wavelengths (bright or dark lines in the spectrum) because electrons do not move randomly away from and toward the nucleus, but may only move between certain fixed, allowed "orbits," each of which is at a definite fixed distance from the nucleus. When an electron moves from one of the fixed orbits to another orbit, the attractive force of the nucleus changes by a definite amount that corresponds to a specific change in energy. The quantity of energy absorbed or emitted by an electron in moving from one allowed orbit to another is called a **quantum (photon)**, and the energy of the particular quantum is indicated by the wavelength (or frequency) of the light emitted or absorbed by the atom. The energy of a photon is given by the Planck equation

$$\Delta E = h\nu = hc/\lambda$$

where ν is the frequency of light emitted or absorbed and λ is the wavelength corresponding to that frequency.

Bohr postulated that the energy of an electron when it is in a particular orbit was given by the formula

$$E_n = -(\text{constant})/n^2$$

where n is the number of the orbit as counted out from the nucleus ($n = 1$ means the first orbit, $n = 2$ means the second orbit, etc.) and is called the **principal quantum number**. The proportionality constant in Bohr's theory is called the Rydberg constant (given the symbol R_H) and has the value 2.18×10^{-18} J. According to Bohr's theory, if an electron were to move from an outer orbit (designated as n_{outer}) to an inner orbit (designated by n_{inner}), a photon of light should be emitted, having energy given by

$$\Delta E = E_{\text{inner}} - E_{\text{outer}} = -R_H[(1/n_{\text{inner}}^2) - (1/n_{\text{outer}}^2)]$$

The wavelength (λ) of this photon would be given by the Planck formula as

$$\lambda = hc/\Delta E$$

Bohr performed calculations of wavelengths for various values of the principal quantum number, n, and found that the predicted wavelengths from theory agreed exactly with experimental wavelengths measured with a spectroscope. Bohr even went so far as to predict emissions by hydrogen atoms in other regions of the electromagnetic spectrum (ultraviolet, infrared) that had not yet been observed experimentally but that were confirmed almost immediately.

Bohr's simple atomic theory of an electron moving between fixed orbits helped greatly to explain observed spectra and formed the basis for the detailed modern atomic theory for more complex atoms with more than one electron. The spectra of larger atoms are considerably more complicated than that of hydrogen, but generally a *characteristic* spectrum is seen. Bohr's theory for hydrogen accounted on a microscopic basis for the macroscopic phenomena of spectral emission lines.

In this experiment, you will measure the wavelengths of the lines in the emission spectrum of hydrogen with a spectroscope and then determine by calculation to which atomic transition (of the electron between the various orbits) each of these spectral lines corresponds. You will also examine the emission spectra of nitrogen, which as a multi-electron atom is considerably more complicated to interpret.

Safety Precautions

- Wear safety glasses at all times in the laboratory.
- In addition to visible light, the hydrogen lamp and nitrogen lamps emit radiation at ultraviolet wavelengths. Ultraviolet radiation is *damaging* to the eyes. Wearing your safety glasses while taking readings with the spectroscope will absorb most of this radiation. Refrain from looking at the source of radiation for any extended period of time.
- The power supply used with the lamps develops a potential of several thousand volts. *Do not touch* any portion of the power supply, wire leads, or lamps unless the power supply is unplugged from the wall outlet.
- Always *unplug* the power supply before adjusting the position of the lamps or any other part of the apparatus.

Apparatus/Reagents Required

Spectroscope with illuminated scale, hydrogen lamp (discharge tube), nitrogen lamp, high-voltage power pack with lamp holder

Procedure

Record all data and observations directly in your notebook in ink.

Check to make sure that the power supply pack is *unplugged.* Remember that this power supply operates at high voltages and is dangerous.

Turn on the illuminated scale of the spectroscope. Look through the eyepiece to make sure that the scale is visible but not so brightly lighted as to obscure the hydrogen spectral lines.

Position the power supply pack containing the hydrogen vapor lamp so that the lamp is directly in front of the slit opening of the spectroscope.

With the instructor's permission, plug in the power supply and turn on the power supply switch to illuminate the hydrogen lamp.

Look through the eyepiece, and adjust the slit opening of the spectroscope so that the hydrogen spectral lines are as bright and as sharp as possible. If necessary, adjust the illuminated scale of the spectroscope so that the numbered scale divisions are easily read but do not obscure the hydrogen spectral lines.

Record the color and location on the numbered scale of the spectroscope for each line in the visible spectrum of hydrogen. You should easily observe red, blue-green, and violet lines. A second very faint violet line may also be visible if the room lighting and illuminated scale lights are not too bright.

Use the calibration plot prepared in Choice I for mercury vapor to determine the wavelengths of the lines in the hydrogen emission spectrum. Look up the true wavelengths of these lines in your textbook or a chemical handbook, and calculate the percent error in the determination of each line's wavelength.

Use the equations provided in the introduction to this choice to calculate the predicted wavelengths in nanometers (according to Bohr's theory) from the electronic transitions in the hydrogen atom corresponding to the following: $n = 3 \rightarrow n = 2$; $n = 4 \rightarrow n = 2$; $n = 5 \rightarrow n = 2$; $n = 6 \rightarrow n = 2$. How do these predicted wavelengths correspond to those you have measured for hydrogen?

Turn off and unplug the power pack containing the hydrogen lamp. Position the power pack containing the nitrogen lamp in front of the slit opening of the spectroscope. Adjust the spectroscope and scale, and record the location of the bright lines in the nitrogen spectrum. A portion of the emission spectrum of nitrogen appears as a band of several colors. Record the position of this band on the spectroscope scale. Using the calibration curve prepared in Choice I, determine the wavelengths of the bright lines in the nitrogen spectrum, and also the approximate wavelength range covered by the band.

Choice III. Emission Lines of Some Metallic Elements

Introduction

A number of common metallic elements emit light strongly in the visible light region when ions of the metals are excited. The spectra can be studied using the same sort of simple spectroscope as has been used in Choices I and II.

A number of metallic elements from Groups IA and IIA have especially bright emission lines in the visible light region. The emissions are so strong and characteristically colored that these elements can often be recognized by the gross color they impart when aspirated into a burner flame, even without use of a spectroscope. For example, lithium ions impart a red color to a burner flame, sodium ions a yellow/orange color, potassium ions a violet color, calcium ions a brick red color, strontium ions a brighter red color, and barium ions a green color. Upon examination with the spectroscope, it is noted that the spectra of these ions contain several additional lines, but generally the *brightest* line in the spectrum when viewed through the spectroscope corresponds to the gross color imparted to a flame when the spectroscope is not used. Naturally, when several of these ions are present together in a sample (as happens with real samples very commonly), one color may mask another so that direct visual identification of the ions may not be possible. In this case, only a calibrated spectroscope can determine what elements are present. This is done by comparing the position of lines in the spectrum of the unknown mixture with the position of lines in known single samples of the ions in question.

In this Choice, you will excite the ions of each of the elements listed earlier, using a Bunsen burner. You will observe both the gross color imparted to the flame and the spectral lines emitted by the elements as viewed through the spectroscope. You will then determine what elements of those tested are present in an unknown mixture.

Two alternative procedures may be available for introducing the metal ion samples into the burner flame. In the first method, a wire loop is used to pick up a drop of metal solution, and the drop is then placed into the flame for vaporization. This method is simple and requires little equipment, but it only produces a brief burst of color before the sample evaporates completely. This makes it difficult for one person to both introduce the sample and record the spectrum. The second method uses a spray bottle to introduce a fine mist of sample solution into the flame. This method allows for a longer lifetime for the color of the ion in the flame but requires separate sprayers for each of the ions (because of the difficulty in cleaning the sprayer between samples).

Safety Precautions

- Wear safety glasses at all times in the laboratory.
- Some of the metal salts used in this experiment are *toxic*. Wash your hands after using them. If the spray method is used, wash down the lab bench to remove all traces of the metal salts.
- Avoid having the burner flame too close to the spectroscope, to avoid damaging the apparatus.
- 6 *M* HCl can burn skin and clothing. Exercise caution in its use, and inform the instructor of any spills.

Apparatus/Reagents Required

Spectroscope with illuminated scale; burner; nichrome wire or spray bottles; 6 *M* HCl; 0.10 *M* solutions of the following metal chlorides: lithium, sodium, potassium, calcium, and strontium; unknown solution containing one of these ions; unknown mixture containing two of these ions

Procedure

Record all data and observations directly in your notebook in ink.

Your instructor will tell you which method for introducing the metal ions into the flame to use. In either case, you will be determining the spectra of solutions of lithium chloride, sodium chloride, potassium chloride, calcium chloride, and strontium chloride.

If the wire method is to be used, obtain several 6-inch lengths of nichrome wire and a small amount of 6 *M* HCl (*Caution!*). Bend the last quarter-inch of each wire into a small loop for picking up the sample solutions. Dip the loops into 6 *M* HCl to remove any oxides present, rinse in distilled water, and then heat in the oxidizing portion of a burner flame until no color is imparted to the flame by the wires.

If the spray method is to be used, using the sink, check to make sure that the sprayers deliver a very fine mist. If the sprayer nozzle is adjustable, try adjusting the nozzle to improve the character of the spray. If the nozzle cannot be adjusted, consult your instructor for a method of cleaning the sprayer.

Set up a laboratory burner directly in front of the slit of the spectroscope but far enough away from the spectroscope to avoid damaging the instrument. Adjust the flame of the burner so that it is as hot as possible. Adjust the illuminated scale of the spectroscope so that approximate positions of the spectral lines can be noted (exact measurements will not be made).

Using the wire method, introduce a drop of one of the metal ion solutions into the oxidizing portion of the flame and note the gross color imparted to the flame by the solution. Then introduce a second drop of the same metal ion solution into the flame while looking at the flame through the eyepiece of the spectroscope. Note the color, intensity, and approximate scale position of the brightest few lines in the metal ion's spectrum. It may be necessary to repeat the introduction of a drop of the metal ion solution to the flame to permit recording of all the spectral lines shown by the metal ion.

Using the spray method, spray a fine mist of one of the metal ion solutions into the flame and note the gross color imparted to the flame by the metal ion. Then while looking through the eyepiece of the spectroscope, spray additional bursts of the same metal ion solution into the flame, noting the color, intensity, and approximate scale position of the brightest few lines in the metal ion's spectrum.

Repeat the determinations (by either method) using the other metal ion samples. If using the wire method, use a new length of wire for each successive sample. If using the spray method, allow the burner to heat for a few minutes between samples to make sure that the previous sample has been evaporated completely.

When each of the known samples of metal ions has been determined, obtain an unknown sample containing just one of the metal ions. Determine the spectrum of the unknown sample, and by matching the colors, intensities, and positions of the spectral lines, identify the unknown sample.

Obtain an unknown sample containing a mixture of two of the metal ions. Determine the spectrum of the mixture, noting the color, intensity, and position of all the brightest lines in the spectrum. Compare the spectrum of the mixture with the spectra of each of the individual known ions, and determine which ions are most likely to be present in the unknown mixture.

Name ______________________________ Section ______________________

Lab Instructor ______________________________ Date ______________________

EXPERIMENT 18 Atomic Spectroscopy

PRE-LABORATORY QUESTIONS

CHOICE I. CALIBRATION OF THE SPECTROSCOPE

1. Given the wavelengths and colors of the major mercury emission lines listed in the introduction to this experiment, construct a graph (on a sheet of graph paper from the end of this manual) which will approximate the spectrum to be determined. Lay out the horizontal axis of the graph in terms of the wavelength of the lines (in nanometers), and make the vertical axis about 2 inches high to approximate the appearance of the lines to be measured. Using colored pens that correspond to those colors listed, sketch in the emission lines. Compare your sketch of the spectrum to the actual spectrum recorded. Attach your sketch to this page.

2. What are the approximate wavelengths (in nanometers) in the spectrum of visible sunlight that correspond to the following colors?

   Red ____________________ Yellow ____________________

   Green ____________________ Blue ____________________

3. What are the frequencies (in Hz) corresponding to the wavelengths you have listed ?

   ____

   ____

CHOICE II. SPECTRA OF ATOMIC HYDROGEN AND NITROGEN

1. The spectral lines observed in the visible spectrum of hydrogen arise from transitions from upper states back to the $n = 2$ principal quantum level. Calculate the predicted wavelengths for the spectral transitions of the hydrogen atom from the $n = 6$ to $n = 2$, for the $n = 5$ to $n = 2$, for the $n = 4$ to $n = 2$, and for the $n = 3$ to $n = 2$ levels in atomic hydrogen.

   ____

   ____

   ____

   ____

   ____

2 Why is it not possible to so easily interpret the spectrum of a polyelectronic atom (such as nitrogen in this experiment)?

CHOICE III. EMISSION LINES OF SOME METALLIC ELEMENTS

1. Use a chemical dictionary or encyclopedia to find the distinction between atomic emission and absorption spectroscopy.

2. Atomic spectroscopy is used in police laboratories for the identification of samples collected at crime scenes—bullet fragments, for example. A bullet typically might consist of an alloy of several metals (copper, zinc, lead, etc.). How would you expect the atomic spectrum of a *mixture* of elements to compare to the individual spectra of the constituent elements in the sample? Would you expect the spectra of a mixture to be a superimposition of the individual spectrum, or would you expect the spectral emissions of one atom to influence the spectrum of another atom? Discuss.

Name ______________________________ Section ____________________

Lab Instructor ______________________________ Date ____________________

EXPERIMENT 18 Atomic Spectroscopy

CHOICE I. CALIBRATION OF THE SPECTROSCOPE

RESULTS/OBSERVATIONS

Description of the spectrum

____

____

____

____

Lines observed:

Color	*Location on spectroscope scale*
____________	____________________
____________	____________________
____________	____________________
____________	____________________
____________	____________________
____________	____________________
____________	____________________
____________	____________________

Attach your graph to this report sheet.

QUESTIONS

1. What is the purpose of the slit in the spectroscope?

2. Why is *mercury* used to calibrate the spectroscope?

3. Why should you *not* look at the emissions from the mercury tube for a prolonged time period?

Name ______________________________ Section ______________________

Lab Instructor ______________________________ Date ______________________

EXPERIMENT 18 Atomic Spectroscopy

CHOICE II. SPECTRA OF ATOMIC HYDROGEN AND NITROGEN

RESULTS/OBSERVATIONS

Hydrogen Emission Spectrum

Description of the spectrum

____

____

Lines observed:

Color	*Location on spectroscope scale*
____________	______________________
____________	______________________
____________	______________________
____________	______________________
____________	______________________
____________	______________________
____________	______________________

Wavelengths of the lines (from calibration graph, Choice I)

Color	*Wavelength*
____________	______________________
____________	______________________
____________	______________________
____________	______________________

Calculate the percent error in the determination of the wavelengths (comparing the wavelengths you have determined with those given in your textbook).

Color *Percent error in wavelength*

Nitrogen Emission Spectrum

Description of the spectrum

Lines observed:

Color *Location on spectroscope scale*

Wavelengths of the lines (from calibration graph, Choice I)

Color	*Wavelength*
________	________
________	________
________	________
________	________
________	________
________	________

QUESTIONS

1. In addition to the visible lines in the hydrogen emission spectrum determined earlier, atomic hydrogen also emits short-wavelength ultraviolet radiation (for example, sunburns are a result of such radiation from the sun). Ultraviolet emissions arise from transitions back to the ground state (n = 1) level. Beginning with the n = 6 level, calculate the wavelengths for the ultraviolet emissions of atomic hydrogen.

2. As you observed the spectrum of atomic nitrogen, you may have noticed that part of the spectrum was *continuous*: that is, in one portion of the spectrum there was a *band* of color rather than a sharp line. What do you think might account for such a continuous spectral region?

Name ______________________ Section ______________________

Lab Instructor ______________________ Date ______________________

EXPERIMENT 18 Atomic Spectroscopy

CHOICE III. EMISSION LINES OF SOME METALLIC ELEMENTS

RESULTS/OBSERVATIONS

Which method did you use for introducing the metal ion samples?

____

Metal ion	*Gross color*	*Lines observed*
Li	____________	______________________
Na	____________	______________________
K	____________	______________________
Ca	____________	______________________
Sr	____________	______________________
Ba	____________	______________________
Unknown 1 (single)	____________	______________________
Unknown 2 (double)	____________	______________________

ID number of Unknown 1 ____________ Identity? ____________

ID number of Unknown 2 ____________ Identity? ____________

QUESTIONS

1. Why were spectra observed only for alkali and alkaline earth metal cation solutions (and not cations of other groups)?

2. Of the metal cations tested, sodium usually gives the brightest and most persistent color to the flame. What problems would this introduce if a real mixture containing both sodium and other cations were to be analyzed by the technique used in this experiment? How could the problems be solved?

EXPERIMENT

19 Molecular Shapes and Structures

Objective

Models will be built according to the predictions of the VSEPR theory to illustrate the regular patterns of molecular shapes.

Introduction

The shapes exhibited by molecules are often very difficult for beginning chemistry students to visualize, especially since most students' training in geometry is limited to *plane* geometry. To understand the geometric shapes exhibited by molecules, a course in solid geometry, which covers the shapes of three-dimensional figures, would be more useful. In this experiment you will encounter some unfamiliar geometrical arrangements that will help you to appreciate the complexity and importance of molecular geometry.

The basic derivation and explanation of molecular shapes arises from the valence shell electron pair repulsion theory, usually known by its abbreviation, **VSEPR**. This theory considers the environment of the most central atom in a molecule and imagines first how the valence electron pairs of that central atom must be arranged in three-dimensional space around the atom to minimize repulsion among the electron pairs. The general principle is as follows: For a given number of pairs of electrons, the pairs will be oriented in three-dimensional space to be as *far away from each other as possible*. For example, if a central atom were to have only two pairs of valence electrons around it, the electron pairs would be expected to be 180° from each other.

The VSEPR theory then also considers which pairs of electron pairs around the central atom are **bonding pairs** (with atoms attached) and which are **nonbonding pairs** (lone pairs). The overall geometric shape of the molecule as a whole is determined by *how many* pairs of electrons are on the central atom and by which of those pairs are used for *bonding* to other atoms.

It is sometimes difficult for students to distinguish between the orientation of the electron pairs of the central atom of a molecule and the overall geometric shape of that molecule. A simple example that clearly makes this distinction concerns the case in which the central atom of the molecule has four valence electron pairs. Consider the Lewis structures of the following four molecules: hydrogen chloride, HCl; water, H_2O; ammonia, NH_3; and methane, CH_4.

```
           ..          ..                              H
                                                       |
       H—Cl:       H—O—H        H—N—H         H—C—H
           ..          ..           |                  |
                                    H                  H
```

The central atom in each of these molecules is surrounded by four pairs of valence electrons. According to the VSEPR theory, these four pairs of electrons will be oriented in three-dimensional space to be as far away from each other as possible. The four pairs of electrons point to the corners of the geometrical figure known as a **tetrahedron**. The four pairs of electrons are said to be tetrahedrally oriented, and are separated by angles of approximately 109.5°.

However, three of the molecules shown are *not* tetrahedral in *overall shape*, because some of the valence electron pairs in the HCl, H_2O, and NH_3 molecules are not *bonding* pairs. The angular position of the bonding pairs (and hence the overall shape of the molecule) is determined by the *total* number of valence electron pairs on the central atom, but the non-bonding electron pairs are *not* included in the description of the molecules' overall shape. For example, the HCl molecule could hardly be said to be tetrahedral in shape, since there are only two atoms in the molecule. HCl is linear even though the valence electron pairs of the chlorine atom are tetrahedrally oriented. Similarly the H_2O molecule cannot be tetrahedral. Water is said to be *V*-shaped (bent, or nonlinear), with the nonlinear shape a result of the tetrahedral orientation of the valence electron pairs of oxygen. Ammonia's overall shape is said to be that of a trigonal (triangular) pyramid. Of the four molecules used as examples, only methane, CH_4, has both tetrahedrally oriented valence electron pairs and an overall geometric shape that can be described as tetrahedral (since all four pairs of electrons about the central atom are bonding pairs).

During this experiment, your instructor will construct a large-scale demonstration model of each of the molecular structures to be studied. At your desk, you will construct a smaller model of the structure, measure the bond angles in the structure with a protractor, sketch the structure on paper, and suggest a real molecule that would be likely to have that structure.

Safety Precautions

- Safety glasses must be worn at all times while in the laboratory.
- Although this experiment does not involve any chemical substances, you should exercise normal caution while in the laboratory.

Apparatus/Reagents Required

Demonstration molecular model kit, student model kit, protractor

Procedure

Record all data and observations directly in your notebook in ink.

Your instructor will build demonstration models of each of the shapes listed in the Table of Geometries on the next page. Examine these large-scale models; then build a similar model with the student kit.

With the protractor, measure all bond angles in your models.

Sketch a representation of the models, and indicate the measured bond angles. Your sketches do not have to be fine artwork, but the overall shape of the molecule, as well as the position of all electron pairs on the central atom (both bonding and nonbonding), must be clear.

For each structure you build and sketch, use your textbook to suggest a real molecule that would be expected to have that shape, based on its Lewis dot electron structure. (Give the Lewis dot formula for each of the molecules you suggest.)

Table of Geometries

	No. Valence pairs on central atom	*Arrangement of valence pairs*	*No. Bonding pairs on central atom*	*Molecular shape*	*Type formula*
(a)	2	linear	2	linear	AB_2
(b)	3	trigonal planar	1	linear	—
(c)	3	trigonal planar	2	bent	—
(d)	3	trigonal planar	3	trigonal planar	AB_3
(e)	4	tetrahedral	1	linear	AB
(f)	4	tetrahedral	2	bent	AB_2
(g)	4	tetrahedral	3	trigonal pyramid	AB_3
(h)	4	tetrahedral	4	tetrahedral	AB_4
(i)	5	trigonal bipyramid	1	linear	—
(j)	5	trigonal bipyramid	2	linear	AB_2
(k)	5	trigonal bipyramid	3	T-shape	AB_3
(l)	5	trigonal bipyramid	4	see-saw	AB_4
(m)	5	trigonal bipyramid	5	trigonal bipyramid	AB_5
(n)	6	octahedral	1	linear	—
(o)	6	octahedral	2	linear	—
(p)	6	octahedral	3	T-shape	—
(q)	6	octahedral	4	square planar	AB_4
(r)	6	octahedral	5	square pyramid	AB_5
(s)	6	octahedral	6	octahedral	AB_6

Note. A "—" in the *Type formula* column indicates that no simple molecules with this structure are known (or likely to be discussed in an introductory general chemistry text).

Name ______________________________ Section ______________________

Lab Instructor ______________________________ Date ______________________

EXPERIMENT 19 Molecular Shapes and Structures

PRE-LABORATORY QUESTIONS

1. Give a short summary of the two main points of VSEPR theory.

____

____

____

____

____

2. Draw Lewis dot structures and predict the bond angles in the following molecules:

PF_3

SF_6

ICl_5

HBr

H_2Se

Name ______________________________ Section ______________________

Lab Instructor ______________________________ Date ______________________

EXPERIMENT 19 Molecular Shapes and Structures

RESULTS/OBSERVATIONS

	Sketch (showing bond angles)	*Lewis dot formula*
(a)		
(b)		
(c)		
(d)		
(e)		
(f)		

	Sketch (showing bond angles)	*Lewis dot formula*
(g)		
(h)		
(i)		
(j)		
(k)		
(l)		
(m)		
(n)		
(o)		

Sketch (showing bond angles) *Lewis dot formula*

(p)

(q)

(r)

(s)

QUESTIONS

1. The models you have built and sketched do not take into account the fact that bonding and nonbonding pairs of electrons do not repel each other to exactly the same extent; repulsion by nonbonding pairs is stronger than by bonding pairs. How will this affect the true geometric shape of the molecules you have drawn? How will the bond angles you have measured be changed by this effect? Use specific examples.

2. The models you have built also do not consider those molecules having double or triple bonds predicted for their Lewis dot structures. What effect might a second or third pair of electrons shared between two individual atoms have on the overall shape of a molecule?

EXPERIMENT

20 Properties of Some Representative Elements

Objective

In this experiment you will examine the properties of some of the more common representative elements.

Introduction

Even in the earliest studies of chemistry, it became evident that certain elemental substances were very much like other substances in their physical and chemical properties. For example, the more common alkali metals (Na, K) were almost indistinguishable to early chemists. Both metals are similar in appearance and undergo reaction with the same reagents (giving formulas of the same stoichiometric ratio). As more and more elemental substances were separated, purified, and identified, more and more similarities between the new elements and previously known elements were detected. Chemists began to wonder why such similarities existed.

In the mid-1800s, Mendeleev and Meyer independently proposed that the properties of the known elements seemed to vary systematically when arranged in the order of their atomic masses. That is, when a list of the elements is made in order of increasing atomic mass, a given set of properties would repeat at regular intervals among the elements. Mendeleev generally receives most credit for the development of the periodic law, because he suggested that some elements were missing from the table. That is, based on the idea that the properties of the elements should repeat at regular intervals, Mendeleev suggested that some elements that should have certain properties had not yet been discovered. Mendeleev went so far as to predict the properties of the as yet undiscovered elements, based on the properties of the already known elements. Mendeleev's periodic law received much acceptance when the missing elements were finally discovered and had the very properties that Mendeleev had predicted for them.

While Mendeleev arranged the elements in order of increasing atomic mass, the modern periodic table is arranged in order of increasing atomic number. At the time of Mendeleev's work, the structure of the atom had not yet been determined. Generally, the arrangement of the elements by atomic mass is very similar to the arrangement by atomic number, with some notable exceptions. Some elements in Mendeleev's arrangement were out of order and did not have the properties expected. When the arrangement is made by atomic number, however, the properties of the elements do fall into a completely regular order.

Group 1A Elements

The Group 1A elements are commonly referred to as the **alkali metals**. These substances are among the most reactive of elements, are never found in nature in the uncombined state, and are relatively difficult and expensive to produce and store. In particular, these elements are very easily oxidized by oxygen in the air; usually they are covered with a layer of oxide coating unless the metal has been freshly cut or cleaned. When obtained from a chemical supply house, these elements are usually stored under a layer of kerosene to keep them from contact with the air.

The elements of Group 1A are very low in density and are soft enough to be easily cut with a knife or spatula. The only real reaction undergone by these elements involves the loss of their single valence electron to some other species:

$$M \rightarrow M^{+} + e^{-}$$

In compounds, the elements of Group 1A are invariably found as the unipositive ion. The least reactive of these elements is lithium (the topmost member, with the valence electron held most tightly by the nuclear charge), whereas the most reactive is cesium (the bottommost member). Cesium is so reactive that the metal may be ionized merely by shining light on it. This property of cesium has been made use of in certain types of photocells (in which the electrons produced by the ionization of cesium are channeled into a wire as electric current). The most common elements of this group, sodium and potassium, are found naturally in great abundance in the combined state on the earth. The elements of Group 1A are rather low in density (for example, they float on the surface of water while reacting) and are very soft (they can be easily cut with a knife).

Group 2A Elements

The elements of Group 2A are commonly referred to as the **alkaline earth elements**. The "earth" in this sense indicates that the elements are not as reactive as the alkali metals of Group 1A and are found (occasionally) in the free state in nature. Beryllium, the topmost element of this group, is not very common. But some of the other members of the group, such as magnesium and calcium, are found in fairly high abundance. As with the Group 1A elements, the relative reactivity of the Group 2A elements increases from top to bottom in the periodic table. For example, metallic magnesium can be kept almost indefinitely without developing an oxide coating (by reaction with air), whereas a freshly prepared sample of pure metallic calcium will develop a layer of oxide in a much shorter period. In compounds, the elements of Group 2A are usually found as dipositive ions (through loss of the two valence electrons).

Group 3A Elements

Whereas all the elements of Group 1A and Group 2A are metals, the first member of Group 3A (boron) is a nonmetal. The dividing line between metallic elements and nonmetallic elements forms the "stairstep" region indicated at the right-hand side of most periodic charts. Because boron is to the right of the stairstep, boron shows many properties that are characteristic of nonmetals. For example, boron is not a very good conductor of heat and electricity, whereas aluminum (which is the element beneath boron in the group) shows the much higher conductivities associated with metallic elements. Boron is generally found covalently bonded in its compounds, which further indicates its nonmetallic nature. Aluminum is the only relatively common element from Group 3A.

In nature, aluminum is generally found as the oxide, and most metallic aluminum is produced by electrolysis of the molten oxide. This is a relatively difficult and expensive process, which is part of the reason the aluminum in cans was commonly recycled even before the present emphasis on preserving the environment. Aluminum is a relatively reactive element; however, this reactivity is sometimes masked under ordinary conditions. Aluminum used for making cooking utensils or cans becomes very quickly coated with a thin layer of aluminum oxide. This oxide layer serves as a protective coating, preventing further oxidation of the aluminum metal underneath. (This is why aluminum pans are seldom as shiny as copper or stainless steel pans.)

Group 4A Elements

Group 4A contains some of the most common, useful, and important elements known. The first member of the group, carbon, is a nonmetal. The second and third members, silicon and germanium, show both some metallic and some nonmetallic properties (and because of this in-between nature, these elements are referred to as **metalloids** or **semimetals**). The last two members of the group, tin and lead, show mostly metallic properties.

The element carbon forms the framework for life. Virtually all biological molecules are, in fact, carbon compounds. Carbon is unique among the elements in that it is able to form long chains of many hundreds or even thousands of similar carbon atoms. Pure elemental carbon itself is usually obtained in either of two forms, graphite or diamond, each of which has a very different structure. Graphite contains flat, two-dimensional layers of covalently bonded carbon atoms, with each layer effectively being a molecule of graphite, independent of the other layers present in the sample. The layers of carbon atoms in graphite are able to move relative to each other. This makes graphite slippery and useful as a lubricant for machinery and locks. Diamond contains three-dimensional extended covalent networks of carbon atoms, making an entire diamond effectively a single molecule. Because all the atoms present in diamond are covalently bonded to each other in three dimensions, diamond is relatively stable and is also the hardest substance known. Natural diamonds used as gemstones are produced deep in the earth under high pressures over eons, but synthetic diamonds are produced in the laboratory by compressing graphite to several thousand atmospheres of pressure. Because of its hardness, diamond is used as an abrasive in industry.

Silicon and germanium are used in the semiconductor industry. Because these elements have some of the properties of nonmetals (brittleness, hardness), as well as some of the properties of metals (slight electrical conductivity), they have been used extensively in transistors.

The last two elements of Group 4A, tin and lead, show mostly metallic properties (shiny luster, conductivity, malleability) and frequently form ionic compounds. However, covalently bonded compounds of these metals are not rare. For example, leaded gasoline contains tetraethyllead, which is effectively an organic lead compound. Tin has been used in the past to make cooking utensils and is used extensively as a liner for steel cans ("tin" cans), because it is less likely to be oxidized than iron and is also less likely to impart a bad taste to foods. Whereas tin is nonpoisonous and is used in the food industry, lead compounds are very dangerous poisons. Formerly, pigments for white paint were made using lead oxide. Lead compounds are remarkably sweet-tasting, and many children have been seriously poisoned by eating flecks of peeling lead paints.

Group 5A Elements

The first few elements of Group 5A are nonmetals, with nitrogen and phosphorus being the two most abundant elements of the group. The earth's atmosphere is nearly 80% N_2 gas by volume. This large percentage of nitrogen is believed to be a remnant of the primordial composition of the earth's atmosphere. Astronomers have determined that several of the planets of the solar system contain large quantities of ammonia, NH_3, in their atmospheres. It is known that ammonia actually comes to equilibrium with its constituent elements

$$2NH_3 \rightarrow N_2 + 3H_2$$

Astronomers theorize that the atmosphere of the early earth was also mostly ammonia, but that the lighter hydrogen gas has escaped the earth's relatively weak gravitational field over the eons, leaving the atmosphere with a large concentration of nitrogen gas. Nitrogen is also an important constituent of many of the molecules synthesized and used by living cells (amino acids, for example).

Phosphorus, the second member of Group 5A, is a typical solid nonmetal. Three allotropic forms of elemental phosphorus exist: red, white, and black phosphorus. White phosphorus is the most interesting of the three forms and is the form that is produced when phosphorus vapor is condensed. White phosphorus is very unstable toward oxidation. Below about 35°C, white phosphorus reacts with oxygen of the air, emitting the energy of the oxidation as light rather than heat (phosphorescence); above 35°C, white phosphorus will spontaneously burst into flame. (For this reason, white phosphorus is stored under water to keep it from contact with air.)

Whereas elemental nitrogen consists of diatomic molecules, white phosphorus contains tetrahedrally shaped P_4 molecules. When white phosphorus is heated to high temperatures in the absence of air, it is converted to the red allotrope. The red allotrope is much less subject to oxidation and is the form usually available in the laboratory. The lower reactivity is due to the fact that the phosphorus atoms have polymerized into one large molecule, which is far less subject to attack by oxygen atoms than the individual P_4 molecules of the white allotrope.

Phosphorus is essential to life, being a component of many biological molecules, most notably the phosphate compounds used by the cell for storing and transferring energy (AMP, ADP, and ATP). Phosphorus also forms part of the structural framework of the body, bones being complex calcium phosphate compounds.

The other members of Group 5A are far less common than are nitrogen and phosphorus. For example, arsenic (As) is known primarily for its severe toxicity and has been used in various rodenticides and weed killers (though As has been supplanted by other agents for most of these purposes).

Group 6A Elements

The most abundant elements of Group 6A (sometimes referred to as the **chalcogens**) are oxygen and sulfur, which, along with selenium, are nonmetallic. Elemental oxygen makes up about 20% of the atmosphere by volume and is vital not only to living creatures but to many common chemical reactions. Most of the oxygen in the atmosphere is produced by green plants, especially by plankton in the oceans of the earth. On a microscopic basis, oxygen is used in living cells for the oxidation of carbohydrates and other nutrients. Oxygen is needed also for the oxidation of petroleum-based fuels for heating and lighting purposes and for myriad other important industrial processes. Oxygen has two allotropic forms: the normal elemental form of oxygen is O_2 (dioxygen), whereas a less stable allotrope, O_3(ozone), is produced in the atmosphere by high-energy electrical discharges.

Sulfur is obtained from the earth in a nearly pure state and needs little or no refining before use. In certain areas of the earth, there are vast underground deposits of pure sulfur. High-pressure steam is pumped into such deposits, liquefying the sulfur and allowing it to be pumped to the surface of the earth for collection. The main use of sulfur is in the manufacture of sulfuric acid, which is the industrial chemical produced each year in the largest amount. When sulfur is burned in air, the sulfur is converted to sulfur dioxide, notable for its choking, irritating odor (similar to a freshly struck match). If sulfur dioxide is further oxidized with a catalyst, it is converted to sulfur trioxide, the anhydride of sulfuric acid.

Sulfur has several allotropic forms. At room temperature, the normal allotrope is orthorhombic sulfur, which consists of S_8 cyclic molecules. At higher temperatures, a slight rearrangement of the shape of these rings occurs, producing a solid allotropic form of sulfur: monoclinic sulfur. This substance has a different crystalline structure than the orthorhombic form. (Monoclinic sulfur slowly changes into orthorhombic sulfur at room temperature.) The most interesting allotrope of sulfur, however, is produced when boiling sulfur is rapidly lowered in temperature. This technique produces a plastic (amorphous) allotrope. The properties of plastic sulfur are very different from those of orthorhombic or monoclinic sulfur. Whereas plastic sulfur is soft and pliable (being able to be stretched and pulled into strings), orthorhombic and monoclinic sulfur are both hard, brittle solids. Sulfur is insoluble in water, but the orthorhombic form is fairly soluble in the nonpolar solvent carbon disulfide, CS_2.

The most common oxygen compound is, of course, water, H_2O. The analogous sulfur compound, hydrogen sulfide, H_2S, is a noxious gas at room temperature. (The gas smells like rotten eggs; eggs contain a protein involving sulfur that releases hydrogen sulfide as the egg spoils.)

Selenium, which occurs in Group 6A just beneath sulfur, is the only other member of the group that occurs in any abundance. Selenium compounds are generally very toxic. Selenium is used in certain prescription dandruff shampoos.

Group 7A Elements

The Group 7A elements are more commonly referred to as the **halogens** ("salt-formers"). The elementary substances are very reactive; these elements are, therefore, usually found in the combined state, as the negative halide ions. Elemental fluorine is the most reactive nonmetal and is seldom encountered in the laboratory because of problems of toxicity and handling of the gaseous element. Most compounds of fluorine are very toxic, although tin(II) fluoride is used in toothpastes and mouthwashes as a decay preventative.

Hydrogen fluoride is very different from the other hydrogen halides, since aqueous solutions of HF are weakly acidic, rather than strongly acidic (HCl, HBr). Hydrogen fluoride is able to attack and dissolve glass and is always stored in plastic bottles.

Gaseous elemental chlorine is somewhat less reactive than elemental fluorine and is used commercially as a disinfectant. Chlorine will oxidize biological molecules present in bacteria (thereby killing the bacteria), while it is reduced itself to chloride ion (which is nontoxic). For example, drinking water is usually chlorinated, either with gaseous elemental chlorine or with some compound of chlorine that is capable of releasing elemental chlorine when needed. Gaseous hydrogen chloride, when dissolved in water, forms the solution known as hydrochloric acid, which is one of the most commonly used acids.

Elemental bromine is a strikingly dark red liquid at room temperature. Elemental bromine is used as a common laboratory test for the presence of double bonds in carbon compounds. Br_2 reacts with the double bond, and its red color disappears, thereby serving as an indicator of reaction.

Elemental iodine is a dark gray solid at room temperature but undergoes sublimation when heated slightly, producing an intensely purple vapor. Elemental iodine is the least reactive of the halogens. Iodine is very important in human metabolism, being a component of various hormones produced by the thyroid gland. Since many people's diets do not contain sufficient iodine from natural sources, commercial table salt usually has a small amount of sodium iodide added as a supplement ("iodized" salt). Iodine is also used sometimes as a topical antiseptic, since it is a weak oxidizing agent that is capable of destroying bacteria. Iodine for this purpose (called "tincture" of iodine) is usually sold in alcohol solution.

Safety Precautions

- Wear safety glasses at all times while in the laboratory.
- The reactions of lithium, sodium, and potassium with water are very dangerous. Your instructor will demonstrate these reactions for you. Do *not* attempt these reactions yourself.
- Hydrogen gas is extremely flammable and forms explosive mixtures with air. Small amounts of hydrogen will be generated in some of the student reactions. Avoid flames and use caution.
- The flames produced by burning magnesium/calcium are intensely bright and can damage the eyes. Do *not* look directly at the flame while these substances are burning.
- Oxides of nitrogen are toxic and extremely irritating to the respiratory system. Generate these gases only in the exhaust hood.
- Hydrochloric and nitric acids will burn the skin. Wash immediately if the acids are spilled and inform the instructor.
- Vapors of methylene chloride and of the halogens are toxic. Keep these substances in the exhaust hood during use. Dispose of methylene chloride as directed by the instructor.
- Hydrogen sulfide and sulfur dioxide are toxic and have noxious odors. Confine them to the exhaust hood.
- Salts of the Group 1A and 2A metals may be toxic; wash after handling these compounds.

Apparatus/Reagents Required

Lithium, sodium, potassium, universal indicator solution, calcium, magnesium ribbon, samples of solid LiCl, NaCl, KCl, $CaCl_2$, $BaCl_2$, and $SrCl_2$, flame test wires, 6 *M* hydrochloric acid, boric acid, aluminum oxide, sodium carbonate, copper wire, concentrated nitric acid, 0.1 *M* NaOH, chlorine water, bromine water, iodine water, methylene chloride, 0.1 *M* NaCl, 0.1 *M* NaBr, 0.1 *M* NaI, ferrous sulfide, solid iodine

Procedure

Record all data and observations directly in your notebook in ink.

A. Some Properties of the Alkali and Alkaline Earth Elements

The reactive metals of Group 1A and Group 2A react with cold water to liberate elemental hydrogen, leaving a solution of the strongly basic metal hydroxide

$$2M(s) + 2H_2O \rightarrow 2M^{+}(aq) + 2OH^{-}(aq) + H_2(g)$$

$$M(s) + 2H_2O \rightarrow M^{2+}(aq) + 2OH^{-}(aq) + H_2(g)$$

Generally the reactivity of the metals increases going from the top of the group toward the bottom of the group. The valence electrons of the metal are less tightly held by the nucleus toward the bottom of the group.

1. Lithium, Sodium, and Potassium (Instructor Demonstration)

The reaction of the Group 1A metals with water is very dangerous and will be performed by your instructor as demonstrations.

In the exhaust hood, behind a safety shield, the instructor will drop small pellets of lithium, sodium, and potassium into beakers containing a small amount of cold water. Compare the speed and vigor of the reactions.

When the reactions have subsided, obtain small portions of the water from the beakers in which the reactions were conducted in separate clean test tubes. Add 2–3 drops of universal indicator to each test tube; refer to the color chart provided with the indicator to determine the pH of the solution. Write equations for the reactions of the metals with water.

2. Reactions of Magnesium and Calcium (Student Procedure)

Repeat the procedure that you have seen in the demonstration, using pieces of the Group 2A metals magnesium and calcium in place of lithium/sodium/potassium. Work in the exhaust hood, with the safety shield pulled down. Account for differences in reactivity between these two Group 2A metals. Test the pH of the water with which the metals were reacted.

Add a single turning of magnesium to 10 drops of 6 *M* HCl in a small test tube. Why does magnesium react with *acid* but not with water?

3. Flame Tests

Obtain a 6–8-inch length of nichrome wire for use as a flame test wire. Form one end of the wire into a small loop that is no more than a few millimeters in diameter.

Obtain about 20 mL of 6 *M* HCl in a small beaker, and immerse the loop end of the wire for 2–3 minutes to clean the wire.

Ignite a Bunsen burner flame, and heat the loop end of the wire in the flame until it no longer imparts a color to the flame.

Obtain a few crystals each of LiCl, NaCl, KCl, $CaCl_2$, $BaCl_2$, and $SrCl_2$. Dip the loop of the flame test wire into one of the salts, and then into the oxidizing portion of the burner flame. Record the color imparted to the flame by the salt.

Clean the wire in 6 *M* HCl, heat in the flame until no color is imparted, and repeat the flame test with the other salts. The colors imparted to the flame by the metal ions are so intense and so characteristic that they are frequently used as a test for the presence of these elements in a sample.

B. Oxides/Sulfides of Some Elements

Oxygen compounds of most elements are known, and are generally referred to as **oxides**. However, there are great differences between the oxides of metallic elements and those of nonmetals. The oxides of metallic elements are *ionic* in nature: they contain the oxide ion, O^{2-}. Ionic metallic oxides form *basic* solutions when dissolved in water

$$O^{2-} + H_2O \rightarrow 2OH^-$$

Oxides of nonmetals are generally *covalently* bonded and form *acidic* solutions when dissolved in water.

Sulfur compounds of both metallic and nonmetallic substances are known. Sulfur compounds of nonmetallic compounds are generally covalently bonded, whereas with metallic ions, sulfur is usually present as the sulfide ion, S^{2-}. Most sulfur compounds have characteristic unpleasant odors, or decompose into compounds that have the characteristic odor.

1. Metallic Oxides

Place a small amount of distilled water in a beaker and add 2–3 drops of universal indicator. Refer to the color chart provided with the indicator, and record the pH of the water.

Obtain a piece of magnesium ribbon about 1 inch in length. Hold the magnesium ribbon with tongs above the beaker of water and ignite the metal with a burner flame. Do *not* look directly at the flame produced by burning magnesium. It is intensely bright and *damaging* to the eyes. When the flame has expired, stir the liquid in the beaker for several minutes and record the pH of the solution. Write an equation for the reactions that have taken place.

Place approximately 10 drops of distilled water in each of three test tubes, and add a drop of universal indicator to each. Add a *very* small quantity of sodium (per)oxide to one test tube, calcium oxide (lime) to a second test tube, and aluminum oxide (alumina) to the third test tube. Stir each test tube with a clean glass rod, record the pH of the solution, and write an equation to account for the pH.

2. Nonmetallic Oxides

Place approximately 10 drops of water into each of two test tubes, and add 1 drop of universal indicator to each. In one test tube, dissolve a small amount of boric acid. To the second test tube, add a small chunk of dry ice (solid carbon dioxide). Stir the test tubes with a clean glass rod, record the pH, and write an equation to account for the pH of the solutions.

3. Sulfur Compounds

Have ready a beaker of cold water. In the exhaust hood, ignite a *tiny* amount of powdered sulfur on the tip of a spatula in the burner flame and *cautiously* note the odor. Extinguish the sulfur in the cold water to prevent too much SO_2 from getting into the room air. Determine the pH of the water in which you extinguished the sulfur. Why is it acidic?

Obtain a *tiny* portion of iron(II) sulfide in a test tube, and add 2 drops of dilute hydrochloric acid. *Cautiously* note the odor of the hydrogen sulfide generated. Transfer the test tube to the exhaust hood to dispose of the hydrogen sulfide.

C. Some Properties of the Halogen Family (Group 7)

All of the members of Group 7A are nonmetallic. The halogen elements tend to gain electrons in their reactions, and the attraction for electrons is stronger if the electrons are closer to the nucleus of the atom. The activity of the halogen elements therefore decreases from top to bottom in Group 7 of the periodic table. As a result of this, elemental chlorine is able to replace bromide and iodide ion from compounds:

$$Cl_2 + 2NaBr \rightarrow Br_2 + 2NaCl$$

$$Cl_2 + 2NaI \rightarrow I_2 + 2NaCl$$

Similarly, elemental bromine is able to replace iodide ion from compounds.

1. Identification of the Elemental Halogens by Color

The elemental halogens are nonpolar and are not very soluble in water. Solutions of the halogens in water are not very brightly colored. However, if an aqueous solution of an elemental halogen is shaken with a nonpolar solvent, the halogen is preferentially extracted into the nonpolar solvent, and imparts a characteristic, relatively bright color to the nonpolar solvent. Since the nonpolar solvent is not miscible with water, the halogen color is evident as a separate colored layer.

In the following tests, dispose of the methylene chloride samples as directed by the instructor.

Obtain about 1 mL of chlorine water in a small test tube. Note the color. In the exhaust hood, add 10 drops of methylene chloride to the test tube. Stopper and shake.

Allow the solvent layers to separate and note the color of the lower (methylene chloride) layer. Elemental chlorine is not very intensely colored and imparts only a pale yellow/green color.

Repeat the procedure using 1 mL of bromine water in place of the chlorine water. Bromine imparts a bright red color to the lower layer.

Repeat procedure using 1 mL of iodine water. Iodine imparts a purple color to the methylene chloride layer.

2. Relative Reactivity of the Halogens

Add 2 mL of chlorine water to each of two test tubes. Add 2 mL of 0.1 *M* NaBr to one test tube and 2 mL of 0.1 *M* NaI to the second tube.

In the exhaust hood, add 10 drops of methylene chloride to each test tube. Stopper, shake, and allow the solvents to separate. Record the colors of the lower layers and identify which elemental halogens have been produced.

Add 2 mL of bromine water to each of two test tubes. Add 2 mL of 0.1 *M* NaI to one test tube and 2 mL of 0.1 *M* NaCl to the other test tube.

In the exhaust hood, add 10 drops of methylene chloride to each test tube. Stopper, shake, and allow the layers to separate. Record the colors of the lower layers. Indicate where a reaction has taken place.

Add 2 mL of iodine water to each of two test tubes. Add 2 mL of 0.1 *M* NaBr to one test tube and 2 mL of 0.1 *M* NaCl to the other test tube.

In the exhaust hood, add 10 drops of methylene chloride to each test tube. Stopper, shake, and allow the layers to separate. Record the colors of the lower layers. Did any reaction take place?

3. Sublimation of Iodine

Using forceps, place a few crystals of elemental iodine in a small beaker. Set the beaker on a ring stand in the exhaust hood, and cover with a watch glass containing some ice. Heat the iodine crystals in the beaker with a small flame for 2–3 minutes, and watch the sublimation of the iodine. Extinguish the flame, and examine the iodine crystals that have formed on the lower surface of the watch glass.

Name ____________________ Section ____________________

Lab Instructor ____________________ Date ____________________

EXPERIMENT 20 Properties of Some Representative Elements

PRE-LABORATORY QUESTIONS

1. For each of the following elements, list the atomic number, average atomic mass, in which group (vertical column) of the periodic table the element can be found, and also in which period (horizontal row) the element is located:

K ____________________

Ba ____________________

I ____________________

Ne ____________________

Se ____________________

Cs ____________________

Ra ____________________

Kr ____________________

2. Why do members of the same vertical group of the periodic chart tend to show similar chemical and physical properties?

____

____

____

____

____

____

Name ______________________ Section ______________

Lab Instructor ______________________ Date ______________

EXPERIMENT 20 Properties of Some Representative Elements

RESULTS/OBSERVATIONS

A. Some Properties of the Alkali and Alkaline Earth Elements

1. Reaction with water

Metal	Evidence of reaction	Relative vigor	Indicator color	Equation for reaction
Li				
Na				
K				
Mg				
Ca				

2. Reaction of Mg with acid ______________________

3. Flame tests

Metal	Color observed

B. Oxides/Sulfides of Some Elements

Substance	Results/equations – universal indicator test
Magnesium oxide	
Sodium oxide	
Calcium oxide	
Aluminum oxide	
Boric acid	
Dry ice	
Sulfur dioxide	
FeS + HCl	

C. Some Properties of the Halogen Family

1. Identification of the elemental halogens

Halogen	Color of water solution	Color of CH_2Cl_2 extract
Cl_2		
Br_2		
I_2		

2. Relative reactivity of the halogens

Elemental halogen	Halide ion	Color of lower layer	Which halogen is in lower layer?	Did a reaction take place?
Cl_2	Br^-			
Cl_2	I^-			
Br_2	I^-			
Br_2	Cl^-			
I_2	Br^-			
I_2	Cl^-			

Equations for reactions that occurred

____

____

____

____

Sublimation of iodine observation ______________________________

QUESTIONS

1. In comparing the *vigor* of the reactions of lithium, sodium, and potassium with cold water, does there appear to be a correlation between the location of these elements in the periodic table and the vigor of reaction?

____

____

____

____

2. You tested solutions of several oxides with universal indicator to determine their acidity/basicity. Can a *general* statement be made concerning the acidity of metal oxides compared to oxides of nonmetallic elements?

____

____

____

____

3. Elemental fluorine, F_2, was not used in the halogens experiment because of difficulties in handling it, and elemental astatine, At_2, was not used because of its radioactivity and short half-life. How would you expect the reactivity of F_2 and At_2 to compare with those of the elemental halogens tested in this experiment?

____

____

____

EXPERIMENT

21 Classes of Chemical Reactions

Objective

Chemical reactions may be classified as precipitation, acid/base, complexation, or oxidation/reduction. Several examples of each will be examined in this experiment.

Introduction

There are over 100 known elements and millions of known compounds. Throughout history, chemists have sought to organize the wealth of data and observations that have been recorded for these substances. In this experiment, you will perform one or more examples of each of the reaction types mentioned in the objective.

A. Precipitation Reactions

Certain substances are not very soluble in water. Frequently such substances are generated *in situ* in a reaction vessel by the addition of various other substances that are themselves very soluble. For example, silver chloride is not soluble in water. If an aqueous solution of silver nitrate (very soluble) is mixed with an aqueous solution of sodium chloride (very soluble), the combination of silver ions from one solution and chloride ions from the other solution generates silver chloride, which then forms a precipitate that settles to the bottom of the container. The solution that remains above the precipitate of silver chloride effectively becomes a solution of sodium nitrate. Silver ions and sodium ions have switched partners, ending up in a compound with the negative ion that originally came from the opposite substance:

$$AgNO_3(aq) + NaCl(aq) \rightarrow AgCl(s) + NaNO_3(aq)$$

The silver ion and sodium ion have replaced each other in this process; this sort of reaction is sometimes referred to as a **displacement reaction**.

B. Acid/Base Reactions

There are many theories and definitions that attempt to explain what constitutes an acid or a base. An early, and still useful theory, developed by Arrhenius in the late 1800s, defines an **acid** as a substance that produces **hydrogen ions**, H^+, when dissolved in water. A **base**, according to Arrhenius, is a species that produces **hydroxide ions**, OH^-, when dissolved in water. For example, hydrogen chloride is an acid in the Arrhenius theory, because hydrogen chloride ionizes when dissolved in water

$$HCl \rightarrow H^+ + Cl^-$$

producing hydrogen ions. In particular, hydrogen chloride is called a **strong acid** because virtually every HCl molecule ionizes when dissolved. Other substances, although they do in fact produce hydrogen ions when dissolved, do not completely ionize when dissolved, and are called **weak acids**. For example, the acidic component of vinegar (acetic acid) is a weak acid

$$CH_3COOH = H^+ + CH_3COO^-$$

The equation for the ionization of acetic acid indicates that this substance reaches an **equilibrium** when dissolved in water, at which point a certain fixed concentration of hydrogen ion is present. The concentration of hydrogen ion produced by dissolving a given amount of weak acid is typically several orders of magnitude less than if the same amount of strong acid is dissolved.

In a similar manner, there are both strong and weak bases in the Arrhenius scheme. Sodium hydroxide, for example, is a strong base:

$$NaOH(s) \rightarrow Na^+(aq) + OH^-(aq)$$

For every mole of NaOH that might be dissolved in water, an equivalent number of moles of OH^- is produced. In contrast, ammonia, NH_3, is a weak base:

$$NH_3(g) + H_2O = NH_4^+(aq) + OH^-(aq)$$

Ammonia in effect *reacts* with water and comes to equilibrium, at which point a certain fixed concentration of OH^- exists. As with weak acids, the concentration of hydroxide ion in a solution of a weak base is much smaller than if a strong base had been used.

The most important reaction of acids and bases is **neutralization**. The hydrogen ion from an aqueous acid combines with the hydroxide ion from an aqueous base, producing water. For example,

$$HCl + NaOH \rightarrow NaCl + H_2O$$

$$HNO_3 + NaOH \rightarrow NaNO_3 + H_2O$$

The net reaction in each of these is the same and is typical of the reaction between acids and bases in aqueous solution

$$H^+ + OH^- \rightarrow H_2O$$

Although the Arrhenius definitions of acids and bases have proved very useful, the theory is restricted to the situation of aqueous solutions. Aqueous solutions are most common, but it is reasonable to question if HCl and NaOH will still behave as an acid or base when dissolved in some other solvent. The Brönsted/Lowry theory of acids and bases extends the Arrhenius definitions to more general situations. An acid is defined in the Brönsted/Lowry theory as any species that provides hydrogen ions, H^+, in a reaction. Since a hydrogen ion is nothing more than a simple proton, acids are often referred to as being proton donors in the Brönsted/Lowry scheme. Bases in the Brönsted/Lowry system are defined as any species that receives hydrogen ions from an acid. Bases are referred to then as proton acceptors. The similarity between the definitions of an acid in the Arrhenius and Brönsted/Lowry schemes is obvious. On first glance, however, it would seem that the definition of what constitutes a base differs between the two systems. Realize, however, that hydroxide ion in aqueous solution (Arrhenius definition) will very readily accept protons from any acid that might be added. Therefore, hydroxide ion is in fact a base in the Brönsted/Lowry system as well.

C. Complexation Reactions

Metal ions, especially in solution, react with certain neutral molecules and many negatively charged ions to form species known as **coordination complexes**. Such compounds are particularly common with ions of the **transition metals**. The neutral or anionic species that reacts with the metal ion is called a **ligand**. In order for a molecule or anion to be able to function as a ligand, the species must have at least one pair of nonbonding valence electrons. Most metal ions have empty, relatively low-energy atomic *d*-orbitals. When a suitable ligand is added to a solution of a metal ion, a coordinate covalent bond can form between the metal ion and the ligand molecule or ion. The nonbonding valence electron pair of the ligand provides the electron pair needed for the covalent bond.

In their complex ions, many metals characteristically bind with the same number of ligand species, regardless of the particular identity of the ligands. The number of ligands bound to a metal ion is called the **coordination number** of the metal. For example, copper(II) ion usually exhibits a coordination number of four, whereas nickel(II) usually is found bound with six ligands.

Coordination complexes of transition metals are often very striking in color and are used as pigments in paints for this reason. These colors arise from transitions of electrons among the *d*-orbitals of the coordinated metal ion. In the presence of a set of ligands, the *d*-orbitals of the metal ion are displaced somewhat from their normal energies. These energy differences between the displaced *d*-orbitals correspond to the absorption/emission of visible light. For example, an aqueous solution of copper(II) ion is pale blue [because of the presence of water molecules coordinated to the copper(II) ion]. If ammonia is added to aqueous copper(II) ion solution, ammonia molecules displace coordinated water molecules and themselves coordinate with the copper(II) ion.

$$Cu(H_2O)_4^{2+} + 4NH_3 \rightarrow Cu(NH_3)_4^{2+} + 4H_2O$$

The resulting copper/ammonia complex is a striking dark blue.

Coordination complexes are very important in biological systems. Important molecules such as hemoglobin and chlorophyll contain coordinated metal ions.

The formation of coordination complexes actually represents a special case of acid/base chemistry. G. N. Lewis extended the study of acid/base processes beyond the theories of Arrhenius and Brönsted/Lowry to include species that had no protons, but that demonstrated reactions very similar or suggestive of other acid/base reactions. Lewis defined a base as a species capable of providing a valence electron pair for formation of a bond between atoms. According to this definition, ligand molecules in coordination compounds would be bases. Similarly, Lewis defined an acid as a species capable of receiving an electron pair in the formation of a bond. Lewis's definitions did not contradict earlier definitions of what constituted an acid or base. Any species that could receive a proton in the Brönsted/Lowry scheme would have to have at least one nonbonding pair of valence electrons for donating to a coordinate covalent bond. However, in the Lewis definition, a species would not have to be able to donate a proton to be an acid. Metal ions that receive pairs of electrons in the formation of a coordination compound could be looked upon as Lewis acids.

D. Oxidation/Reduction Reactions

A large and important class of chemical reactions can be classified as oxidation/reduction (or **redox**) processes. Oxidation/reduction processes involve the transfer of electrons from one species to another. For example, in the reaction

$$Zn(s) + CuSO_4(aq) \rightarrow ZnSO_4(aq) + Cu(s)$$

$Zn(s)$ represents uncharged elemental zinc atoms, whereas in $ZnSO_4(aq)$, zinc exists in solution as Zn^{2+} ions. Each zinc atom has effectively lost two electrons in the process and is said to have been oxidized:

$$Zn(s) \rightarrow Zn^{2+}(aq) + 2e^- \qquad \textit{oxidation}$$

Oxidation is defined as a loss of electrons by an atom or ion, or better, as a transfer of electrons from one species to another. Similarly, in the preceding reaction, Cu^{2+} ions in solution have each gained two electrons, becoming uncharged elemental copper atoms:

$$Cu^{2+}(aq) + 2e^- \rightarrow Cu(s) \qquad \textit{reduction}$$

Copper ions are said to have been reduced in the process. Reduction is defined as a gain of electrons by a species, or better, as the transfer of electrons to a species from another species.

To clarify the transfer of electrons between the species undergoing oxidation and the species undergoing reduction, redox reactions are usually divided into two half-reactions, one for each process. This was done for the zinc/copper reaction discussed earlier. Some redox reactions and their corresponding half-reactions follow:

$$2HgO \rightarrow 2Hg + O_2$$

$$2Hg^{2+} + 4e^- \rightarrow 2Hg \qquad \textit{reduction}$$

$$2O^{2-} \rightarrow O_2 + 4e^- \qquad \textit{oxidation}$$

$$Cl_2 + 2I^- \rightarrow 2Cl^- + I_2$$

$$Cl_2 + 2e^- \rightarrow 2Cl^- \qquad \textit{reduction}$$

$$2I^- \rightarrow I_2 + 2e^- \qquad \textit{oxidation}$$

Be careful to distinguish between oxidation/reduction reactions (in which electrons are completely transferred from one species to another) and complexation reactions (in which one species provides both electrons for a covalent bond between itself and another species).

Safety Precautions

- Wear safety glasses at all times while in the laboratory.
- Chromium compounds are toxic, mutagenic, and can cause severe burns to the skin. Wash after use.
- Lead and barium compounds are toxic if ingested.
- Silver nitrate will discolor the skin if spilled. The discoloration is not harmful, but requires several days to wear off.
- All acids/bases should be assumed to be damaging to the skin if spilled. Wash immediately and inform the instructor if any spills occur.
- Copper, nickel, and thiocyanate ion solutions are toxic. Wash after use.
- Concentrated ammonia solution is a severe respiratory irritant and cardiac stimulant. Confine all use of ammonia to the exhaust hood.
- Ethylene diamine is toxic and its vapor is harmful. Confine the use of ethylene diamine to the exhaust hood.
- The dimethylglyoxime (DMG) solution is flammable. No flames are permitted during its use.
- Hydrogen gas is extremely flammable and forms explosive mixtures with air.
- The flame produced by burning magnesium is intensely bright and can damage the retina. Do *not* look directly at the flame of burning magnesium.
- Hydrogen peroxide solutions are unstable and will burn the skin if spilled.
- Potassium permanganate will stain or burn the skin if spilled.
- "Chlorine water" may burn the skin if spilled and may emit toxic chlorine gas. Wash hands after use, and confine the use to the exhaust hood.
- Methylene chloride is toxic if inhaled or absorbed through the skin, and is a possible mutagen. Wash after use and confine its use to the exhaust hood.

Apparatus/Reagents Required

0.1 *M* solutions of the following: silver nitrate, sodium chloride, lead acetate, potassium chromate, barium chloride, sulfuric acid, hydrochloric acid, acetic acid, sodium hydroxide, ammonia, sodium thiocyanate; conductivity tester; universal indicator; 0.5 *M* solutions of copper(II) and nickel(II) sulfate; concentrated ammonia; 1 *M* ammonium nitrate; 0.5 *M* iron(III) nitrate; 1 *M* sodium chloride; 1 *M* hydrochloric acid; 5% 1,0-phenanthroline; 1% dimethylglyoxime; hydrogen gas; magnesium ribbon; steel wool; oxygen gas; 3% hydrogen peroxide; 1 *M* lead acetate; metallic zinc strips; metallic copper strips; 10% potassium iodide; 10% potassium bromide; chlorine water; methylene chloride; 0.5 *M* NaCl; 0.5 *M* NaBr; 10% ethylene diamine

Procedure

Record all data and observations directly in your notebook in ink.

A. Precipitation Reactions

Combine approximately 5-drop portions of the 0.1 *M* aqueous solutions in the list that follows. Use the table of solubilities from your textbook or Appendix I of this manual to predict *what precipitate* forms in each double displacement reaction. Write equations for the reactions:

silver nitrate and sodium chloride →

lead acetate and potassium chromate →

barium chloride and dilute sulfuric acid →

B. Acid/Base Reactions

Obtain 10-drop samples of 0.1 *M* HCl and 0.1 *M* acetic acid in separate clean semimicro test tubes. Using pH paper, determine the pH of each solution. How do you account for the fact that the pH of the acetic acid solution is higher than that of the HCl solution, although both are at the same molar concentration (0.1 *M*)?

Obtain 10-drop mL samples of 0.1 *M* NaOH and 0.1 *M* ammonia in separate clean test tubes. Using pH paper, determine the pH of each solution. How do you account for the fact that the pH of the ammonia solution is lower than that of the NaOH solution, although both are at the same molar concentration (0.1 *M*)?

One easy method for demonstrating the difference in the degree of ionization between strong acid and weak acid solutions (or of other electrolytes) is to measure the relative electrical conductivity of the solutions. Solutions of strong acids (or other strongly ionized electrolytes) conduct electrical currents well, whereas solutions of weak acids (or other weak electrolytes) conduct electricity poorly. Your instructor has set up a light bulb conductivity tester and will demonstrate the electrical conductivity of the 0.1 *M* acid and base solutions. The light bulb will glow brightly when the electrodes of the conductivity tester are immersed in strong acid or base solution but will glow only dimly when immersed in solutions of weak acid or base. Record which of the acids/bases are strong electrolytes, and which are weak.

Obtain 2.0 ± 0.1 mL of 0.1 *M* HCl in a clean test tube, and add 1 drop of universal indicator. Record the color of the indicator and the pH of the solution. This indicator changes color gradually with pH and can be used to monitor the pH of a solution during a neutralization reaction. Keep handy the color chart provided with the indicator. It shows the color of the indicator under different pH conditions.

Obtain a sample of 0.1 *M* NaOH in a small beaker, and begin adding NaOH to the sample of HCl and indicator one drop at a time with a medicine dropper. Record the color and pH after 5 drops of NaOH have been added.

Continue adding NaOH dropwise, recording the color and pH at 5-drop increments until the pH has risen to pH greater than 10. Approximately how many drops of NaOH were required to reach pH 7 (neutral)? Given that 1 mL of liquid is approximately equivalent to 20 average drops, approximately how many mL of 0.1 *M* NaOH were required to neutralize the 2-mL sample of 0.1 *M* HCl? Repeat the process, using 2 mL of 0.1 *M* acetic acid in place of HCl. How does the change in pH during the addition of NaOH *differ* when a weak acid is neutralized?

C. Complexation Reactions

1. Ammonia Complexes

Obtain approximately 10 drops of 0.5 *M* copper(II) sulfate and 0.5 *M* nickel(II) sulfate solutions in separate clean semimicro test tubes. Record the colors of the solutions.

In the *exhaust hood*, add concentrated ammonia solution *dropwise* to each of the metal ion solutions. Initially, a precipitate of the metal hydroxide may form, but on further addition of ammonia, brightly colored complex ions will form. Record your observations. Write equations for the complexation reactions that occurred.

Repeat the tests on nickel ion and copper ion, using a solution of 1 *M* ammonium sulfate in place of the concentrated ammonia solution. Why is there no change in color when the ammonium ion is added?

2. Iron Complexes

Obtain 10-drop samples of 0.5 *M* $Fe(NO_3)_3$ in each of four clean semimicro test tubes. Record the color of the solution.

Add 1 *M* NaCl solution dropwise to one iron(III) solution until the color of the resulting complex is evident.

Add 1 *M* HCl to the other iron solution until the color of the complex is evident.

Which ion (Na^+, H^+, or Cl^-) in the preceding two tests is the ligand that reacts with iron(III) ion?

Add a few drops of 0.1 *M* NaSCN to the third iron(III) sample. Record the color of the resulting iron/thiocyanate complex. The intense color of the iron/thiocyanate complex is used commonly as a test to detect the presence of iron in a sample.

Add a few drops of 5% 1,10-phenanthroline to the fourth iron(III) sample. 1,10-Phenanthroline is an organic ligand frequently used as a chromogenic agent in the determination of iron by spectrophotometry.

3. An Insoluble Nickel Complex

Obtain 10 drops of 0.5 *M* nickel(II) sulfate solution in a clean semimicro test tube. In the exhaust hood, add one drop of concentrated ammonia solution and 5 drops of 1% dimethylglyoxime reagent. Record the appearance of the nickel/dimethylglyoxime complex. Dimethylglyoxime is an organic ligand used as a standard qualitative and quantitative test for the presence of nickel ion (see Experiment 20).

4. Copper Complexes

Add approximately 10 drops of 0.5 *M* $CuSO_4$ solution to each of three semimicro test tubes. To one of the test tubes, add 0.5 *M* NaCl solution dropwise until the color of the complex is evident. To the second copper(II) sample, add 0.5 *M* NaBr dropwise until the complex has formed. In the exhaust hood, add 10% ethylenediamine solution dropwise to the third sample of copper(II) ion until the color of the complex is evident. How do these complexes compare to the copper/ammonia complex prepared in Part 1?

D. Oxidation/Reduction Reactions

Oxidation/reduction reactions are very common and take many forms. Reactions in which elemental substances combine to form a compound or in which compounds are decomposed to elemental substances (or simpler compounds) are redox reactions. The reactions of metallic substances with acids, or with corrosive gases, and the reaction of many species in solution are all examples of oxidation/reduction processes.

1. Oxidation of Hydrogen

Your instructor will *demonstrate* (in the exhaust hood) the burning of hydrogen gas in air. A dish full of ice water will be held above the flame of the burning hydrogen. Water vapor produced by the reaction will condense as liquid water on the bottom of the dish. The unbalanced reaction is simply

$$H_2 + O_2 \rightarrow H_2O$$

In this process, elemental hydrogen changes from an oxidation number of 0 to +1, whereas elemental oxygen is reduced from the 0 to the –2 oxidation state.

2. Oxidation of Magnesium

Obtain a small strip of magnesium ribbon. When magnesium is ignited, the flame produced is dangerous to the eyes. *Caution*! Do *not* look directly at burning magnesium. Hold the magnesium ribbon with tongs, and have ready a beaker of distilled water to catch the product of the reaction as it is formed. Ignite the magnesium ribbon in a burner flame, and hold the burning ribbon over the beaker of water. The unbalanced reaction is

$$Mg + O_2 \rightarrow MgO$$

Allow the beaker of water containing the magnesium oxide produced by the reaction to stand for 5–10 minutes; then test the water with pH paper. Why is the solution basic? Magnesium is oxidized in the reaction, whereas oxygen is reduced.

3. Oxidation of Iron

Your instructor will demonstrate (in the exhaust hood) the burning of iron (steel wool) in a stream of pure oxygen. Iron oxidizes only slowly in air at room temperature (forming rust), but the reaction is considerably faster in pure oxygen at elevated temperatures. Iron has two common oxidation states. The unbalanced reactions are

$$Fe + O_2 \rightarrow FeO \quad \text{and} \quad Fe + O_2 \rightarrow Fe_2O_3$$

4. Oxidation/Reduction of Hydrogen Peroxide

Place about 20 mL of 3% hydrogen peroxide in a large test tube. Obtain a single crystal of potassium permanganate. Ignite a wooden splint, then blow out the flame so that the splint is still glowing. Add the crystal of permanganate to the hydrogen peroxide sample to catalyze the decomposition of the hydrogen peroxide. The glowing wood splint should immediately burst into flame as it comes in contact with the pure oxygen being generated by the decomposition reaction

$$H_2O_2 \rightarrow H_2O + O_2$$

5. Oxidation/Reduction of Copper and Zinc

Obtain about 2 mL of 0.5 *M* copper(II) sulfate solution in a small test tube. Add a small strip or turning of metallic zinc and allow the solution to stand for 15–20 minutes. Then remove the zinc and examine the coating that has formed on the strip. The reaction is

$$Zn(s) + CuSO_4(aq) \rightarrow ZnSO_4(aq) + Cu(s)$$

6. Oxidation/Reduction of Lead and Zinc

Obtain about 2 mL of 1 *M* lead acetate solution in a small test tube, and place the test tube in a place where it will not be disturbed. Add a small strip of metallic zinc to the solution. Allow the solution to stand for 10–15 minutes. After this period, examine the "tree" of metallic lead that has formed. The reaction is

$$Zn(s) + Pb(CH_3COO)_2(aq) \rightarrow Pb(s) + Zn(CH_3COO)_2(aq)$$

7. Oxidation/Reduction of the Halogens

Obtain about 10 drops each of 10% potassium iodide and 10% potassium bromide in separate semimicro test tubes. Add 10 drops of "chlorine water" to each test tube; stopper and shake briefly. Record any color changes. Add 5 drops of methylene chloride to each test tube, stopper, and shake. Elemental iodine and elemental bromine have been produced by replacement; these substances are more soluble in methylene chloride than in water and are preferentially extracted into this solvent. The colors of the methylene chloride layer are characteristic of these halogens. The reactions are

$$Cl_2 + KI \rightarrow KCl + I_2 \quad \text{and} \quad Cl_2 + KBr \rightarrow KCl + Br_2$$

8. Oxidation/Reduction of Copper and Silver

Scrub a 2-inch length of copper wire with steel wool and rinse with water. Place 2 mL of 0.1 *M* silver nitrate solution in a test tube, and add the copper wire. Allow the test tube to stand for 15–20 minutes, then examine the copper wire. The reaction is

$$Cu(s) + AgNO_3(aq) \rightarrow Cu(NO_3)_2(aq) + Ag(s)$$

Name ____________________ Section ____________________

Lab Instructor ____________________ Date ____________________

EXPERIMENT 21 Classes of Chemical Reactions

PRE-LABORATORY QUESTIONS

1. Give simple explanations of the four types of chemical reactions considered in this experiment, indicating the nature of what happens during each type of reaction, and how the reaction types differ.

2. Find three examples in your textbook or in a chemical encyclopedia of balanced chemical equations for each of the four types of chemical reactions discussed in this experiment (precipitation, acid/base, complexation, oxidation/reduction). Do not repeat any of the examples given in the experiment.

Name ______________________ Section ______________________

Lab Instructor ______________________ Date ______________________

EXPERIMENT 21 Classes of Chemical Reactions

RESULTS/OBSERVATIONS

A. Precipitation Reactions

Silver nitrate + sodium chloride

____

____

Lead acetate + potassium chromate

____

____

Barium chloride + sulfuric acid

____

____

B. Acid/Base Reactions

pH of 0.1 *M* HCl ______________ pH of 0.1 *M* acetic acid ______________

Explanation ____

pH of 0.1 *M* NaOH ______________ pH of 0.1 *M* ammonia ______________

Explanation ____

Conductivity tests:

0.1 *M* HCl ____

0.1 *M* acetic acid ____

0.1 *M* NaOH ____

0.1 *M* ammonia ____

Universal indicator test:

HCl + NaOH

Acetic acid + NaOH

C. Complexation Reactions

1. Ammonia complexes

Copper sulfate + ammonia

Nickel sulfate + ammonia

Copper sulfate + ammonium sulfate

Nickel sulfate + ammonium sulfate

2. Iron complexes

Iron nitrate + NaCl

Iron nitrate + HCl

Iron nitrate + NaSCN

Iron nitrate + 1,10-phenanthroline

3. Insoluble nickel complex

Nickel nitrate + DMG

4. Copper complexes

Copper(II) sulfate + NaCl

Copper(II) sulfate + NaBr

Copper(II) sulfate + ethylene diamine

Comparison with the preceding $CuSO_4/NH_3$ reaction

D. Oxidation/Reduction Reactions

1. Hydrogen + oxygen

____

2. Magnesium + oxygen

____

3. Iron + oxygen

____

4. Hydrogen peroxide decomposition

____

5. Zinc strip + copper sulfate solution

____

6. Zinc strip + lead acetate solution

____

7. Potassium iodide + chlorine water

____

Potassium bromide + chlorine water

____

8. Copper strip + silver nitrate solution

____

QUESTIONS

1. Balance all the reactions given in the introduction and procedure sections of this experiment.

2. What is the purpose of the single crystal of potassium permanganate used in the decomposition of hydrogen peroxide?

3. Use your textbook or a chemical encyclopedia to determine the *coordination number* of the metal ions in each of the coordination complexes studied in Part C.

EXPERIMENT

22 Water and Water Pollutants

Objective

In this experiment, three common procedures applied to water resources, to test their fitness for human consumption, will be examined: determination of total dissolved oxygen content, determination of chloride ion content, and determination of calcium ion content.

Introduction

One of humanity's most precious resources is pure water. Most of the mass of the human body consists of water or substances derived from water. An average human can survive without food for several days, or even weeks, but cannot live more than a few days without water. Water, of course, is also needed for many other purposes.

Agriculture, for example, is very dependent on large sources of potable water. If rain does not provide enough water for the growing of crops, water must be provided from some other source through irrigation. The state of California grows much of the fresh vegetables and fruits sold in the United States. Many growing regions of California were originally desert, however, and have been converted into arable farms and orchards by massive irrigation projects.

Industry consumes the major portion of the fresh water used in the United States each day. Oftentimes water used for industrial processes must meet even higher standards than those for human consumption. For example, water is used as a coolant in many situations. Water has a high heat capacity and can absorb and remove a great deal of heat from its surroundings. Power plants use millions of gallons of fresh water daily for this purpose.

To be considered fit for human consumption or for industrial uses, water must meet very high standards. Water must be virtually free of bacterial and viral organisms. For this reason, most drinking water is treated with elemental chlorine (or some chlorine compound that generates Cl_2 *in situ*) to kill such species. Water must be free of suspended solids (sand, silt, or waste material). Water treatment plants contain large filters and settling tanks to remove such solids.

Water must be free of both naturally occurring chemical contaminants or those that have been introduced by humans. For example, water from wells or rivers may contain a large concentration of calcium ion from the leaching of natural minerals. Calcium ion is easily precipitated as calcium carbonate by reaction with carbon dioxide in the air. If water containing large levels of calcium ion were used as a coolant in an industrial process, there is a good chance that the precipitated calcium carbonate would eventually block the pipes used to deliver the water. This happens also in the home, especially in the boilers of home heating systems, or in hot water heaters. Water that contains excessive amounts of calcium ion is said to be *hard*.

Natural sources of water often contain large amounts of sodium chloride. This frequently results from salt, used to treat roadways during the winter, leaching into the groundwater supply. Small amounts of sodium chloride are not harmful, but if the level gets too high, the water becomes unfit for human consumption. Excessive sodium chloride can also be detrimental to industrial processes.

In order to support aquatic life (fish and plants) water must contain not less than a particular minimum amount of dissolved oxygen gas. Fish and underwater plants must obtain oxygen to live. Many pollutants cause the amount of dissolved oxygen gas to decrease substantially below the level needed to support aquatic life. For example, organic wastes (such as sewage) are oxidized by dissolved oxygen (and thereby consume the oxygen).

In this experiment, water samples will be analyzed to determine (1) the amount of available oxygen contained, (2) the concentration of chloride ion they contain, and (3) the level of calcium ion they contain. Although samples of naturally occurring water could be used for this experiment, the samples you will analyze have been prepared so that they contain a known amount of the species under analysis. The methods to be used, however, are essentially the same as would be applied to a sample of water from a well or stream.

"Available" oxygen will be determined by an iodine/thiosulfate method. An excess of potassium iodide will be added to the acidified water sample. Oxygen (or other oxidizing agents) will oxidize a portion of the iodide ion to elemental iodine. The elemental iodine (actually, triiodide ion, I_3^-) thus produced will be determined by titration with standard sodium thiosulfate solution, using starch as indicator.

Chloride ion concentration will be determined by Mohr precipitation titration with standard silver nitrate solution, to a silver chromate endpoint. A small quantity of sparingly soluble calcium carbonate is added to adjust the pH to slightly basic.

Calcium ion concentration will be determined in the sample by titration with standard disodium EDTA solution. A small quantity of the magnesium/EDTA complex is added to sharpen the Eriochrome Black T endpoint, and the sample is buffered at pH 10 with an ammonia/ammonium chloride system.

Safety Precautions

- Wear safety glasses at all times while in the laboratory.
- Use a rubber safety bulb when transferring solutions with a pipet. Do *not* pipet by mouth.
- Sulfuric acid is dangerous to the skin. Wash immediately if spilled and inform the instructor.
- Silver nitrate and iodine/iodide solutions will stain the skin. Nitrates are strong oxidizing agents and are toxic. Wash after handling.
- Chromium compounds are toxic. Wash hands after use. Dispose as directed by the instructor.
- Ammonia is a respiratory irritant and heart stimulant. Confine the pouring of the ammonia buffer to the exhaust hood.

Apparatus/Reagents Required

Burets, 25-mL pipets and safety bulb, 0.0500 M standard silver nitrate solution, 0.0500 M standard disodium EDTA (Na_2EDTA) solution, 0.0500 M standard sodium thiosulfate solution, potassium iodide, starch indicator solution, 0.035 M sodium chromate indicator solution, pH 10 6.0 M ammonia/ammonium chloride buffer, Eriochrome Black T indicator solution, calcium carbonate, 0.01 M Mg/EDTA solution, 6 M sulfuric acid

Procedure

Record all data and observations directly in your notebook in ink.

A. Determination of Dissolved Oxygen in Water

Clean out two 250-mL Erlenmeyer flasks and rinse with distilled water. Label the flasks as samples 1 and 2.

Clean out the buret and the 25-mL pipet with soap and water. Rinse with several portions of tap water, followed by several portions of distilled water. Rinse the buret with several portions of standard 0.0500 M sodium thiosulfate solution (discard the rinsings), and then fill the buret with the sodium thiosulfate solution.

Obtain a sample of water for oxygen analysis. Record the identification number of the sample. Using a rubber safety bulb, pipet 25 mL of the water sample into each of the two Erlenmeyer flasks.

Add 2 g (roughly measured) of potassium iodide and 10 mL of 6 M sulfuric acid (*Caution!*) to sample 1. Iodide ion will reduce any oxidizing agents present (available oxygen) and will itself be oxidized to elemental iodine, I_2. A brown solution (of iodine) will result if the water sample contains available oxygen. Elemental iodine can be titrated with standard sodium thiosulfate solution.

Record the initial reading of sodium thiosulfate in the buret to the nearest 0.02 mL. Begin adding sodium thiosulfate to sample 1, swirling the flask during the addition. The brown color of the elemental iodine will begin to fade. Stop adding sodium thiosulfate when the sample reaches a *light yellow* color.

The disappearance of brown color from an iodine solution gives an indistinct endpoint. Starch forms an intense blue/black color with even a small amount of iodine. It is much easier to see the sudden complete disappearence of the blue/black color of the starch/iodine complex.

Add 2–3 mL of starch solution to the sample. A deep blue or black color will result. Continue adding sodium thiosulfate from the buret drop by drop until the starch/iodine color just disappears. Record the final buret volume to the nearest 0.02 mL.

Repeat the titration with sample 2, adding potassium iodide and sulfuric acid before titrating, and adding starch just before the endpoint is reached. From the concentration of the sodium thiosulfate solution and from the volume used in the titration, calculate the concentration of available oxygen in the water sample.

B. Determination of Chloride Ion in Water

Clean out two 250-mL Erlenmeyer flasks and rinse with distilled water. Label the flasks as samples 1 and 2. Clean out the buret and the 25-mL pipet with soap and water. Rinse with several portions of tap water, followed by several portions of distilled water. Rinse the buret with several portions of standard 0.0500 *M* silver nitrate solution (*Caution!*) and discard the rinsings. Then fill the buret with the silver nitrate solution.

Obtain a sample of water for chloride analysis. Record the identification number of the sample. Using a rubber safety bulb, pipet 25 mL of the water sample into each of the two Erlenmeyer flasks. Add 3–5 drops of sodium chromate indicator solution (*Caution!*) to each of the sample flasks. The sample will be yellow at this point. Silver chromate, which has a red color, will form during the titration. The endpoint is the *first* appearance of the red color.

Record the initial volume of silver nitrate in the buret to the nearest 0.02 mL. Add an amount of solid calcium carbonate about the size of a pea to sample 1. Start adding silver nitrate from the buret to sample 1 slowly, swirling the flask during the addition. A white precipitate of silver chloride will begin to form as the silver nitrate is added. (The precipitate may appear yellow because of the chromate ion present in the sample.) Eventually, red streaks will begin to appear as silver chromate begins to form (as the concentration of chloride ion decreases). Add silver nitrate solution from the buret until one drop causes a permanent red color of silver chromate to appear.

Record the final volume of silver nitrate solution in the buret to the nearest 0.02 mL.

Repeat the titration with sample 2. From the volume of standard silver nitrate used to titrate the sample, and the concentration of the silver nitrate, calculate the concentration of chloride ion present in the water sample.

C. Determination of Calcium Ion in a Water Sample

Clean out two 250-mL Erlenmeyer flasks and rinse with distilled water. Label the flasks as samples 1 and 2. Clean out the buret and the 25-mL pipet with soap and water. Rinse with several portions of tap water, followed by several portions of distilled water. Rinse the buret with several portions of standard 0.0500 *M* Na_2EDTA solution (discard the rinsings); then fill the buret with the Na_2EDTA solution.

Obtain a sample of water for calcium analysis. Record the identification number of the sample. Using a rubber safety bulb, pipet 25 mL of the water sample into each of the two Erlenmeyer flasks. Take the two flasks to the exhaust hood, and add 20 mL of ammonia/ammonium chloride buffer (*Caution!*) to each flask.

Record the initial volume of standard Na_2EDTA solution in the buret to the nearest 0.02 mL. To sample 1, add 10 mL of the Mg/EDTA solution (which promotes a sharper color change at the endpoint) and 5–6 drops of Eriochrome Black T indicator solution. The sample should be red at this point.

Begin titrating with standard EDTA solution from the buret. The red color of the initial solution will gradually change to purple/gray as the blue calcium/EDTA complex forms. The endpoint is signaled by the disappearance of the initial red/purple color, and the appearance of the *pure blue* color of the calcium/EDTA complex.

Record the final volume of Na_2EDTA in the buret to the nearest 0.02 mL.

Repeat the process with sample 2. From the volume of Na_2EDTA used to titrate the sample and the concentration of the Na_2EDTA, calculate the concentration of calcium ion in the water sample.

Name ______________________ Section ______________________

Lab Instructor ______________________ Date ______________________

EXPERIMENT 22 Water and Water Pollutants

PRE-LABORATORY QUESTIONS

1. Use your textbook or a chemical encyclopedia to find a definition of *available oxygen,* as the term is applied to natural fresh water sources.

2. Use your textbook or a chemical encyclopedia to find a definition of *hard water,* as well as several methods for its treatment.

3. A 25.0-mL sample of well water for chloride determination requires 34.32 mL of 0.05012 *M* $AgNO_3$ solution to reach a sodium chromate endpoint. Calculate the concentration of chloride ion in the water sample.

Name ____________________ Section ____________________

Lab Instructor ____________________ Date ____________________

EXPERIMENT 22 Water and Water Pollutants

RESULTS/OBSERVATIONS

A. Available Oxygen Determination

Concentration of $Na_2S_2O_3$ standard solution ____________________

	Sample 1	Sample 2
Initial buret reading	______	______
Final buret reading	______	______
Volume thiosulfate used	______	______
Moles thiosulfate used	______	______
Volume of water sample	______	______
Concentration of sample	______	______

Mean concentration of sample ____________ Code number ____________

B. Chloride Determination

Concentration of standard $AgNO_3$ solution ____________________

	Sample 1	Sample 2
Initial buret reading	______	______
Final buret reading	______	______
Volume $AgNO_3$ used	______	______
Moles $AgNO_3$ used	______	______
Volume of water sample	______	______
Concentration of sample	______	______

Mean concentration of sample ____________ Code number ____________

C. Calcium Determination

Concentration of Na_2EDTA solution ______________________________

	Sample 1	Sample 2
Initial buret reading	____________	____________
Final buret reading	____________	____________
Volume Na_2EDTA used	____________	____________
Moles Na_2EDTA used	____________	____________
Volume of water sample	____________	____________
Concentration of sample	____________	____________

Mean concentration of sample ____________ Code number ____________

QUESTIONS

1. Elemental iodine is volatile: It undergoes substantial sublimation even at room temperature. For this reason, the iodine in Part A was generated by adding KI to the samples one at a time. What error in the determination of available oxygen would be introduced if a portion of the iodine generated had vaporized prior to the titration?

2. Why was it necessary to add a small amount of magnesium/EDTA complex to the calcium samples before titrating?

3. How do pollutants such as sewage lower the dissolved oxygen content of water sources?

EXPERIMENT

23 Colligative Properties of Solutions

Objective

Colligative properties of solutions depend on the *quantity* of solute dissolved in the solvent rather than the *identity* of the solute. The phenomenon of freezing point lowering will be examined quantitatively as an example of a colligative property in Choice I. The phenomena of osmosis and dialysis will be investigated qualitatively in Choice II.

Choice I. Freezing Point Depression and Molar Mass Determination

Introduction

When a solute is dissolved in a solvent, the properties of the solvent are *changed* by the presence of the solute. The magnitude of the change generally is *proportional to* the amount of solute added. Some properties of the solvent are changed only by the *number* of solute particles present, without regard to the particular *nature* of the solute. Such properties are called **colligative properties** of the solution. Colligative properties include changes in vapor pressure, boiling point, freezing point, and osmotic pressure.

For example, if a nonvolatile, nonionizing solute is added to a volatile solvent (such as water), the amount of solvent that can escape from the surface of the liquid at a given temperature is lowered, relative to the case where only the pure solvent is present. The vapor pressure above such a solution will be lower than the vapor pressure above a sample of the pure solvent under the same conditions. Molecules of nonvolatile solute physically *block* the surface of the solvent, thereby preventing as many molecules from evaporating at a given temperature. (See Figure 23-1.) As shown in the figure, if the vapor pressure of the solution is lowered, there is an increase in the boiling point of the solution as well as a decrease in the freezing point.

In this experiment, you will determine the freezing points of a pure solvent (naphthalene), a solution of a known solute (1,4-dichlorobenzene, $C_6H_4Cl_2$) dissolved in naphthalene, and an unknown solution of sulfur in naphthalene. The sulfur is considered to be an unknown solute because you will be determining the molar mass of sulfur (and the formula of elemental sulfur). It may turn out to be rather surprising.

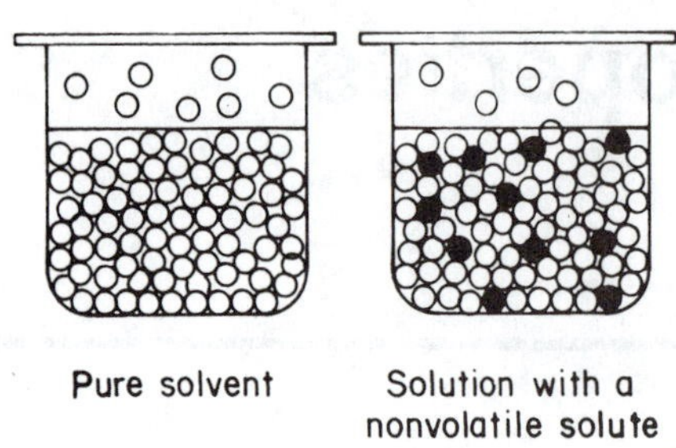

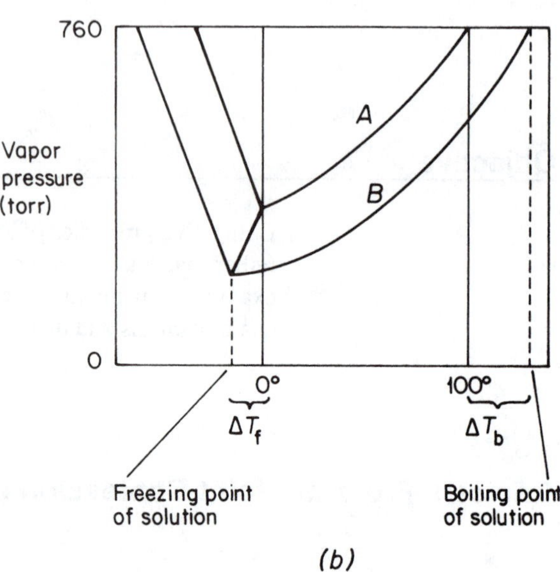

Figure 23-1. (a) The presence of a nonvolatile solute lowers the vapor pressure of the solvent by blocking the surface. (b) Plot of vapor pressure vs. temperature for a pure solvent and a solution. Note that the presence of a solute changes the properties of the solvent.

The decrease in freezing point (ΔT_f) when a nonvolatile, nonionizing solute is dissolved in a solvent is proportional to the molal concentration (m) of solute present in the solvent:

$$\Delta T_f = K_f m \quad (23\text{-}1)$$

K_f is a constant for a given solvent (called the molal freezing point depression constant) and represents by how many degrees the freezing point will change when 1.00 mol of solute is dissolved per kilogram of solvent. For example, K_f for water is 1.86°C kg/mol, whereas K_f for the solvent benzene is 5.12°C kg/mol.

The molal concentration of a solution represents how many moles of solute are dissolved per kilogram of the solvent. For example, if 0.50 mol of the sugar sucrose were dissolved in 100 g of water, this would be equivalent to 5.0 mol of sugar per kilogram of solvent, and the solution would be 5.0 molal. The molality of a solution is defined as

$$\text{molality, } m = \frac{\textit{moles of solute}}{\textit{kilograms of solvent}} \quad (23\text{-}2)$$

The measurement of freezing point lowering is routinely used for the determination of the molar masses of unknown solutes. If Equations 23-1 and 23-2 are combined, it can be derived that the molar mass of a solute is related to the freezing point lowering experienced by the solution and to the composition of the solution. Consider the following example:

The molal freezing point depression constant K_f for the solvent benzene is 5.12°C kg/mol. A solution of 1.08 g of an unknown in 10.02 g of benzene freezes 4.60°C lower than pure benzene. Calculate the molar mass of the unknown.

The molality of the solution can be obtained using Equation 23-1:

$$m = \Delta T_f/K_f = 4.60°C/5.12°C \text{ kg/mol} = 0.898 \text{ mol/kg}$$

Using the definition of molality from Equation 23-2, the number of moles of unknown present can be calculated

$$m = 0.898 \text{ mol/kg} = \text{mol unknown}/0.01002 \text{ kg solvent}$$

$$\text{moles of unknown} = 0.00898 \text{ mol}$$

Thus, 0.00898 mol of the unknown weighs 1.08 g, and the molar mass of the unknown is given by

$$\text{molar mass} = 1.08 \text{ g}/0.00898 \text{ mol} = 120. \text{ g/mol}$$

In the preceding discussion, we considered the effect of a *nonionizing* solute on the freezing point of a solution. If the solute does indeed ionize, the effect on the freezing point will be *larger.* The depression of the freezing point of a solvent is related to the number of particles of solute present in the solvent. If the solute ionizes as it dissolves, the total number of moles of all particles present in the solvent will be larger than the formal concentration indicates. For example, a 0.1 m solution of NaCl is effectively 0.1 m in *both* Na^+ and Cl^- ions.

Safety Precautions

- Wear safety glasses at all times while in the laboratory.
- Naphthalene and 1,4-dichlorobenzene are toxic, and their vapors are harmful. Do not handle these substances. Work in an exhaust hood. Wear disposable gloves.
- Use glycerine when inserting the thermometer through the rubber stopper. Protect your hands with a towel.
- Acetone is highly flammable. Use acetone only in the exhaust hood.
- Dispose of the naphthalene solutions as directed by the instructor.

Apparatus/Reagents Required

8-inch test tube with two-hole stopper to fit, thermometer, stirring wire, beaker tongs, naphthalene, 1,4-dichlorobenzene, sulfur, acetone, disposable gloves

Procedure

Record all data and observations directly in your notebook in ink.

A. Determination of the Freezing Point of Naphthalene

Set up a 600-mL beaker about three-fourths full of water in an exhaust hood, and heat the water to boiling.

Using glycerine as a lubricant and protecting your hands with a towel, insert your thermometer through a two-hole stopper so that the temperature can still be read from 60–90°C. Equip the second hole of the stopper with a length of copper wire, coiled at the bottom into a ring that will fit around the thermometer. The copper wire will be used to *stir* the solution. (See Figure 23-2.)

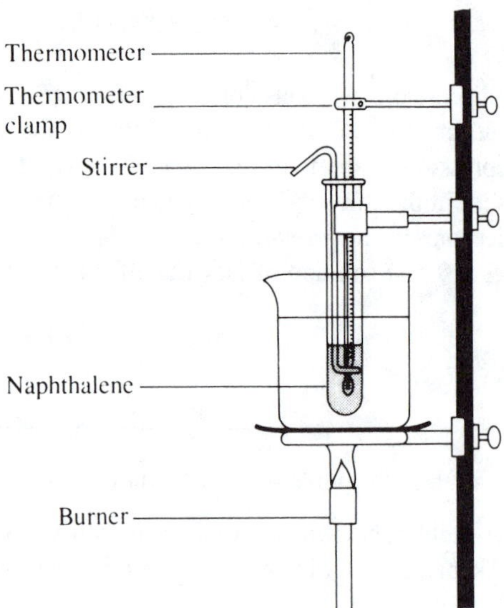

Figure 23-2. Apparatus for freezing point determination. Be certain the thermometer can be read from 60–90°C.

Weigh a clean, dry 8-inch test tube to the nearest 0.01 g.

Add approximately 10 g of naphthalene to the test tube, and reweigh precisely, to the nearest 0.01 g.

Use a clamp to set the test tube vertically in the boiling water bath. Allow the naphthalene to melt completely.

When the naphthalene has melted completely, insert the rubber stopper containing the thermometer and copper wire into the test tube. See Figure 23-2. Make certain that the bulb of the thermometer is immersed in the molten naphthalene and that the copper stirring wire can be agitated freely.

Continue to heat the test tube to remelt any naphthalene that might have crystallized on the thermometer or stirring wire. Move the stirring wire up and down to make certain the temperature is uniform throughout the naphthalene sample.

When the naphthalene has remelted completely (generally the temperature will be above 95°C), remove the beaker of hot water from under the test tube (*Caution!*) using tongs or a towel to protect your hands from the heat.

Ensure that the thermometer is immersed in the center of the molten naphthalene and does *not* touch the walls of the test tube. When the temperature of the molten naphthalene has dropped to 90°C, begin recording the temperature of the naphthalene every 30 seconds. Record the temperatures to the nearest 0.1°C.

Continue taking the temperature of the naphthalene until the temperature has dropped to around 60–65°C. The naphthalene will freeze in this temperature range.

On graph paper from the back of this manual, plot a **cooling curve** for naphthalene: graph the *temperature* of the naphthalene as it cools versus the *time* in minutes. You will notice that the naphthalene *remains* at the same temperature for several minutes; this results in a horizontal, flat region in the cooling curve. This temperature represents the freezing point of pure naphthalene. (See Figure 23-3.)

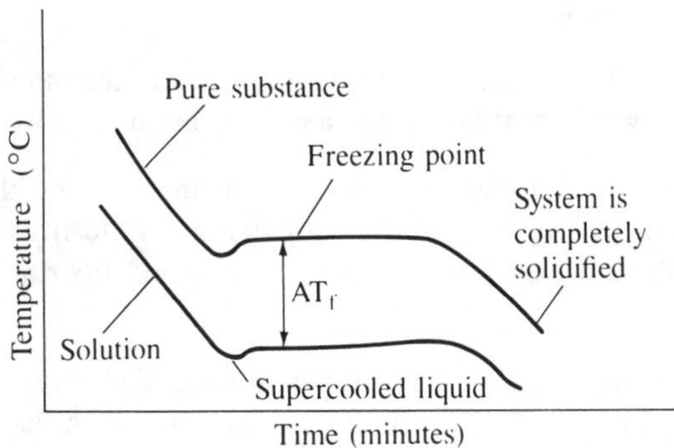

Figure 23-3. Cooling curve for a pure solvent and of a solution involving the solvent. The flat portions of the curves represent freezing.

Use the boiling water bath to remelt the naphthalene in the test tube, and remove the thermometer and wire stirrer. In the exhaust hood, rinse any adhering naphthalene from the thermometer/stirrer with acetone (*Caution: Flammable!*). Allow the thermometer/stirrer to remain in the hood until all acetone has evaporated.

B. Determination of K_f for Naphthalene

Reweigh the test tube containing the naphthalene to the nearest 0.01 g. Some of the naphthalene may have been lost on the thermometer/stirrer apparatus or vaporized.

Add approximately 1 g of 1,4-dichlorobenzene crystals to the test tube containing the naphthalene, and reweigh the test tube to the nearest 0.01 g.

Melt the naphthalene/dichlorobenzene mixture in boiling water.

Make certain that all acetone has evaporated from the thermometer/stirrer apparatus (any remaining acetone would *also* lower the freezing point of naphthalene).

Insert the thermometer/stirrer apparatus into the test tube containing the molten naphthalene/dicholorobenzene mixture. Stir the mixture for several minutes while it is in the boiling water bath to make certain that the solution is completely homogeneous.

Following the same procedure as in Part A, determine the cooling curve and the freezing point of the 1,4dichlorobenzene/naphthalene solution.

From the freezing point depression of the naphthalene/dichlorobenzene solution and the composition of the solution, calculate the molal freezing point depression constant, K_f, for naphthalene. The molecular formula of 1,4-dichlorobenzene is $C_6H_4Cl_2$.

Use the boiling water bath to remelt the naphthalene mixture in the test tube, and remove the thermometer and wire stirrer. In the exhaust hood, rinse any adhering naphthalene from the thermometer/stirrer with acetone (*Caution: Flammable!*). Allow the thermometer/stirrer to remain in the hood until all acetone has evaporated. Dispose of the naphthalene/dichlorobenzene mixture as directed by the Instructor.

C. Determination of the Freezing Point of a Sulfur Solution

Clean and dry an 8-inch test tube, and weigh it to the nearest 0.01 g.

Add approximately 10 g of naphthalene to the test tube, and reweigh the test tube and contents to the nearest 0.01 g.

Add approximately 0.5 g of powdered sulfur to the naphthalene in the test tube and reweigh again to the nearest 0.01 g.

Make certain that all acetone has evaporated from the thermometer/stirrer apparatus (any remaining acetone would also lower the freezing point of naphthalene).

Melt the mixture in the boiling water bath (following the procedure described earlier) and insert the thermometer/stirrer apparatus. Stir thoroughly to make certain that the sulfur is completely dissolved. Determine the cooling curve and freezing point of the sulfur/naphthalene solution.

From the freezing point depression of the sulfur/naphthalene solution and K_f for naphthalene as determined in Part B, calculate the molar mass of sulfur. Determine the number of sulfur atoms that must be present in a molecule of elemental sulfur to give rise to this molar mass

Attach your three graphs to the laboratory report.

Choice II. Osmosis and Dialysis

Introduction

Plant and animal cell membranes must provide some mechanism for the transfer of materials across the membrane without the wholesale loss of the cell's fluid contents or microstructures. Although there are several mechanisms by which molecules are passed through cell walls, perhaps the simplest to understand and demonstrate are the processes of *osmosis* and *dialysis.*

Cell membranes are said to be **semipermeable**: such membranes contain pores that are of a sufficient size that they permit the passage of ions and small molecules, while *not* allowing the passage of most macromolecules or cell organelles. Osmosis represents the passage of *solvent* through a semipermeable membrane, whereas dialysis refers to the passage of *solute* molecules as well. Both osmosis and dialysis occur whenever a concentration gradient (difference) exists between the solutions on either side of a membrane. Osmosis and dialysis occur in an attempt to *equalize* the concentrations of the solutions on either side of the membrane.

For example, suppose you had just eaten a sugary snack after a few hours without food. After digestion of the snack takes place, your bloodstream would have a relatively high concentration of carbohydrates, whereas in the interior of your cells, the concentration of carbohydrates would be expected to be lower (since whatever sugars had been present would have been metabolized during the period without food). Dialysis of sugar molecules from the region of high concentration in the bloodstream to the region of low concentration inside the cells would take place until the concentrations on either side of the cell membrane were similar. Osmosis of water from the cell interior into the bloodstream will also take place.

Osmosis and dialysis are special examples of the process of diffusion: Molecules of solute and solvent diffuse across a membrane separating two solutions in an effort to equalize the concentrations of the solutions, which is very nearly the same as would happen between the two solutions if the membrane were not present.

The *force* with which osmosis/dialysis takes place—that is, how strongly solvent or solute flows through the membrane—depends on the degree of difference in concentration across the membrane. If the two solutions separated by the membrane are of very nearly the same concentration, the osmosis/dialysis that takes place will be very gentle. If a *large* difference in concentration exists across the membrane, however, the osmosis/dialysis will be very strong (perhaps so forceful as to rupture the membrane). For this reason, solutions to be injected into the bloodstream, such as the dextrose solutions used for intravenous feeding in hospitals, have their concentrations adjusted to be very nearly that of normal bodily fluids: Such solutions are said to be *isotonic* with the normal fluids of the body.

The force with which osmosis takes place is measured in terms of the **osmotic pressure** that exists between a solution and a sample of the pure solvent. The osmotic pressure represents the pressure that would have to be applied to the solution to just *prevent* osmosis of the solvent across a membrane. Osmotic pressure (Π) is a colligative property of solutions and is dependent on the molarity of solute present in the solution:

$$\Pi = MRT \qquad (23\text{-}3)$$

Accurate osmotic pressures are not conveniently measurable in the general chemistry laboratory, and so this experiment will only investigate the processes of osmosis and dialysis on a qualitative basis. Osmosis into both a concentrated and a dilute solution of the same solute will be examined to determine into which solution osmosis occurs with the greater force. Dialysis of sodium chloride (an ionic solute), dextrose (a molecular carbohydrate solute), and starch (a colloidal macromolecular solute) will be investigated, and the relative *time* required for dialysis determined.

Safety Precautions

- Wear safety glasses at all times while in the laboratory.
- Silver nitrate solutions will discolor the skin if spilled. Nitrates are strong oxidizing agents and are toxic. Wash your hands after handling.
- Benedict's and Fehling's solutions are toxic. Wash your hands after handling.

Apparatus/Reagents Required

Cellophane dialysis tubing, paper punch, 5% NaCl, 25% NaCl, 5% glucose, 1% starch solution, 0.1 *M* silver nitrate, Benedict's or Fehling's reagents, 0.1 *M* I_2/KI reagent, 600 mL beakers, graduated cylinder

Procedure

A. Demonstration of Osmotic Pressure

Cut two strips of cellophane dialysis tubing each 10–12 inches long. With a paper punch, make a small hole through each piece of tubing about 1 inch from each end. See Figure 23-4.

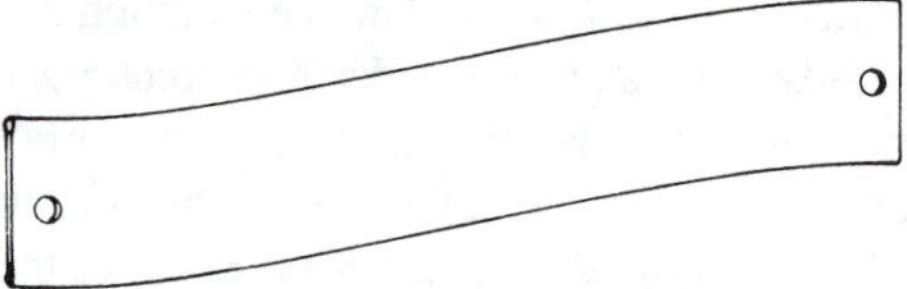

Figure 23-4. Preparation of dialysis tubing. Note the location of the punch holes.

Soak the strips of tubing in water until they become soft and the tubing walls can be spread apart.

Set up two 600-mL beakers in a location where they will not be a disturbed. Place 450–500 mL of distilled water in each beaker.

Obtain 30–40 mL of 5% NaCl solution in a clean beaker, and then transfer exactly 25 mL of this solution to a graduated cylinder.

Holding one of the pieces of dialysis tubing by both ends to prevent spillage, transfer the 25 mL of 5% NaCl solution to the tubing. See Figure 23-5(a).

Insert a stirring rod through the holes in the end of the dialysis tubing, and suspend the tubing containing 5% NaCl solution in one of the beakers of distilled water. If the solution in the tubing is not covered by the water in the beaker, add additional water to the beaker. See Figure 23-5(b).

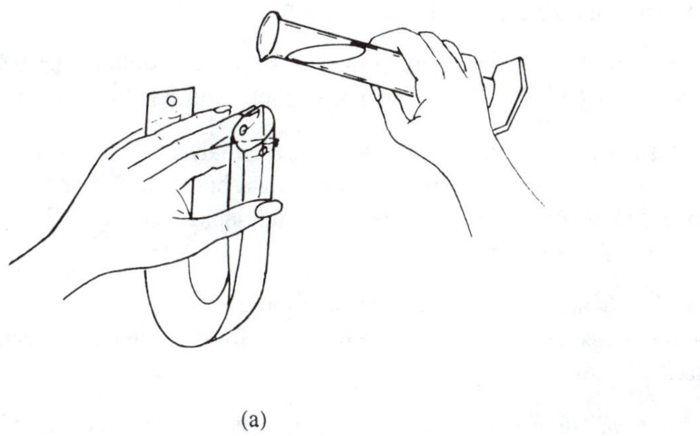

(a)

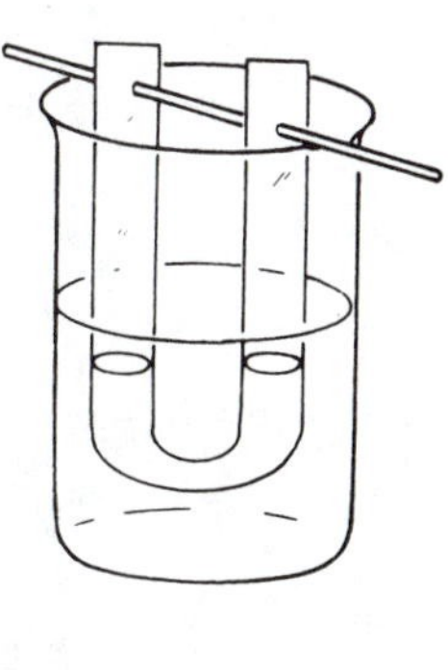

(b)

Figure 23-5. (a) Make a "sling" with a length of dialysis tubing containing the NaCl solution. (b) Suspend the solution in the tubing so that it lies beneath the water's surface.

Repeat the process using the second length of dialysis tubing and the other beaker, using 25% NaCl in place of the 5% solution.

Allow the two beakers to stand for approximately one hour.

At the end of this time, carefully determine the volume of the solutions in the dialysis tubing lengths by pouring the solutions separately into a graduated cylinder.

Determine the volume of water that has entered each length of tubing by osmosis. Did more water diffuse into the concentrated or the dilute NaCl solution?

B. Dialysis

In separate clean beakers, obtain 15–18 mL each of 5% NaCl solution, 5% dextrose solution, and 1% starch.

Transfer about 1 mL of each solution to separate clean test tubes. These test-tube samples will be used for a *preliminary* test for each of the three solutes to be used for dialysis.

To the test tube containing 5% NaCl solution, add 5–6 drops of 0.1 *M* silver nitrate (*Caution!*). The appearance of a white precipitate of AgCl upon treatment with silver nitrate during the dialysis experiment will confirm the presence of chloride ion.

To the test tube containing starch, add 5–6 drops of 0.1 *M* I_2/KI reagent. The appearance of a blue-black color upon treatment with iodine reagent during the dialysis experiment will confirm the presence of starch.

To the test tube containing dextrose, add approximately 5 mL of Benedict's or Fehling's reagent, and heat briefly in a boiling water bath. The appearance of a green or orange precipitate upon similar treatment will confirm the presence of dextrose during the dialysis.

Obtain a 10–12-inch length of dialysis tubing and punch a hole about 1 inch from either end. Soak the tubing in water until the walls can be spread apart.

Set up a 600-mL beaker containing 450–500 mL of distilled water in a place where the beaker will not be disturbed.

Holding the dialysis tubing by both ends to prevent spillage, transfer approximately 10 mL of the 5% NaCl, 5% dextrose, and 1% starch solutions to the tubing.

Insert a stirring rod through the holes in the dialysis tubing, rinse the outside of the tubing with a stream of distilled water from a wash bottle, and suspend the tubing in the beaker of distilled water [see Figure 23-5(b)]. Add additional water if necessary to cover the solution in the dialysis tubing.

Immediately after placing the dialysis tubing in the distilled water, remove approximately 3–4 mL of water from the beaker and perform the tests for NaCl, dextrose, and starch described earlier.

At *10-minute intervals* for the next hour, remove small portions of distilled water from the beaker and repeat the tests for the three solutes.

In what *order* were the solutes able to dialyze through the membrane? How is this order related to the size of the particles involved?

Name ______________________ Section ______________________

Lab Instructor ______________________ Date ______________________

EXPERIMENT 23 Colligative Properties of Solutions

PRE-LABORATORY QUESTIONS

CHOICE I. FREEZING POINT DEPRESSION AND MOLAR MASS DETERMINATION

1. Define the term *colligative property* (you may refer to your textbook).

____

____

____

____

2. Suppose the freezing point of a solution of 2.00 g of an unknown molecular substance in 10.00 g of the solvent benzene is measured. If the solution freezes at a temperature of 6.33°C lower than pure benzene itself, calculate the molar mass of the unknown substance.

____

____

____

____

____

3. For a given number of moles of solute, why do *ionic* substances have a larger effect on the freezing and boiling points of solvents than do *non*ionic substances?

____

____

____

____

____

____

CHOICE II. OSMOSIS AND DIALYSIS

1. As mentioned in the introduction to this choice, solutions intended for injection into the bloodstream are prepared to have the same osmotic pressure as the body's own fluids. The typical intravenous feeding solution used in hospitals consists of 5% (weight/volume) dextrose, $C_6H_{12}O_6$. Calculate the molarity of such a dextrose solution, and its osmotic pressure at 37°C.

2. Ionic solutes produce more particles when dissolved in water than is indicated by their formal concentration. For example, when one mole of NaCl is dissolved in a large amount of water, effectively two moles of solute particles are introduced into the solvent. Saline solution for injection into the human body is approximately 0.9% NaCl. Is such a solution isotonic with the 5% dextrose solution described in Question 1?

Name ______________________________ Section ______________________

Lab Instructor ______________________________ Date ______________________

EXPERIMENT 23 Colligative Properties of Solutions

CHOICE I. FREEZING POINT DEPRESSION AND MOLAR MASS

RESULTS/OBSERVATIONS

A. Determination of the Freezing Point of Naphthalene

Mass empty test tube ______________________________

Mass test tube plus naphthalene ______________________________

Mass naphthalene taken ______________________________

Freezing point of naphthalene (from graph) ______________________________

B. Determination of K_f for Naphthalene

Mass test tube plus naphthalene ______________________________

Mass naphthalene in test tube ______________________________

Mass test tube/naphthalene plus dichlorobenzene ______________________________

Mass 1,4-dichlorobenzene taken ______________________________

Freezing point naphthalene/dichlorobenzene solution ______________________________

Molality of dichlorobenzene solution ______________________________

K_f for naphthalene ______________________________

C. Determination of the Freezing Point of a Sulfur Solution

Mass empty test tube ______________________________

Mass test tube with naphthalene ______________________________

Mass of naphthalene taken ______________________________

Mass of test tube after adding sulfur ______________________________

Mass of sulfur added ______________________________

Freezing point of naphthalene/sulfur solution ____________________

Molality of sulfur in the solution ____________________

Molar mass of sulfur ____________________

Number of sulfur atoms per molecule ____________________

QUESTIONS

1. A phenomenon called **supercooling** is frequently encountered in this experiment. In supercooling, a solution momentarily drops below its freezing point, and then warms up again, before solidification begins. What event is likely to give rise to supercooling?

2. The molal freezing point constant, K_f, is a property of the solvent, not the solute. What does this say about the fact that freezing point depends on the amount of solute, rather than on the solute's nature?

3. Look up the freezing point constant, K_f, for naphthalene in a handbook. How closely does your value for K_f compare? What might have led to your obtaining a different value?

4. A phenomenon that happens sometimes during freezing point depression experiments is that the solute is affected in some manner by the solvent. One common occurrence is for a solute to dimerize; that is, two solvent molecules combine to produce a single double molecule (a dimer). What effect would there be on a molar mass determination if the solute were to dimerize?

Name ______________________ Section ______________

Lab Instructor ______________________ Date ______________

EXPERIMENT 23 Colligative Properties of Solutions

CHOICE II. OSMOSIS AND DIALYSIS

RESULTS/OBSERVATIONS

A. Demonstration of Osmotic Pressure

Volume of 5% sodium chloride solution taken ______________________

Volume measured after osmosis ______________________

Volume of water that passed through membrane ______________________

Volume of 25% sodium chloride solution taken ______________________

Volume measured after osmosis ______________________

Volume of water that passed through membrane ______________________

Into which NaCl solution was the osmosis stronger? ______________________

Why?

____

____

____

B. Dialysis

Observation of $AgNO_3$ test for Cl^- ______________________

Observation of I_2 test for starch ______________________

Observation of Benedict's test for dextrose ______________________

After approximately what time intervals were the solutes detectable by dialysis from the tube into the beaker of water?

NaCl ______________ dextrose ______________ starch ______________

QUESTIONS

1. How were the *time intervals* determined for dialysis related to the *sizes* of the molecules involved?

2. If the experiment was performed correctly, you should *not* have been able to detect starch in the beaker even after a full hour. What is different about a starch solution that causes this?

3. If red blood cells are placed in a solution which is not *isotonic* with the liquids inside the cells, the cells will undergo either *hemolysis* or *crenation* depending on the difference in concentrations. Use a dictionary or science encyclopedia to define these terms.

EXPERIMENT

24 Colloids

Objective

In this experiment, some of the properties and characteristics of *colloidal dispersions* and gels will be investigated.

Introduction

In a true **solution**, particles of solute are homogeneously dispersed among molecules of solvent on an individual *molecular* or *ionic* basis. That is, in a solution, the solute exists as *single*, relatively independent particles. Particles of solute in a solution move freely and randomly throughout the solvent matrix and do not settle out of solution on standing in a closed system. Solute particles in solution have diameters on the order of one nanometer (10^{-9} m).

An opposite situation exists in the type of mixture known as a **suspension**. In a suspension, grossly *large* particles of a nonsoluble substance are mixed with the solvent and shaken to homogenize the mixture *temporarily.* Particles in a suspension are macroscopic and consist of many billions of molecules or ions, with the diameter of the particles greater than one micrometer (10^{-6} m). The particles of a suspension settle out of solution fairly rapidly; most suspensions for use in the laboratory or in the home are labeled "shake well before using" for this reason.

Colloidal dispersions represent an intermediate situation between true solutions and suspensions. The solute particles in a colloid are *not* individual molecules or ions; rather, they are microscopically sized *groups* of molecules or ions, or large intertangled polymeric molecules, with sizes on the order of 10–100 nanometers. Although the solute particles in a colloid do not consist of individual molecules or ions, the particles do *not* settle out of the mixture on standing because of an effect known as **Brownian motion**. The solute particles in a colloid are small enough that irregular, random collisions with rapidly moving solvent molecules are strong enough to *support* the solute particles and prevent them from settling out of the mixture. In suspensions, the particles are too big for collisions with solvent molecules to have much of an effect. The particles of suspension are generally large enough that such collisions are equally probable from all directions (thereby canceling out with time any effect from the collisions).

Colloids generally can be distinguished from true solutions by their interaction with light (**Tyndall effect**). Although the particles in colloids are microscopically small, they are still large enough to scatter light. A concentrated colloid appears *cloudy* because of its ability to scatter light. A very dilute colloid may appear fully clear and transparent to the naked eye under room lighting, but if an intense beam of light is passed through a colloid at right angles to the viewer, the beam of light is clearly visible. This does *not* happen with true solutions because the molecules or ions of a solution are too small to scatter light. The Tyndall effect is observed quite commonly in everyday life. For example, the air in most rooms contains colloidally dispersed dust. A beam of sunlight passing through such dust is clearly visible.

Under the correct conditions, some colloidal dispersions can be made to form semisolids called **gels**. This is especially true if the colloidal solute is a polymeric species whose long-chain molecules can become tangled and interconnected. For example, common gelatin is a colloidal material that forms a gel when a hot mixture is cooled. In a gel, the long-chain molecules of the colloidal substance form an extended three-dimensional network, in which molecules of the solvent become trapped among the chains of the dispersed substance. In this experiment, you will prepare several gels.

Starch and gelatin form colloidal dispersions in boiling water. The resulting dispersions typically appear cloudy when the mixture is concentrated, but are considerably more clear in dilute mixtures. Starch and gelatin mixtures may exhibit the Tyndall effect when a beam of light is passed through the sample at right angles to the observer. The light beam is scattered by the relatively large particles of the colloid, and the beam of light is clearly visible as it passes through the sample. If the light beam is intense enough, the actual individual particles of the colloid may become visible (but only as small, bright flashes of light). The Brownian motion imparted to the colloidal particles by collision with solvent molecules may thus be observed. The properties of starch and gelatin colloids will be contrasted with the behavior of distilled water and some true solutions when treated and observed in the same manner.

Safety Precautions

- Wear safety glasses at all times in the laboratory.
- Use tongs or insulated gloves when handling hot solutions.

Apparatus/Reagents Required

Soluble starch, gelatin powder, glucose, sodium chloride, high-intensity light source, distilled water, square jar, magnifying glass

Procedure

Record all data and observations directly in your notebook in ink.

Heat a 600-mL beaker about half full of water to boiling.

Measure out around 5 g of soluble starch. Transfer the starch to a beaker, add 10–15 mL of water, and stir to make a paste.

When the water in the 600-mL beaker is boiling, stir the starch paste and pour it into the boiling water. Continue boiling the mixture with constant stirring until the mixture *clarifies*; then remove it from the heat and allow it to cool to room temperature.

Set up a second 600-mL beaker about half full of water and heat to boiling.

Measure out about 5 g of gelatin. Transfer to a beaker and add 10–15 mL of cold water. Allow the mixture to stand for 5–10 minutes to soften the gelatin.

When the water in the 600-mL beaker is boiling, stir the gelatin mixture and pour into the boiling water. Stir the boiling mixture for around 30 seconds, then remove it from the heat and allow it to cool to room temperature.

Obtain about 600 mL of distilled water and filter through a gravity funnel and filter paper to remove any dust that might be present.

Dissolve 5 g of glucose in about 200 mL of the filtered distilled water.

Dissolve 5 g of sodium chloride in a second 200-mL portion of the filtered distilled water.

Set up the high-intensity lamp source in a convenient location so that the light beam from the source will be 2–3 inches above the lab bench.

Obtain a square glass jar or tall-form beaker, and transfer into it some of the starch colloid already prepared. Move the jar or beaker so that the beam of light from the high-intensity lamp passes through the starch. View the starch solution at right angles to the beam of light. The light beam should be clearly visible in the starch. Examine the starch solution with a magnifying glass while it is illuminated by the beam of light; look for evidence of the Brownian motion of the colloid particles (flashes of light).

Clean out the jar or beaker, rinse with distilled water, and repeat the examination of the Tyndall effect in the light beam with the gelatin colloid and with the solutions of glucose and sodium chloride.

Clean out the jar, and fill it with the remaining portion of filtered distilled water. Examine with the light beam for evidence of the Tyndall effect.

Replace the filtered distilled water with samples of tap water and unfiltered distilled water. Examine each sample in the light beam for evidence of the Tyndall effect.

Transfer the beakers containing the remaining portions of starch and gelatin to an ice bath; they will set into gels. Examine the gels, and test with the light source for the Tyndall effect.

Name ______________________________ Section ______________________

Lab Instructor ______________________________ Date ______________________

EXPERIMENT 24 Colloids

PRE-LABORATORY QUESTIONS

1. The terms *Brownian motion* and *Tyndall effect* are frequently confused. Use your textbook or an encyclopedia to find and write a specific definition for each of these terms.

2. Compare the relative particle size and resulting properties of "true" solutions, colloidal dispersions, and suspensions.

3. Write a specific definition of *gel colloid* (you may use your textbook or a chemical dictionary).

Name ______________________ Section ______________________

Lab Instructor ______________________ Date ______________________

EXPERIMENT 24 Colloids

RESULTS/OBSERVATIONS

Gross appearance of starch colloid ______________________

Gross appearance of gelatin colloid ______________________

Appearance of starch colloid with high-intensity lamp ______________________

____

With magnifying glass ______________________

____

Appearance of gelatin colloid with high-intensity lamp ______________________

____

With magnifying glass ______________________

____

Appearance of glucose solution with high-intensity lamp ______________________

____

Appearance of sodium chloride solution with high-intensity lamp ______________________

____

Appearance of filtered distilled water with high-intensity lamp ______________________

____

Appearance of tap water with high-intensity lamp ______________________

____

Observations of starch and gelatin gels ______________________

____

QUESTIONS

1. Why was it necessary to use *filtered* distilled water in preparing the solutions for this experiment?

2. Give two examples of suspensions as used in medicine, the home, or the chemistry laboratory. Indicate the active ingredient(s) in each suspension, and the approximate concentration.

EXPERIMENT

25 Rates of Chemical Reactions

Objective

The study of the *speed* with which a chemical process takes place is crucial. Chemists or chemical engineers want reactions to take place quickly enough that they will be useful, but not so quickly that the reaction cannot be studied or controlled. Biologists use the study of reaction rates as an indication of the mechanism by which a biochemical process takes place. In this experiment, the effect of varying the concentration of the reactants in a process will be investigated as to how this affects the measured rate of the reaction. The effect of temperature in speeding up or slowing down reactions will also be examined briefly.

Introduction

The rate at which a chemical reaction occurs depends on several factors: the *nature* of the reaction, the *concentrations* of the reactants, the *temperature*, and the presence of possible *catalysts*. Each of these factors can influence markedly the observed speed of the reaction.

Some reactions at room temperature are very slow. For example, although wood is quickly oxidized in a fireplace at elevated temperatures, the oxidation of wood at room temperature is negligible. Many other reactions are essentially instantaneous. The precipitation of silver chloride when solutions containing silver ions and chloride ions are mixed is an extremely rapid reaction, for example.

For a given reaction, the rate typically *increases* with an increase in the concentrations of the reactants. The relation between rate and concentration is a remarkably simple one in many cases. For example, for the reaction

$$aA + bB \rightarrow \text{products}$$

the rate can usually be expressed by the relationship

$$\text{Rate} = k\,[A]^m[B]^n$$

in which m and n are usually small whole numbers. In this expression, called a **rate law**, [A] and [B] represent, respectively, the concentration of substances A and B, and k is called the **specific rate constant** for the reaction (which provides the correct numerical proportionality). The exponents m and n are called the **orders** of the reaction with respect to the concentrations of substances A and B, respectively. For example, if $m = 1$, the reaction is said to be *first order* with respect to the concentration of A. If $n = 2$, the reaction would be *second order* with respect to the concentration of B. The so-called *overall order* of the reaction is represented by the *sum* of the individual orders of reaction. For the examples just mentioned, the reaction would have overall order

$$1 + 2 = 3 \text{ (third order).}$$

The rate of a reaction is also significantly dependent on the *temperature* at which the reaction occurs. An increase in temperature increases the rate. A rule of thumb (which does have a theoretical basis) states that an increase in temperature of 10 Celsius degrees will double the rate of reaction. While this rule is only approximate, it is clear that a rise in temperature of 100°C would affect the rate of reaction appreciably. As with concentration, there is a quantitative relationship between reaction rate and temperature; but here the relationship is less straightforward. The relationship is based on the idea that, in order to react, the reactant species must possess a certain minimum amount of energy at the time the reactant molecules actually collide during the rate-determining step of the reaction. This minimum amount of energy is called the **activation energy** for the reaction and generally reflects the kinetic energies of the molecules at the temperature of the experiment. The relationship between the specific rate constant (k) for the reaction, the absolute temperature (T), and the activation energy (E_a) is

$$\log k = -E_a/2.3RT + \text{constant}$$

In this relationship, R is the ideal gas constant, which has value $R = 8.31$ J/mol K. The equation therefore gives the activation energy, E_a, in units of *joules*. By experimentally determining k at various temperatures, the activation energy can be calculated from the *slope* of a plot of log k versus $1/T$. The slope of such a plot would be ($-E_a/2.3R$).

In this experiment, you will study a reaction called the "iodine clock." In this reaction, potassium iodate (KIO_3) and sodium hydrogen sulfite ($NaHSO_3$) react with each other, producing elemental iodine

$$5HSO_3^- + 2IO_3^- \rightarrow I_2 + 5SO_4^{2-} + H_2O + 3H^+$$

This is an oxidation/reduction process, in which iodine(V) is reduced to iodine(0), and sulfur(IV) is oxidized to sulfur(VI). Because elemental iodine is *colored* (whereas the other species are colorless), the rate of reaction can be monitored simply by determining the *time required* for the appearance of the *color* of the iodine. As usual with other reactions in which elemental iodine is produced, a small quantity of starch is added to heighten the color of the iodine. Starch forms an intensely colored blue/black complex with iodine. While it would be difficult to detect the first appearance of iodine itself (since the solution would be colored only a very pale yellow), if starch is present, the first few molecules of iodine produced will react with the starch present to give a much sharper color change.

The rate law for this reaction would be expected to have the general form

$$\text{Rate} = k\,[HSO_3^-]^m[IO_3^-]^n$$

in which m is the order of the reaction with respect to the concentration of bisulfite ion, and n is the order of the reaction with respect to the concentration of iodate ion. Notice that even though the stoichiometric coefficients of the reaction are known, these are *not* the exponents in the rate law. The order of the reaction must be determined *experimentally*, and may bear no relationship to the stoichiometric coefficients of the balanced chemical equation. The rate law for a reaction reflects what happens in the slowest, or *rate-determining,* step of the reaction mechanism. A chemical reaction generally occurs as a series of discrete microscopic steps, called the **mechanism** of the reaction, in which only one or two molecules are involved at a time. For example, in the bisulfite/iodate reaction, it would be statistically almost impossible for five bisulfite ions and two iodate ions to all come together in the same place at the same time for reaction. It is much more likely that one or two of these molecules first interact with each other, forming some sort of *intermediate* perhaps, and then this intermediate reacts with the rest of the ions at some later time. By careful experimental determination of the rate law for a process, information is obtained about exactly what molecules react during the slowest step in the reaction, and frequently this information can be extended to suggest what happens in all the various steps of the reaction's mechanism.

In this experiment, you will determine the order of the reaction with respect to the concentration of potassium iodate. You will perform several runs of the reaction, each time using the *same* concentration of all other reagents, but *varying* the concentration of potassium iodate in a systematic manner. By measuring the time required for reaction to occur with different concentrations of potassium iodate, and realizing that the time required for reaction is inversely proportional to the rate of the reaction, the exponent of iodate ion in the rate law should be readily evident.

Safety Precautions

- Wear safety glasses at all times while in the laboratory.
- Sodium hydrogen sulfite (sodium bisulfite) is harmful to the skin and releases noxious SO_2 gas if acidified. Use with adequate ventilation in the room.
- Potassium iodate is a strong oxidizing agent and can damage skin. Wash after using. Do not expose KIO_3 to any organic chemical substance or an uncontrolled oxidation may result.
- Elemental iodine may stain the skin if spilled. The stains are generally not harmful at the concentrations to be used in this experiment, but will require several days to wear off.

Apparatus/Reagents Required

Laboratory timer (or watch with second hand), Solution 1 (containing potassium iodate at 0.024 *M* concentration), Solution 2 (containing sodium hydrogen sulfite at 0.016 *M* concentration and starch)

Procedure

Record all data and observations directly in your notebook in ink.

A. Solutions to Be Studied

Two solutions have been prepared for your use in this experiment. It is essential that the two solutions are *not mixed* in any way before the actual kinetic run is made. Be certain that graduated cylinders used in obtaining and transferring the solutions are *rinsed* with distilled water between solutions. Also rinse thermometers and stirring rods before transferring between solutions.

Solution 1 is 0.024 *M* potassium iodate. Solution 2 is a mixture containing two different solutes, sodium hydrogen sulfite and starch. Solution 2 has been prepared so that the solution contains hydrogen sulfite ion at 0.016 *M* concentration. The presence of starch in Solution 2 may make the mixture appear somewhat cloudy.

Obtain about 400 mL of Solution 1 in a clean, dry 600-mL beaker. Keep covered with a watch glass to minimize evaporation. Obtain about 150 mL of Solution 2 in a clean, dry 250-mL beaker. Cover with a watch glass.

B. Kinetic Runs

Clean out several graduated cylinders and beakers for the reactions. It is important that Solutions 1 and 2 do not mix until the reaction time is to be measured. Use separate graduated cylinders for the measurement of each solution.

The general procedure for the kinetic runs is as follows (specific amounts of reagents to be used in the actual runs are given in a table further on in the procedure):

Measure out the appropriate amount of Solution 2 in a graduated cylinder. Take the temperature of Solution 2 while it is in the graduated cylinder, being sure to rinse and dry the thermometer to avoid mixing the solutions prematurely. Transfer the measured quantity of Solution 2 from the graduated cylinder to a clean, dry 250-mL beaker.

Measure out the appropriate amount of Solution 1 using a clean graduated cylinder and transfer to a clean, dry small beaker. Again using a clean graduated cylinder, add the appropriate amount of distilled water for the run and stir to mix. Measure the temperature of Solution 1, being sure to rinse and dry the thermometer to avoid mixing the solutions prematurely.

If the temperatures of Solutions 1 and 2 differ by more than one degree, wait until the two solutions come to the same temperature.

When the two solutions have come to the same temperature, prepare to mix them. Have ready a clean stirring rod for use after mixing the solutions.

Noting the time (to the nearest second), pour Solution 1 into the beaker containing Solution 2 and stir for 15–30 seconds. Watch the mixture carefully, and record the time the blue/black color of the starch/iodine mixture appears.

Repeat the same run, using the same amounts of Solutions 1 and 2 before going on to the next run.

The table that follows indicates the amounts of Solution 1, distilled water, and Solution 2 to be mixed for each run. Distilled water is added in varying amounts to keep the total volume the same for all the runs.

Table of Kinetic Runs

Run	*mL Solution 1*	*mL distilled water*	*mL Solution 2*
A	10.0	80.0	10.0
B	20.0	70.0	10.0
C	30.0	60.0	10.0
D	40.0	50.0	10.0
E	50.0	40.0	10.0

From the time required for the appearance of the blue/black color, and the concentration of potassium iodate used in each kinetic run, determine the *order* of the reaction with respect to iodate ion concentration.

Construct a plot of *time required for reaction* (vertical axis) versus *concentration of iodate ion* (horizontal axis). Construct a second plot, in which you plot the *reciprocal* of the time required for reaction on the vertical axis versus the concentration of iodate ion. Why is the reciprocal graph a straight line?

C. Temperature Dependence

A rule of thumb indicates that the rate of reaction is *doubled* for each 10°C temperature increase. Confirm this qualitatively by performing two determinations of run A in the table at a temperature approximately 10°C *higher* than the temperature used previously. Repeat the procedure, making two determinations of run A, at a temperature approximately 10°C *lower* than the original determinations.

Name ______________________________ Section ______________________

Lab Instructor ______________________________ Date ______________________

EXPERIMENT 25 Rates of Chemical Reactions

PRE-LABORATORY QUESTIONS

1. Given the following data, determine the *orders* with respect to the concentrations of substances A and B in the reaction

$$A + B \rightarrow \text{products}$$

$[A]_{initial}$	$[B]_{initial}$	Time for reaction
0.10 *M*	0.10 *M*	262 s
0.20 *M*	0.10 *M*	131 s
0.30 *M*	0.10 *M*	87 s
0.20 *M*	0.20 *M*	66 s
0.10 *M*	0.20 *M*	131 s

____

____

____

____

____

____

2. What must be the *units* of the specific rate constant in Pre-Laboratory Question 1, given that the rate of reaction would be measured in *M*/s?

____

____

3. Why is it not possible to predict the *form* of the rate law for a reaction from a knowledge of the overall stoichiometry of a reaction?

____

____

____

Name ______________________________ Section ______________________

Lab Instructor ______________________________ Date ______________________

EXPERIMENT 25 Rates of Chemical Reactions

RESULTS/OBSERVATIONS

Kinetic Runs

Time required for I_2 color to appear	First trial	Second trial
Run A (10 mL Solution 1)	______________	______________
Run B (20 mL Solution 1)	______________	______________
Run C (30 mL Solution 1)	______________	______________
Run D (40 mL Solution 1)	______________	______________
Run E (50 mL Solution 1)	______________	______________

Based on your results, what is the *order* of the reaction with respect to potassium iodate concentration? Explain your reasoning.

____

____

____

____

____

Temperature Dependence

What higher temperature did you use? ______________________________

What time was required for reaction? ______________________________

What lower temperature did you use? ______________________________

What time was required for reaction? ______________________________

Do these times confirm the rule of thumb? ______________________________

QUESTIONS

1. In this experiment you determined the dependence of the reaction rate on the concentration of potassium iodate. Devise an experiment for determining the dependence of the rate on the concentration of sodium sulfite.

2. Why was it necessary to keep the total volume of the reagents after mixing constant in all the kinetic runs (i.e., why was it necessary to add distilled water in inverse proportion to the quantity of Solution 1 that was required)?

3. Why was it necessary that the two solutions to be mixed be at the same temperature before mixing? What error would have been introduced if the solutions were not at the same temperature?

4. A term that is often used by kineticists is the **molecularity** of a reaction. Use a scientific dictionary or chemical encyclopedia to write a definition of molecularity and explain how this term differs from the order of a reaction as discussed earlier.

5. Determination of the rate law for a reaction is often the first step in elucidating the *mechanism* of the reaction. Write a specific definition for what is meant by a reaction mechanism.

EXPERIMENT

26 Chemical Equilibrium

Objective

Many chemical reactions, especially those of organic substances, do *not* go to *completion*. Rather, they come to a point of **chemical equilibrium** before the reactants are fully converted to products. At the point of equilibrium, the concentrations of all reactants remain *constant* with time. The *position* of this equilibrium is described by a function called the **equilibrium constant**, K_{eq}, which is a *ratio* of the amount of product present to the amount of reactant remaining once the point of equilibrium has been reached. In Choice I of this experiment, you will determine the equilibrium constant for an esterification reaction. In Choice II, you will investigate how outside forces acting on a system at equilibrium provoke changes within the system (Le Châtelier's principle).

Introduction

Early in the study of chemical reactions, it was noted that many chemical reactions do not produce as much product as might be expected, based on the amounts of reactants taken originally. These reactions appeared to have *stopped* before the reaction was complete. Closer examination of these systems (after the reaction had seemed to stop) indicated that there were still significant amounts of all the original *reactants* present. Quite naturally, chemists wondered why the reaction had seemed to stop, when all the necessary ingredients for *further* reaction were still present.

Some reactions appear to stop because the products produced by the original reaction *themselves begin to react*, in the reverse direction to the original process. As the concentration of products begins to build up, product molecules will react more and more frequently. Eventually, as the speed of the forward reaction decreases while the speed of the reverse reaction increases, the forward and reverse processes will be going on at exactly the *same* rate. Once the forward and reverse rates of reaction are identical, there can be no further net change in the concentrations of any of the species in the system. At this point, a dynamic state of *equilibrium* has been reached. The original reaction is still taking place but is opposed by the reverse of the original reaction also taking place.

The point of chemical equilibrium for a reaction is most commonly described numerically by the **equilibrium constant**, K_{eq}. The equilibrium constant represents a *ratio* of the concentrations of all product substances present at the point of equilibrium to the concentrations of all original reactant substances (all concentrations are raised to the power indicated by the coefficient of each substance in the balanced chemical equation for the reaction). For example, for the general reaction

$$aA + bB = cC + dD$$

the equilibrium constant would have the form

$$K_{eq} = \frac{[C]^c [D]^d}{[A]^a [B]^b}$$

A ratio is used to describe the point of equilibrium for a particular chemical reaction scheme because such a ratio will be independent of specific amounts of substance that might have been used initially in a particular experiment. The equilibrium constant K_{eq} is a constant for a given reaction at a given temperature.

Choice I. Determination of an Equilibrium Constant

Introduction

In Choice I of this experiment, you will determine the equilibrium constant for the esterification reaction between *n*-propyl alcohol and acetic acid.

$$CH_3COOH + CH_3CH_2CH_2OH = CH_3COOCH_2CH_2CH_3 + H_2O$$

acetic acid	*n-propyl alcohol*	*n-propyl acetate (an ester)*	*water*
HAc	PrOH	PrAc	H_2O

Using the abbreviations indicated for each of the substances in this system, the form of the equilibrium constant would be

$$K_{eq} = \frac{[PrAc][H_2O]}{[HAc][PrOH]}$$

You will set up the reaction in such a way that the *initial* concentrations of HAc and PrOH are *known*. The reaction will then be allowed to stand for one week to come to equilibrium. As HAc reacts with PrOH, the acidity of the mixture will *decrease*, reaching a minimum once the system reaches equilibrium. The quantity of acid present in the system will be determined by titration with standard sodium hydroxide solution. Esterification reactions are typically catalyzed by the addition of a strong mineral acid. In this experiment a small amount of sulfuric acid will be added as a catalyst. The amount of catalyst added will have to be determined, since this amount of catalyst will also be present in the equilibrium mixture and will contribute to the total acid content of the equilibrium mixture.

By analysis of the amounts of each reagent used this week in setting up the reaction and by determining the amount of acetic acid (HAc) that will be present next week once the system has reached equilibrium, you will be able to calculate the concentration of all species present in the equilibrium mixture. From this, the value of the equilibrium constant can be determined.

The concentration of acetic acid in the mixture is determined by the technique of *titration.* Acetic acid reacts with sodium hydroxide (NaOH) on a 1:1 stoichiometric basis

$$HAc + NaOH \rightarrow NaAc + H_2O$$

A precise volume of the reaction mixture is removed with a pipet, and a standard NaOH solution is added slowly from a buret until the acetic acid has been completely neutralized (this is signaled by an indicator, which changes color). From the volume and concentration of the NaOH used and the volume of reaction mixture taken, the concentration of acetic acid in the reaction mixture may be calculated.

Safety Precautions

- Wear safety glasses at all times while in the laboratory.
- Glacial (pure) acetic acid and sulfuric acid burn skin badly if spilled. Wash immediately if either acid is spilled, and inform the instructor at once.
- Acetic acid, *n*-propyl alcohol, and *n*-propyl acetate are all highly flammable and their vapors may be toxic or irritating if inhaled. Absolutely no flames are permitted in the laboratory.
- The NaOH solution used in this experiment is very dilute but will concentrate by evaporation on the skin if spilled. Wash after use.
- When pipeting solutions, use a rubber safety bulb. Do *not* pipet by mouth.

Apparatus/Reagents Required

50-mL buret, plastic wrap, standard 0.10 *M* NaOH solution, phenolphthalein indicator solution, 1-mL pipet and rubber safety bulb, *n*-propyl alcohol (1-propanol), glacial acetic acid, 6 *M* sulfuric acid

Procedure

Record all data and observations directly in your notebook in ink.

A. First Week: Setup of the Initial Reaction Mixture

Clean a buret with soap and water until water does not bead up on the inside of the buret. Clean a 1-mL volumetric transfer pipet. Rinse the buret and pipet with tap water several times to remove all soap. Follow by rinsing with small portions of distilled water.

Obtain approximately 100 mL of standard 0.10 *M* NaOH solution in a clean dry beaker. Rinse the buret several times with small portions of NaOH solution (discard the rinsings); then fill the buret with NaOH solution. Keep the remainder of the NaOH solution in the beaker covered until it is needed.

Clean two 250-mL Erlenmeyer flasks for use as titration vessels. Label the flasks as 1 and 2. Rinse the flasks with tap water; follow with small portions of distilled water. Place approximately 25 mL of distilled water in each flask, and set aside until needed.

Since glacial acetic acid and *n*-propyl alcohol are both liquids, it is more convenient to measure them out by *volume* than by mass.

Clean and dry a 125-mL Erlenmeyer flask. Label the flask as *reaction mixture*. Cover a rubber stopper that securely fits the flask with plastic wrap (this prevents the stopper from being attacked by the vapors of the reaction mixture).

Clean and dry a small graduated cylinder. Using the graduate, obtain 14 ± 0.2 mL of glacial acetic acid (0.25 mol) and transfer the acetic acid to the clean, dry reaction mixture Erlenmeyer flask.

Rinse the graduated cylinder with water and redry. Obtain 19 ± 0.2 mL of *n*-propyl alcohol (0.25 mol) and add to the acetic acid in the reaction mixture Erlenmeyer flask. Stopper the flask and swirl the flask for several minutes to mix the reagents.

Using the 1-mL volumetric pipet and safety bulb, transfer 1.00 mL of the reaction mixture to each of the two 250-mL Erlenmeyer flasks (1 and 2). Restopper the flask containing the *n*-propyl alcohol/acetic acid reaction mixture to prevent evaporation,

Add 3–4 drops of phenolphthalein indicator to each of the two samples to be titrated.

Record the initial level of the NaOH solution in the buret. Place Erlenmeyer flask 1 under the tip of the buret, and slowly begin adding NaOH solution to the sample. Swirl the flask during the addition of NaOH. As NaOH is added, red streaks will begin to appear in the sample due to the phenolphthalein, but the red streaks will disappear as the flask is swirled. The endpoint of the titration is when a single additional drop of NaOH causes a faint, permanent pink color to appear. Record the level of NaOH in the buret.

Repeat the titration using Erlenmeyer flask 2, recording initial and final levels of NaOH.

Discard the samples in flasks 1 and 2.

From the average volume of NaOH used to titrate 1.00 mL of the reaction mixture and the concentration of the NaOH, calculate the concentration (in mol/L) of acetic acid in the reaction mixture:

$$\text{moles of NaOH used} = (\text{concentration of NaOH, } M) \times (\text{volume used to titrate, L})$$

$$\text{moles of HAc} = \text{moles of NaOH at the color change of the indicator}$$

$$\text{molarity of HAc} = \frac{\text{moles of HAc}}{\text{liters of reactionmixturetaken with pipet}}$$

Since the reaction was begun using equal molar amounts of acetic acid and *n*-propyl alcohol (i.e., 0.25 mol of each), the concentration of acetic acid calculated also represents the concentration of *n*-propyl alcohol in the original mixture.

B. First Week: Determination of Sulfuric Acid Catalyst

The reaction between *n*-propyl alcohol and acetic acid is slow unless the reaction is catalyzed. Mineral acids speed up the reaction considerably, but the presence of the mineral acid catalyst must be considered in determining the remaining concentration of acetic acid in the system once equilibrium has been reached. Next week, you will again titrate 1.00-mL samples of the reaction mixture with NaOH, to determine what concentration of acetic acid remains in the mixture at equilibrium. However, since NaOH reacts with both the acetic acid of the reaction and also with the mineral acid catalyst, some method must be found to determine the concentration of the mineral acid in the reaction mixture. Sulfuric acid will be used as the catalyst.

Refill the buret (if needed) with standard NaOH and record the initial level. Clean out Erlenmeyer flasks 1 and 2, rinse, and fill with approximately 25 mL of distilled water. Clean out and have handy the 1-mL pipet and rubber safety bulb.

Add, with swirling, 10 drops of 6 *M* sulfuric acid catalyst to the acetic acid/propyl alcohol reaction mixture. *Immediately* pipet a 1.00-mL sample of the catalyzed reaction mixture into both flasks. Do not delay pipeting, or the concentration of acetic acid will begin to change as the reaction occurs.

Recording initial and final NaOH levels in the buret, titrate the catalyzed reaction mixture in flasks 1 and 2, using 3–4 drops of phenolphthalein indicator to signal the endpoint.

Since the samples of catalyzed reaction mixture contain the same quantity of acetic acid as the samples of uncatalyzed mixture, the increase in volume of NaOH required to titrate the second set of 1-mL samples represents a measure of the amount of sulfuric acid present.

By subtracting the average volume of standard NaOH used in Part A (acetic acid only) from the average volume of NaOH used in Part B (acetic acid + sulfuric acid), calculate how many mL of the standard NaOH solution are required to titrate the sulfuric acid catalyst present in 1 mL of the reaction mixture. This volume represents a correction that can be applied to the volume of NaOH that will be required to titrate the samples next week, after equilibrium has been reached.

Stopper the 125-mL flask containing the acetic acid/*n*-propyl alcohol mixture. Place the reaction mixture in your locker in a safe place until next week.

C. Second Week: Determination of the Equilibrium Mixture

After standing for a week, the reaction system of *n*-propyl alcohol and acetic acid will have come to equilibrium.

Clean a buret and 1-mL pipet. Rinse and fill the buret with the standard 0.10 *M* NaOH solution.

Clean and rinse two 250-mL Erlenmeyer flasks (samples 1 and 2). Place approximately 25 mL of distilled water in each of the Erlenmeyer flasks.

Uncover the acetic acid/*n*-propyl alcohol reaction mixture. Using the rubber safety bulb, pipet 1.00- mL samples into each of the two Erlenmeyer flasks. You may notice that the *odor* of the reaction mixture has changed markedly from the sharp vinegar odor of acetic acid that was present last week.

Add 3–4 drops of phenolphthalein to each sample, and titrate the samples to the pale pink endpoint with the standard NaOH solution; record the initial and final levels of NaOH in the buret.

Calculate the mean volume of standard NaOH required to titrate 1.00-mL of the equilibrium mixture. Using the volume of NaOH required to titrate the sulfuric acid catalyst present in the mixture (from Part B), calculate the volume of NaOH that was used in titrating the acetic acid component remaining in the equilibrium mixture.

From the volume and concentration of NaOH used to titrate the acetic acid in 1.00-mL of the equilibrium mixture, calculate the concentration of acetic acid in the equilibrium mixture in moles per liter. Since the reaction was begun with equal molar amounts of acetic acid and *n*-propyl alcohol (0.25 mol of each), the concentration of *n*-propyl alcohol in the equilibrium mixture is the same as that calculated for acetic acid.

Calculate the *change* in the concentration of acetic acid between the initial and the equilibrium mixtures. From the change in the concentration of acetic acid, calculate the concentrations of the two products of the reaction (*n*-propyl acetate, water) present in the equilibrium mixture.

From the concentrations of each of the four components of the system at equilibrium, calculate the equilibrium constant for the reaction.

Choice II. Stresses Applied to Equilibrium Systems

Introduction

Many reactions come to equilibrium. The reaction in Choice I seemed to have stopped before the full amount of product expected had been formed. When equilibrium had been reached in this system, there were significant amounts of both products as well as original reactants still present. In this choice, you will study changes made in a system *already in equilibrium*.

Le Châtelier's principle states that, if we *disturb* a system that is already in equilibrium, then the system will *react* so as to minimize the effect of the disturbance. This is most easily demonstrated in cases where additional reagent is added to a system in equilibrium, or when one of the reagents is removed from the system in equilibrium.

Solubility Equilibria

Suppose we have a solution that has been saturated with a solute: This means that the solution has already dissolved as much solute as possible. If we try to dissolve additional solute, no more will dissolve, because the saturated solution is in equilibrium with the solute:

$$\text{Solute} + \text{Solvent} \;=\; \text{Solution}$$

Le Châtelier's principle is most easily seen when an ionic solute is used: Suppose we have a saturated solution of sodium chloride, NaCl. Then

$$NaCl(s) \;=\; Na^{+}(aq) + Cl^{-}(aq)$$

will describe the equilibrium that exists. Suppose we then try adding an *additional amount* of one of the ions involved in the equilibrium. For example, suppose we added several drops of HCl solution (which contains the chloride ion). According to Le Châtelier's principle, the equilibrium would shift so as to consume some of the added chloride ion. This would result in a net decrease in the amount of NaCl that could dissolve. If we watched the saturated NaCl solution as the HCl was added, we should see some of the NaCl precipitate as a solid.

Complex Ion Equilibria

Oftentimes, dissolved metal ions will react with certain substances to produce brightly colored species called **complex ions**. For example, iron(III) reacts with the thiocyanate ion (SCN^{-}) to produce a bright red complex ion:

$$Fe^{3+} + SCN^{-} \;=\; [FeNCS^{2+}]$$

This is an equilibrium process that is easy to study, because we can monitor the bright red color of $[FeNCS^{2+}]$ as an indication of the position of the equilibrium: If the solution is very red, there is a lot of $[FeNCS^{2+}]$ present; if the solution is not very red, then there must be very little $[FeNCS^{2+}]$ present.

Using this equilibrium, we can try adding additional Fe^{3+} or additional SCN^{-} to see what effect this has on the red color according to Le Châtelier's principle. We will also add a reagent (silver ion) that removes SCN^{-} from the system to see what effect this has on the red color.

Acid/Base Equilibria

Many acids and bases exist in solution in equilibrium sorts of conditions: This is particularly true for the weak acids and bases. For example, the weak base ammonia is involved in an equilibrium in aqueous solution

$$NH_3 + H^{+} \;=\; NH_4^{+}$$

Once again, we can use Le Châtelier's principle to play around with this equilibrium: we will try adding more ammonium ion or hydrogen ion to see what happens. Since none of the components of this system is itself colored, we will be adding an acid/base indicator that changes color with pH, to have an index of the position of the ammonia equilibrium. The indicator we will use is the same used in Choice I, phenolphthalein, which is pink in basic solution and colorless in acidic solution.

Safety Precautions

- Wear safety glasses at all times while in the laboratory.
- Concentrated ammonia is a strong respiratory and cardiac stimulant; use concentrated ammonia *only in the exhaust hood.*
- Concentrated hydrochloric acid is *severely damaging to skin* and its *vapor is highly toxic*; use concentrated HCl in the *exhaust hood*; wear disposable gloves while handling the acid to protect your hands; if HCl is spilled on the skin, wash immediately and inform the instructor.
- Iron(III) chloride and potassium thiocyanate are toxic; wash hands after use.

Apparatus/Reagents Required

Saturated sodium chloride solution, 12 M HCl, 1 M HCl, 0.1 M $FeCl_3$, 0.1 M KSCN, concentrated ammonia solution, ammonium chloride, 0.1 M $AgNO_3$, phenolphthalein indicator

Procedure

Record all data and observations directly in your notebook in ink.

A. Solubility Equilibria

Obtain 5 mL of saturated sodium chloride solution in each of two test tubes. This solution was prepared by adding solid NaCl to water until no more would dissolve. Then the clear solution was filtered from any undissolved solid NaCl.

Add 10 drops of 12 M HCl (*Caution!*) to one test tube of saturated NaCl solution. A small amount of solid NaCl should form and precipitate out of the solution. The crystals may form slowly, and may be very small. Examine the test tube carefully.

Add 10 drops of 1 M HCl to the other test tube of saturated NaCl solution. Why does no precipitate form in this instance?

On the lab report sheet, describe what happens in terms of Le Châtelier's principle.

B. Complex Ion Equilibria

Prepare a stock sample of the bright red complex ion [$FeNCS^{2+}$] by mixing 2 mL of 0.1 M iron(III) chloride and 2 mL of 0.1 M KSCN solutions. The color of this mixture is too intense to use as it is, so dilute this mixture with 100 mL of water.

Pour about 5 mL of the diluted red stock solution into each of four test tubes. Label the test tubes as 1, 2, 3, and 4.

Test tube 1 will have *no change* made in it, so that you can use it to compare color with what will be happening in the other test tubes.

To test tube 2, add about 1 mL of 0.1 *M* $FeCl_3$ solution. Watch very carefully as the $FeCl_3$ solution is poured into the red solution.

To test tube 3, add about 1 mL of 0.1 *M* KSCN solution. Watch very carefully as the KSCN solution is poured into the red solution.

To test tube 4, add 0.1 *M* $AgNO_3$ solution dropwise until a change becomes evident. Ag^+ ion removes SCN^- ion from solution as a solid (silver thiocyanate, AgSCN).

Describe the intensification or fading of the red color in each test tube in terms of Le Châtelier's principle.

C. Acid/Base Equilibria

In the exhaust hood, prepare a dilute ammonia solution by adding 4 drops of concentrated ammonia to 100 mL of water.

Add 3 drops of phenolphthalein to the dilute ammonia solution, which will turn pink (ammonia is a base, and phenolphthalein is pink in basic solution).

Place about 5 mL of the pink dilute ammonia solution into each of three test tubes.

To one of the test tubes, add several small crystals of ammonium chloride (which contains the ammonium ion, NH_4^+).

To a second test tube, add a few drops of 12 *M* HCl (*Caution!*).

To the third test tube, add 1 drop of concentrated ammonia (*Caution!*).

Describe what happens to the pink color in terms of how Le Châtelier's principle is affecting the ammonia equilibrium.

Name ______________________________ Section ______________________

Lab Instructor ______________________________ Date ______________________

EXPERIMENT 26 Chemical Equilibrium

PRE-LABORATORY QUESTIONS

CHOICE I. DETERMINATION OF AN EQUILIBRIUM CONSTANT

1. Write the expression for the equilibrium constants for the following reactions:

$N_2(g) + 3H_2(g) = 2NH_3(g)$

$SO_2(g) + O_2(g) = SO_3(g)$

$2CO(g) + O_2(g) = 2CO_2(g)$

____

____

____

2. A 1.00-mL pipet sample of dilute acetic acid required 23.4 mL of 0.10 *M* NaOH to titrate the sample to a phenolphthalein endpoint. Calculate the concentration of acetic acid in the sample. How many significant figures are justified in your answer? Why?

____

____

____

____

CHOICE II. STRESSES APPLIED TO EQUILIBRIUM SYSTEMS

1. Given the gas-phase reaction

$$N_2 + 3\,H_2 = 2\,NH_3.$$

Suppose that the reaction has already taken place, and the system has come to equilibrium. If the following changes are made to the system in equilibrium, tell what effect these changes will have on the system (whether the equilibrium is shifted to the left, the right, or is not shifted).

a. Additional N_2 is added to the system.

___

b. Ammonia is removed from the system as soon as it forms.

___

c. The pressure of the system is increased (all components are gases).

___

d. A very efficient catalyst is used for the reaction.

___

Name ____________________ Section ____________________

Lab Instructor ____________________ Date ____________________

EXPERIMENT 26 Chemical Equilibrium

CHOICE I. DETERMINATION OF AN EQUILIBRIUM CONSTANT

RESULTS/OBSERVATIONS

Concentration of standard NaOH solution ____________________

	Sample 1	Sample 2
Volume of NaOH to titrate 1 mL initial uncatalyzed mixture (*first week*)	________	________
Mean volume		________
Concentration of acetic acid, *M*, in original mixture	________	________
Volume of NaOH to titrate 1 mL catalyzed reaction mixture (*first week*)	________	________
Mean volume		________
Volume correction for sulfuric acid (*to be applied next week*)	________	
Volume of NaOH to titrate 1 mL of equilibrium mixture (*second week*)	________	________
Mean volume		________
Volume (*corrected*) of NaOH to titrate acetic acid in equilibrium mixture	________	________
Concentration of acetic acid, *M*, in equilibrium mixture	________	________
Change in concentration of acetic acid in reaching equilibrium	________	________

Complete the following table:

	Concentrations, M	
	Initial mixture	Equilibrium mixture
Acetic acid	____________	____________
n-Propyl alcohol	____________	____________
n-Propyl acetate	____________	____________
Water	____________	____________

Calculate the equilibrium constant for the reaction:

____

____

____

____

QUESTIONS

1. Sulfuric acid was used as a catalyst in this reaction. Would the presence of a catalyst affect the position of the equilibrium (i.e., the relative amounts of substances present once equilibrium was reached)? Why?

____

____

2. Could some other acid have been used as the catalyst? Why?

____

____

____

3. A common misconception among beginning students is that "you leave liquids and solids out of equilibrium constants." *All* the species involved in this reaction were liquids, and *all* of their concentrations were included in the calculation of the equilibrium constant. Explain.

____

____

____

____

Name ______________________________ Section ______________________

Lab Instructor ______________________________ Date ______________________

EXPERIMENT 26 Chemical Equilibrium

CHOICE II. STRESSES APPLIED TO EQUILIBRIUM SYSTEMS

RESULTS/OBSERVATIONS

A. Solubility Equilibria

Effect of adding HCl to saturated NaCl observation

____

____

Explanation

____

____

B. Complex Ion Equilibria

Effect of adding Fe^{3+} to [$FeNCS^{2+}$] observation

____

____

Explanation

____

____

Effect of adding SCN^- to [$FeNCS^{2+}$] observation

____

____

Explanation

Effect of adding Ag^+ to $[FeNCS^{2+}]$ observation

Explanation

C. Acid/Base Equilibria

Effect of adding $NH4^+$ observation

Explanation

Effect of adding HCl observation

Explanation

QUESTIONS

1. Explain how a saturated solution represents an equilibrium between the solution and any undissolved solute present.

2. Addition of 12 *M* (concentrated) HCl to saturated NaCl results in precipitation of NaCl. Yet when 1 *M* HCl is added to the same NaCl solution, *no* precipitate forms. Explain.

3. How do you explain why the addition of silver nitrate to the Fe/SCN equilibrium had an effect on the equilibrium, even though neither silver ion nor nitrate ion is written as part of the equilibrium reaction?

4. How do you explain why the addition of HCl affected the ammonia equilibrium, even though neither hydrogen ion nor chloride ion is written as part of the ammonia equilibrium reaction?

EXPERIMENT

27 The Solubility Product of Silver Acetate

Objective

The extent to which a sparingly soluble salt dissolves in water is frequently indicated in terms of the salt's solubility product equilibrium constant, K_{sp}. In this experiment, K_{sp} for the salt silver acetate will be determined, and the common ion effect will be demonstrated.

Introduction

A previous experiment has discussed the concept of solubility. Solubility was defined to be the number of grams of a substance that will dissolve in 100 mL of solvent. For substances that dissolve to a reasonable extent in a solvent, the solubility is a useful method of describing how much solute is present in a solution. However, for substances that are only very sparingly soluble in a solvent, the solubility of the salt is not a very convenient means for describing a saturated solution. When the solute is only very sparingly soluble in the solvent, the solution is more conveniently (and correctly) described by the equilibrium constant for the dissolving process.

Consider the salt silver chloride, AgCl, which is very sparingly soluble in water. When a portion of solid silver chloride is placed in a quantity of pure water, Ag^+ ions and Cl^- ions begin to dissolve from the crystals of solid and enter the water. As the number of Ag^+ and Cl^- ions present in the water increases, the likelihood of the Ag^+ and Cl^- ions *re*entering the solid increases. Eventually, an equilibrium is established between the solution and the undissolved solute crystals remaining: Ag^+ and Cl^- ions will be dissolving from the crystals, while elsewhere Ag^+ and Cl^- ions will "undissolve" at the same rate. Beyond this point, there will be no net increase in the number of Ag^+ and Cl^- ions to be found in the solution at any future time. The solution has become **saturated**, which means that the ions that are dissolved have reached a state of equilibrium with the remaining undissolved solute.

$$AgCl(s) = Ag^+(aq) + Cl^-(aq)$$

In writing the equilibrium constant expression for this equilibrium, we might at first be tempted to write

$$K = [Ag^+(aq)][Cl^-(aq)]/[AgCl(s)]$$

in which square brackets [ ] refer as usual to concentrations in moles per liter. However, because AgCl(*s*) is a pure solid, its concentration is a *constant* value (a function of the density of the solid) and is included in the measured value of the equilibrium constant. For this reason, a new equilibrium constant, K_{sp}, called the **solubility product constant** is defined for the dissolving process outlined earlier:

$$K_{sp} = [Ag^+(aq)][Cl^-(aq)]$$

The fact that the concentration of the remaining undissolved solid AgCl does not enter into this equilibrium is easily observed experimentally because the measured value of the equilibrium constant is completely independent of the quantity of solid AgCl that remains. Whether there is one single crystal of undissolved AgCl remaining in contact with the saturated solution or 1000 g, the value of K_{sp} is a fixed number (varying only with temperature). The solubility product, K_{sp}, is generally preferred over the actual solubility (in g solute/100 mL solvent) because it is independent of a specific amount of solvent.

The solubility product, K_{sp}, is defined in terms of molar concentrations. However, the solubility in moles or grams per liter can be calculated from the solubility product. Suppose the solubility product for AgCl in pure water has been determined from experiment to be

$$K_{sp} = [Ag^+(aq)][Cl^-(aq)] = 1.6 \times 10^{-10}$$

If AgCl(*s*) has been dissolved in pure water, then

$$[Ag^+(aq)] = [Cl^-(aq)] = 1.3 \times 10^{-5}\ M$$

and 1.3×10^{-5} mol/L of AgCl(*s*) must have dissolved to produce these concentrations of the individual ions. The solubility of AgCl in grams per liter can then be simply calculated using the formula weight of AgCl(*s*).

In the preceding discussion, we considered the case in which the salt AgCl was dissolved in pure water. If we try to dissolve salt AgCl in a solution that already contains either Ag^+ ion or Cl^- ion (from some other source), we will have to consider the effect that these already present ions will have on the dissolving of AgCl. Suppose again that K_{sp} for AgCl has been measured experimentally and was found to be given by

$$K_{sp} = [Ag^+(aq)][Cl^-(aq)] = 1.6 \times 10^{-10}$$

Suppose we try to dissolve AgCl(*s*) in a solution that contains 0.10 *M* Cl^- ion from another source (perhaps 0.10 *M* sodium chloride solution). On a microscopic basis, there is no distinction among Cl^- ions, whether they were produced by AgCl(*s*) dissolving or were provided from some other source. The solubility product equilibrium constant must be satisfied for AgCl(*s*) regardless of what else might be present in the solution. However, the $[Cl^-(aq)]$ in the solution was already at the level of 0.10 *M* before AgCl(*s*) was added to the solution. Substituting this value into the expression for K_{sp},

$$K_{sp} = [Ag^+(aq)][0.10] = 1.6 \times 10^{-10}$$

gives us that the maximum concentration of Ag^+ ion that can be present in the 0.10 *M* NaCl solution is

$$[Ag^+(aq)] = (1.6 \times 10^{-10})/0.10 = 1.6 \times 10^{-9}\ M$$

Since the only source of Ag^+ ion is from the dissolving of AgCl(*s*), then 1.6×10^{-9} mol/L of AgCl(*s*) must have dissolved. This is approximately 10,000 times less AgCl(*s*) than dissolved per liter of pure water. The presence of the common ion, $Cl^-(aq)$, already in the solvent has decreased the solubility of the sparingly soluble salt $AgCl(s)$.

In general, the presence of a common ion from another source will decrease the solubility of a sparingly soluble substance. The common ion effect is merely a demonstration of Le Châtelier's principle in action. If additional $Cl^-(aq)$ is added to an equilibrium system involving $Cl^-(aq)$, then the equilibrium will shift so as to minimize the effect of the added $Cl^-(aq)$, resulting in a net decrease in the amount of $AgCl(s)$ that can dissolve.

In this experiment, you will determine K_{sp} for the salt silver acetate. Silver acetate is not truly a "very sparingly soluble substance," but its concentration in a saturated solution is much less than for many common salts. However, because the solubility of silver acetate is appreciable, you may not get quite the agreement among results in this experiment that is usually expected. K_{sp} for silver acetate will be determined in water, in solutions in which a common ion is present, and in a solution in which ions that are not part of the solubility equilibrium are present. The concentration of silver ion in the silver acetate solutions will be determined by a precipitation titration of a standard potassium chloride solution (Mohr method), using potassium chromate as indicator. Silver acetate and potassium chloride react according to

$$AgC_2H_3O_2(aq) + KCl(aq) \rightarrow AgCl(s) + KC_2H_3O_2(aq)$$

From the volume of silver acetate solution required to titrate a given sample of standard KCl, the concentration of the silver acetate solution may be calculated by usual solution stoichiometric methods.

Safety Precautions

- Wear safety glasses at all times while in the laboratory.
- Silver compounds can stain the skin if spilled. The stain is due to the reduction of silver ion to metallic silver on the skin and is not dangerous. The stain will wear off in 2–3 days.
- Chromium and nitrate compounds are toxic. Wash hands after use.
- Dispose of all reagents as directed by the instructor.

Apparatus/Reagents Required

10- and 25-mL pipets, buret and clamp, saturated silver acetate solution (in distilled water), saturated silver acetate solution (in 0.100 M potassium nitrate), saturated silver acetate solution (in 0.100 M silver nitrate), saturated silver acetate solution (in 0.100 M sodium acetate), 5% potassium chromate indicator, standard 0.0500 M potassium chloride

Procedure

Record all data and observations directly in your notebook in ink.

Clean the 10- and 25-mL pipets and the buret by soaking with soap and water, followed by rinses with tap water and distilled water. Set up the buret in the buret clamp.

In order to be certain that the silver acetate solutions are saturated with silver acetate, an excess quantity of solid silver acetate has been added to each solution, which must be removed before the solutions are used in the titrations that follow.

A. Determination of Silver Acetate in Distilled Water

Obtain approximately 70 mL of saturated silver acetate (in distilled water) and filter through a gravity funnel/filter paper to remove any cloudiness/solid. Rinse the buret with small portions of the filtered silver acetate solution; then fill the buret and record the initial reading to the nearest 0.02 mL (reading the bottom of the meniscus).

Pipet 10 mL of the standard 0.0500 *M* KCl solution into a clean Erlenmeyer flask. Then add 15 mL of distilled water.

Add approximately 1 mL of 5% potassium chromate solution. The color change of this indicator is rather subtle, from bright lemon yellow to a "dirty" yellow/brown color.

Titrate the KCl solution with the silver acetate until the indicator changes color. Record the final volume of silver acetate solution in the buret.

Repeat the determination twice for consistent results.

From the concentration of the KCl and from the pipet and buret volumes, calculate the concentration, in moles per liter, of silver ion in the saturated solution of silver acetate in distilled water. Using the mean concentration of silver ion for your three determinations, calculate the solubility product of silver acetate.

B. Determination of Silver Acetate in 0.100 *M* Potassium Nitrate

Obtain approximately 70 mL of saturated silver acetate solution containing 0.100 *M* potassium nitrate. Potassium nitrate is an ionic substance, but is not involved in the solubility equilibrium of silver acetate. Filter the solution to remove cloudiness/solid. Rinse the buret with the filtered solution; then fill the buret and record the initial volume to the nearest 0.02 mL.

Pipet 10 mL of the standard 0.0500 *M* KCl solution into an Erlenmeyer flask. Add 1 mL of the 5% potassium chromate indicator.

Titrate the potassium chloride sample to the indicator endpoint, and record the final volume of silver acetate to the nearest 0.02 mL.

Repeat the determination twice for consistent results.

From the concentration of the standard potassium chloride solutions, and from the pipet and buret volumes, calculate the concentration of silver ions, in moles per liter, in the saturated silver acetate solution that also contains 0.100 *M* potassium nitrate. Using the mean concentration of silver ion, calculate the solubility product of silver acetate.

C. Determination of Silver Acetate in 0.100 *M* Silver Nitrate

Obtain approximately 70 mL of saturated silver acetate solution containing 0.100 *M* silver nitrate. Silver nitrate is an ionic solute that contains an ion common to the silver acetate solubility equilibrium. Filter the solution to remove cloudiness/solid. Rinse the buret with the filtered solution; then fill the buret and record the initial volume to the nearest 0.02 mL.

Pipet a 25-mL portion of the standard 0.0500 *M* KCl solution into an Erlenmeyer flask and add approximately 1 mL of the 5% potassium chromate solution.

Titrate the KCl sample with silver acetate solution until the indicator changes from lemon yellow to "dirty" yellow/brown.

Repeat the determination twice for consistent results.

From the concentration of the standard potassium chloride solution, and from the pipet and buret volumes, calculate the concentration of silver ion, in moles per liter, in the saturated silver acetate solution that also contains 0.100 *M* silver nitrate. Using the mean concentration of silver ion for your three determinations, calculate the solubility product of silver acetate.

D. Determination of Silver Acetate in 0.100 *M* Sodium Acetate

Obtain approximately 70 mL of saturated silver acetate solution containing 0.100 *M* sodium acetate. Sodium acetate is an ionic solute that contains an ion common to the silver acetate solubility equilibrium. Filter to remove cloudiness/solid. Rinse the buret with the filtered solution, and then fill the buret. Record the initial buret reading to the nearest 0.02 mL.

Pipet 10 mL of the standard 0.0500 *M* KCl solution into an Erlenmeyer flask, and add 15 mL of distilled water. Add approximately 1 mL of the 5% potassium chromate solution.

Titrate the KCl sample with the silver acetate solution to the indicator endpoint. Record the final volume of silver acetate to the nearest 0.02 mL.

Repeat the determination twice for consistent results.

From the concentration of the standard potassium chloride solution, and from the pipet and buret volumes, calculate the concentration of silver ion, in moles per liter, in the saturated silver acetate solution that also contains 0.100 *M* sodium acetate. Using the mean concentration of silver ion for your three determinations, calculate the solubility product for silver acetate.

Name ______________________ Section ______________________

Lab Instructor ______________________ Date ______________________

EXPERIMENT 27 The Solubility Product of Silver Acetate

PRE-LABORATORY QUESTIONS

1. Write the specific definition of K_{sp} found in your textbook.

____

____

____

____

2. Using a handbook of chemical data, look up the values of K_{sp} for the following substances at 25°C:

Silver chloride ______________________ Calcium carbonate ______________________

Silver bromide ______________________ Silver acetate ______________________

Barium sulfate ______________________ Calcium hydroxide ______________________

3. The solubility product of lead chromate is 2.0×10^{-16}. Calculate the solubility in moles per liter of lead chromate in each of the following solutions:

a. Saturated lead chromate in water

____

____

b. Saturated lead chromate in 0.10 M Na_2CrO_4 solution

____

____

c. Saturated lead chromate in 0.001 M $Pb(NO_3)_2$ solution

____

____

Name ______________________________ Section ______________________

Lab Instructor ______________________________ Date ______________________

EXPERIMENT 27 The Solubility Product of Silver Acetate

RESULTS/OBSERVATIONS

A. Silver Acetate/Water

mL standard KCl taken ______________________________

Titration volumes	Sample 1	Sample 2	Sample 3
Initial volume AgAc	________	________	________
Final volume AgAc	________	________	________
Volume AgAc used	________	________	________
Molarity of Ag^+ ion	________	________	________

Mean molarity of Ag^+ ion ______________________________

K_{sp} for silver acetate ______________________________

B. Silver Acetate/Potassium Nitrate

mL standard KCl taken ______________________________

Titration volumes	Sample 1	Sample 2	Sample 3
Initial volume AgAc	________	________	________
Final volume AgAc	________	________	________
Volume AgAc used	________	________	________
Molarity of Ag^+ ion	________	________	________

Mean molarity of Ag^+ ion ______________________________

K_{sp} for silver acetate ______________________________

C. Silver Acetate/Silver Nitrate

mL standard KCl taken ______________________________

Titration volumes	Sample 1	Sample 2	Sample 3
Initial volume AgAc	______	______	______
Final volume AgAc	______	______	______
Volume AgAc used	______	______	______
Molarity of Ag^+ ion	______	______	______

Mean molarity of Ag^+ ion ______

K_{sp} for silver acetate ______

D. Silver Acetate/Sodium Acetate

mL standard KCl taken ______

Titration volumes	Sample 1	Sample 2	Sample 3
Initial volume AgAc	______	______	______
Final volume AgAc	______	______	______
Volume AgAc used	______	______	______
Molarity of Ag^+ ion	______	______	______

Mean molarity of Ag^+ ion ______

K_{sp} for silver acetate ______

Based on Parts A, B, C, and D, calculate a mean value and the standard deviation for K_{sp} for silver acetate.

______

QUESTIONS

1. In theory, the solubility product equilibrium constant should be, in fact, constant throughout your four sets of determinations. What factors may not have been considered that might have led to different measured values for K_{sp} under the various conditions used in the experiment?

______

______

2. Compare your mean K_{sp} value with the literature value. Calculate the percent error.

______

3. How does K_{sp} vary with temperature?

______

EXPERIMENT

28 Acids, Bases, and Buffered Systems

Objective

In this experiment, a study of the properties of acidic and basic substances will be made, using indicators and a pH meter to determine pH. A buffered system will be prepared by half-neutralization, and the properties of the buffered system will be compared to those of a nonbuffered medium.

Introduction

One of the most important properties of aqueous solutions is the concentration of hydrogen (hydronium) ion present. The H^+ [H_3O^+, $H_2O{\cdot}H^+$] ion greatly affects the solubility of many organic and inorganic substances; the nature of complex metal cations found in solution; and the rate of many chemical reactions; and may profoundly affect the reactions of bio-chemical species.

The pH Scale

Typically the concentration of H^+ ion in aqueous solutions may be small, requiring the use of scientific notation to describe the concentration numerically. For example, an acidic solution may have $[H^+] = 2.3 \times 10^{-5}\ M$, whereas a basic solution may have $[H^+] = 4.1 \times 10^{-10}\ M$. Because scientific notation may be difficult to deal with, especially when you make *comparisons* of numbers, a mathematical simplification, called pH, has been defined to describe the concentration of hydrogen ion in aqueous solutions:

$$pH = -\log_{10}[H^+]$$

By use of a logarithm, the power of ten of the scientific notation is converted to a "regular" number, and use of the minus sign in the definition of pH produces a positive value for the pH. For example, if $[H^+] = 1.0 \times 10^{-4}\ M$, then

$$pH = -\log[1.0 \times 10^{-4}] = -[-4.00] = 4.00$$

and similarly, if $[H^+] = 5.0 \times 10^{-2}\ M$, then

$$pH = -\log[5.0 \times 10^{-2}] = -[-1.30] = 1.30$$

Although basic solutions are usually considered to be solutions of *hydroxide* ion, OH^-, basic solutions also must contain a certain concentration of *hydrogen* ion, H^+, because of the equilibrium that exists in aqueous solutions due to the autoionization of water. The concentration of hydrogen ion in a basic solution can be determined by reference to the equilibrium constant for the autoionization of water, K_W:

$$K_W = [H^+][OH^-] = 1.0 \times 10^{-14} \quad \text{(at 25°C)}$$

For example, if a basic solution contains $[OH^-]$ at the level of 2.0×10^{-3} *M*, then the solution also contains hydrogen ion at the level of

$$[H^+] = 1.0 \times 10^{-14}/[OH^-]$$

$$[H^+] = 1.0 \times 10^{-14}/[2.0 \times 10^{-3}] = 5 \times 10^{-12}\ M$$

In pure water, there must be *equal* concentrations of hydrogen ion and hydroxide ion (one of each ion is produced when a water molecule ionizes). So, in pure water

$$[H^+] = [OH^-] = (1.0 \times 10^{-14})^{1/2} = 1.0 \times 10^{-7}\ M$$

The pH of pure water is thus 7.00; this value serves as the *dividing line* for the pHs of aqueous solutions. A solution in which there is *more* hydrogen ion than hydroxide ion ($[H^+] > 1.0 \times 10^{-7}$ *M*) will be *acidic* and will have pH less than 7.00. Conversely, a solution with *more* hydroxide ion than hydrogen ion ($[OH^-] > 1.0 \times 10^{-7}$ *M*) will be *basic* and will have pH greater than 7.00.

If the pH of a solution is known (or has been measured experimentally), then the hydrogen ion concentration of the solution may be calculated

$$[H^+] = 10^{-pH}$$

For example, a solution with pH = 4.20 would have hydrogen ion concentration given by

$$[H^+] = 10^{-pH} = 10^{-4.20} = 6.3 \times 10^{-5}\ M$$

Experimental Determination of pH

The experimental determination of the pH of a solution is commonly performed by either of two methods. The first of these methods involves the use of chemical dyes called **indicators**. These substances are generally weak acids/bases and can exist in either of two colored forms, depending on whether or not the molecule is protonated or has been deprotonated. For example, let HIn represent the protonated form of the indicator. In aqueous solution, an equilibrium exists:

$$\underset{\textit{first color}}{HIn} \;=\; \underset{\textit{second color}}{H^+ + In^-}$$

Depending on what other acidic/basic substances are present in the solution, the equilibrium of the indicator will shift, and one or the other colored form of the indicator will predominate and impart its color to the entire solution.

Indicators commonly change color over a relatively short pH range (about 2 pH units), and, when properly chosen, can be used to estimate the pH of a solution. Among the most common indicators are litmus (which is usually used in the form of test paper that has been impregnated with the indicator) and phenolphthalein (which is most commonly used as a solution added to samples for titration analysis). Litmus changes color from red to blue as the pH of a solution increases from pH 6 to pH 8. Phenolphthalein goes from a colorless form to red as the pH of a solution increases from pH 8 to pH 10. A given indicator is useful for determining pHs only in the region in which it changes color. Indicators are available for measurement of pH in all the important ranges of acidity and basicity. By matching the color of a suitable indicator in a solution of known pH to the color of the indicator in an unknown solution, the pH of the unknown solution can be estimated.

The second method for the determination of pH involves an instrument called a **pH meter**. The pH meter contains two electrodes, one of which is sensitive to the concentration of hydrogen ion. Typically the pH sensitive electrode is a *glass membrane* electrode, whereas the second electrode may consist of a silver/silver chloride system, or a calomel electrode. Frequently, a glass membrane pH electrode is combined in the same physical chamber as a silver/silver chloride electrode and is referred to as a "combination" pH electrode. In the solution being measured, the electrical potential between the two electrodes is a function of the hydrogen ion concentration of the solution. The pH meter has been designed so that the pH of the solution can be directly read from the scale provided on the face of the meter. It is essential always to calibrate a pH meter before use: a solution of known pH is measured with the electrodes, and the display of the meter is adjusted to read the correct pH for the known pH solution. A properly calibrated pH meter provides for much more precise determinations of pH than does the indicator method, and is ordinarily used when a very accurate determination of pH is needed.

Strong and Weak Acids/Bases

Some acids and bases undergo substantial ionization when dissolved in water and are called **strong acids** or **strong bases**. Strong acids and bases are completely ionized in dilute aqueous solutions. Other acids and bases, because they ionize *in*completely in water (often to the extent of only a few percent), are called **weak acids** or **weak bases**. Hydrochloric acid (HCl) and sodium hydroxide (NaOH) are examples of a typical strong acid and a strong base, respectively. A 0.1 *M* solution of HCl contains hydrogen ion concentration at effectively 0.1 *M* concentration because of the complete ionization of HCl molecules.

$$\mathrm{HCl}(aq) \rightarrow \mathrm{H}^+(aq) + \mathrm{Cl}^-(aq)$$

$$0.1\,M \qquad 0.1\,M \qquad 0.1\,M$$

Similarly, a 1×10^{-2} *M* solution of NaOH contains hydroxide ion concentration at the level of 1×10^{-2} *M* because of the complete ionization of NaOH.

$$\mathrm{NaOH}(aq) \rightarrow \mathrm{Na}^+(aq) + \mathrm{OH}^-(aq)$$

$$1 \times 10^{-2}\,M \qquad 1 \times 10^{-2}\,M \qquad 1 \times 10^{-2}\,M$$

Acetic acid ($HC_2H_3O_2$ or CH_3COOH) and ammonia (NH_3) are typical examples of a weak acid and weak base, respectively. Weak acids and weak bases must be treated by the techniques of chemical equilibrium in determining the concentrations of hydrogen ion or hydroxide ion in their solutions. For example, for the general weak acid HA, the equilibrium reaction would be

$$HA(aq) \;=\; H^+(aq) + A^-(aq)$$

and the equilibrium constant expression would be given by

$$K_a = [H^+][A^-]/[HA]$$

For acetic acid, as an example, the ionization equilibrium reaction and K_a are represented by the following:

$$CH_3COOH(aq) \;=\; H^+(aq) + CH_3COO^-(aq)$$

$$K_a = [H^+][CH_3COO^-]/[CH_3COOH]$$

K_a is a constant and is characteristic of the acid HA. For a given weak acid, the product of the concentrations in the expression will remain constant at equilibrium, regardless of the manner in which the solution of the acid was prepared. For a general weak base B, the equilibrium reaction would be

$$B + H_2O \;=\; BH^+(aq) + OH^-(aq)$$

and the equilibrium constant expression would be given by

$$K_b = [BH^+][OH^-]/[B]$$

As an example, the ionization and K_b for the weak base ammonia are represented by

$$NH_3 + H_2O \;=\; NH_4^+(aq) + OH^-(aq)$$

$$K_b = [NH_4^+][OH^-]/[NH_3]$$

The value of the equilibrium constant K_a for a weak acid (or K_b for a weak base) can be determined experimentally by several methods. One simple procedure involves little calculation, is accurate, and does not require knowledge of the actual concentration of the weak acid (or base) under study. For example, a sample of weak acid (HA) is dissolved in water and is then divided into two equal-volume portions. One portion is then titrated with a sodium hydroxide solution to a phenolphthalein endpoint, thereby converting all HA molecules present into A^- ions:

$$OH^- + HA \rightarrow H_2O + A^-$$

The number of moles of A^- ion produced by the titration is equal, of course, to the number of moles of HA in the original half-portion titrated, and is also equal to the number of moles of HA in the remaining unused portion of weak acid. The two portions are then mixed, and the pH of the combined solution is measured. Because $[HA] = [A^-]$ in the combined solution, these terms (represented below by $[n]$) cancel each other in the equilibrium constant expression for K_a for the acid

$$K_a = [H^+][A^-]/[HA] = [H^+][n]/[n]$$

and therefore the ionization equilibrium constant, K_a, for the weak acid will be given directly from a measurement of the hydrogen ion concentration

$$K_a = [H^+] = 10^{-pH}$$

for the combined solution. In other words, the ionization equilibrium constant for a weak acid can be determined simply by measuring the pH of a half-neutralized sample of the acid.

pH of Salt Solutions

Salts that can be considered to have been formed from the complete neutralization of strong acids with strong bases—such as NaCl (which can be considered to have been produced by the neutralization of HCl with NaOH)—ionize completely in solution. Such strong acid/strong base salts do not react with water molecules to any appreciable degree (do not undergo hydrolysis) when they are dissolved in water. Solutions of such salts are neutral and have pH = 7. Other examples of such salts are KBr (from HBr and KOH) and $NaNO_3$ (from HNO_3 and NaOH).

However, when salts formed by the neutralization of a *weak* acid or base are dissolved in water, these salts furnish ions that tend to react to some extent with water, producing molecules of the weak acid or base and releasing some H^+ or OH^- ion to the solution. Solutions of such salts will not be neutral, but rather will be acidic or basic themselves.

Consider the weak acid HA. If the sodium salt of this acid, Na^+A^- is disolved in water, the A^- ions released to the solution will react with water molecules to some extent

$$A^- + H_2O \quad = \quad HA + OH^-$$

The solution of the salt will be *basic* because hydroxide ion has been released by the reaction of the A^- ions with water. For example, a solution of sodium acetate (a salt of the weak acid acetic acid) is basic because of reaction of the acetate ions with water molecules, releasing hydroxide ions:

$$CH_3COO^- + H_2O \quad = \quad CH_3COOH + OH^-$$

Conversely, solutions of salts of weak bases (such as $NH_4^+Cl^-$, derived from the weak base NH_3) will be *acidic*, because of reaction of the ions of the salt with water. For example,

$$NH_4^+ + H_2O \quad = \quad NH_3 + H_3O^+$$

Most salts of transition metal ions are acidic. A solution of $CuSO_4$ or $FeCl_3$ will typically be of pH 5 or lower. The salts are completely ionized in solution. The acidity comes from the fact that the metal cation is hydrated. For example, $Cu(H_2O)_4^{2+}$ better represents the state of a copper(II) ion in aqueous solution. The large positive charge on the metal cation attracts electrons from the O–H bonds in the water molecules, thereby weakening the bonds and producing some H+ ions in the solution

$$Cu(H_2O)_4^{2+} \quad = \quad Cu(H_2O)_3(OH)^+ + H^+$$

Buffered Solutions

Some solutions, called buffered solutions, are remarkably resistant to pH changes caused by the addition of an acid or base from an outside source. A **buffered solution** in general is a mixture of a weak acid and a weak base. The most common sort of buffered solution consists of a mixture of a conjugate acid/base pair. For example, a mixture of the salt of a weak acid or base, and the weak acid or base itself would constitute a buffered solution. The half-neutralized solution of a weak acid described earlier as a means of determining K_a would therefore represent a buffered system. That solution contained equal amounts of weak acid HA and of the anion A^- present in its salt. If a small amount of strong acid were added to the buffered system from an external source, the H^+ ion introduced by the strong acid would tend to react with the A^- ion of the salt, thereby preventing a change in pH. Similarly, if a strong base were added from an external source to the HA/A^- system, the OH^- introduced by the strong base would tend to be neutralized by the HA present in the buffered system. If similar small amounts of strong acid or strong base were added to a nonbuffered system (such as plain distilled water), the pH would be changed drastically by the addition.

Safety Precautions

- Wear safety glasses at all times while in the laboratory.
- The acids, bases, and salts to be used are in fairly dilute solution, but may be irritating to the skin. Wash if they are spilled and inform the instructor.
- The pH meter uses electricity. Do not spill water or a solution near the meter; electrical shock may result.

Apparatus/Reagents Required

pH meter and electrodes, buret and clamp, unknown sample for indicator determination, known pH 0–6 samples, indicators, unknown acid sample (for half-neutralization and buffered solution study), 0.2 *M* NaOH solution, 0.1 *M* HCl, 0.1 *M* NaOH, salt solutions for pH determination [NaCl, KBr, $NaNO_3$, K_2SO_4, $NaC_2H_3O_2$, NH_4Cl, K_2CO_3, $(NH_4)_2SO_4$, $CuSO_4$, and $FeCl_3$], Universal indicator solution

Procedure

Record all data and observations directly in your notebook in ink.

A. pH Using Indicators

Obtain a solution of unknown pH for indicator determination from your instructor. Place about half an inch of this solution in a small test tube, and add 1–2 drops of one of the indicators in the table of indicators that follows. Record the color of the solution after addition of the indicator.

Indicator	Color Change	pH of Color Change
methyl violet	yellow to violet	–1 to 1.7
malachite green	yellow to green	0.2 to 1.8
cresol red	red to yellow	1 to 2
thymol blue	red to yellow	1.2 to 2.8
bromphenol blue	yellow to blue	3 to 4.7
methyl orange	red to yellow	3.2 to 4.4
bromcresol green	yellow to blue	3.8 to 5.4
methyl red	red to yellow	4.4 to 6.0
methyl purple	purple to green	4.5 to 5.1
bromthymol blue	yellow to blue	6.0 to 7.6
litmus	red to blue	4.7 to 6.8

Repeat the test with each of the indicators listed, using a small amount of the unknown solution and 1–2 drops of indicator.

From the information given in the table, estimate the pH of your unknown solution. Note that the color of an indicator is most indicative of the pH in the region in which the indicator changes color.

Having established the pH of your unknown to within approximately one pH unit, obtain a known solution with a pH about equal to that of your unknown. Stock solutions of standard reference buffered solutions with integral pH values from 0 to 6 have been prepared for you.

Test 1-mL samples of the known pH sample with the indicators most useful in determining the approximate pH of the unknown sample. Compare the colors of the known and unknown solutions. On the basis of the comparisons, decide whether the pH of the unknown is slightly higher or lower than the pH of the known sample.

Select a second known solution, with the aim of bracketing the pH of your unknown solution between two known solutions. Treat the second known solution with the useful indicator(s) and compare the color(s) with the unknown sample(s). Continue testing known pH solutions until you find two that differ by one pH unit, one of which has a higher pH than the unknown, and one of which has a lower pH than the unknown.

For a final estimation of the pH of your unknown, use a 5-mL sample of the unknown and of the two known solutions that bracket its pH. Add 2 drops of the indicator to each solution. View the solutions against a well-lighted sheet of white paper, and estimate the pH of your unknown (as compared to the two known solutions) to ± 0.3 pH units. (See Figure 28-1.)

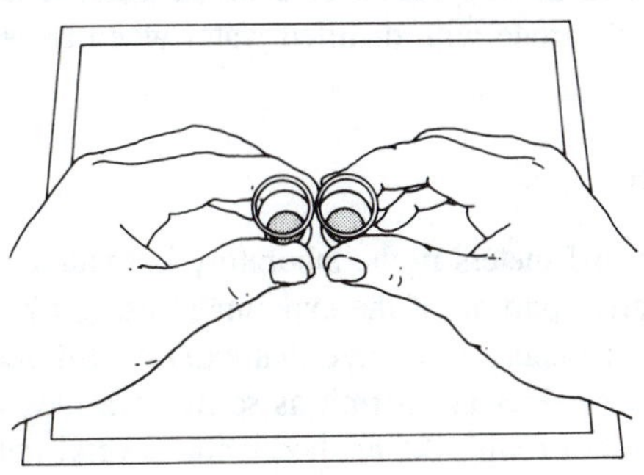

Figure 28-1. Comparison of known and unknown solution colors. Make certain the background is well-lighted.

B. Calibration of the pH Meter

Typically, several models of pH meters are available in the laboratory (see Figure 28-2). Your instructor will provide specific instructions for the operation of the meter you will use.

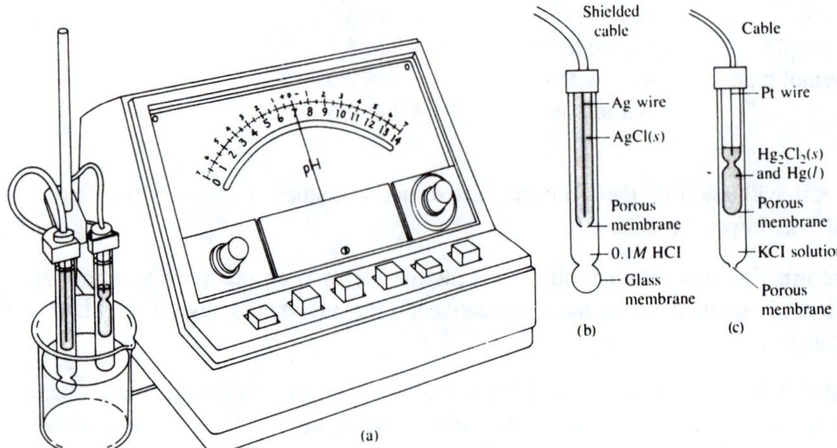

Figure 28-2. (a) A typical pH meter/electrode setup. The two electrodes may be combined in one container to form a combination pH-sensing electrode. (b) pH sensing electrode. The glass membrane measures the difference in $[H^+]$ between the interior of the electrode and the solution being tested. (c) Calomel reference electrode. This completes the electrical circuit in the solution and is very stable and reproducible.

In all cases, the meters must be *calibrated* before being used for pH measurements. Samples of pH standard reference buffered solutions are available for calibrating the meters. Generally, the combination electrode is dipped into one of the reference buffered solutions and allowed to stand for several minutes; this permits the electrode to come to equilibrium with the buffered solution. The set or calibrate knob on the face of the meter is then adjusted until the meter display reads the correct pH for the reference buffered solution. For precise work, a two-point calibration, in which the meter is checked in two different reference buffers, may be necessary. Your instructor will explain the two-point calibration method if you will be using it.

Remember that the pH sensing electrode is made of *glass* and is therefore very *fragile.* (Combination pH electrodes are very expensive, and you may be asked to pay for the electrode if you break it.) Handle the electrode gently, do not stir solutions with the electrode, and keep the electrode in a beaker of distilled water when not in use to keep it from drying out. Rinse the electrode with distilled water when transferring it from one solution to another.

C. pH of Salt Solutions

Note: If the number of pH meters in the laboratory is limited, your instructor may ask you to perform this qualitative portion of the experiment using *Universal indicator*, rather than the pH meter. Universal indicator is a dye that exhibits different colors over the entire pH range. Universal indicator does not permit as sensitive a measurement of pH as would be possible with a pH meter or with the method using several indicators as discussed in Part

A of this experiment. For a quick "ballpark" estimate of pH, however, Universal indicator can be very useful.

In this case, obtain 10 drops of each salt solution, and test for pH with one drop of Universal indicator. Use the color chart provided with the indicator to estimate the pH of each salt solution. Write an equation accounting for the pH observed.

If sufficient pH meters are available in the lab, obtain 20–25 mL (in a small beaker) of one of the 0.1 *M* salt solutions listed in the table that follows. Rinse the pH electrode with distilled water and immerse the electrode in the salt solution.

Salt	Formula
sodium chloride	$NaCl$
potassium bromide	KBr
sodium nitrate	$NaNO_3$
potassium sulfate	K_2SO_4
sodium acetate	$NaC_2H_3O_2$
ammonium chloride	NH_4Cl
potassium carbonate	K_2CO_3
ammonium sulfate	$(NH_4)_2SO_4$
copper(II) sulfate	$CuSO_4$
iron(III) chloride	$FeCl_3$

Allow the electrode to stand in the solution for 2–3 minutes so that it may come to equilibrium with the solution; then record the pH of the salt solution.

Rinse the pH electrode with distilled water, and determine the pH of each of the other salt solutions listed in the table.

Based on the pH measured for each salt solution, write a chemical equation that will explain the observed pH.

D. Determination of K_a for a Weak Acid

Obtain a sample of an unknown solid acid for K_a determination.

If the acid is provided as a solution, use your graduated cylinder to measure out two 50-mL portions into 250-mL Erlenmeyer flasks. If the acid is provided as a solid, place the solid acid sample in 100 mL of distilled water, stir thoroughly to dissolve the acid, and use your graduated cylinder to divide the acid solution into two equal volume portions and place into two 250-mL Erlenmeyer flasks.

Clean, rinse, and fill a buret with 0.2 *M* NaOH solution.

Take one portion of your unknown acid solution, add 3–4 drops of phenolphthalein indicator, and titrate the acid solution until a faint pink color appears and persists. At this point, you have neutralized half the original acid provided and have a solution that contains the salt of the weak acid. It is not necessary in this experiment to determine the exact volume necessary for the titration.

Mix the neutralized solution from the titration with the remaining portion of weak acid solution. Stir well.

Check the calibration of the pH meter with the standard reference buffers provided to make sure that the meter has not drifted in its readings.

Rinse off the electrode of the pH meter with distilled water, and transfer the electrode to the mixture of unknown weak acid and salt. Allow the electrode to stand for 2–3 minutes, and record the pH of the solution.

From the observed pH, calculate $[H^+]$ in the solution and K_a for the weak acid unknown. Save the solution for use later.

E. Properties of a Buffered Solution

The half-neutralized solution prepared in Part D is a buffered system since it contains approximately equal amounts of a free weak acid (HA) and of a salt of the weak acid (Na^+A^-) from the titration. This buffered system should be very resistant to changes in pH, compared to a nonbuffered system, when small quantities of strong acid/strong base are added from an external source.

Check the calibration of the pH meter with the standard reference buffers provided to make sure that the meter has not drifted in its readings.

Place 25 mL of the unknown acid buffered mixture in a small beaker and measure its pH with the meter. Add 5 drops of 0.1 *M* strong acid HCl to the buffered mixture and stir. Allow the electrode to stand for 2–3 minutes and record the pH. The pH should only drop by a small fraction of a pH unit (if at all).

Place another 25-mL sample of the buffered mixture in a small beaker. Add 5 drops of 0.1 *M* strong base NaOH to the buffered mixture and stir. Record the pH after allowing the electrode to stand for 2–3 minutes in the solution. The pH should only have risen by a small fraction of a pH unit (if at all).

For comparison, repeat the addition of small amounts of 0.1 *M* HCl and 0.1 *M* NaOH to 25-mL portions of one of the standard reference buffered solutions available. Determine the pH of the buffered solution itself, as well as the pHs after the addition of strong acid or strong base.

To contrast a buffered system with an unbuffered solution, place 25 mL of distilled water in a small beaker, immerse the electrode, and record the pH. (Do not be surprised if the pH is not exactly 7.0—only pure water that is not in contact with air or glass containers gives pH exactly 7.00.) Add 5 drops of 0.1 *M* HCl to the distilled water, and record the pH. It should decrease by several pH units.

Place another 25 mL of distilled water in a beaker. Add 5 drops of 0.1 *M* NaOH solution and record the pH. It should increase by several pH units.

Write net ionic equations demonstrating why an HA/A^- buffered system resists changes in its pH when strong acid (H^+) or strong base (OH^-) is added.

Name ______________________ Section ______________

Lab Instructor ______________________ Date ______________

EXPERIMENT 28 Acids, Bases, and Buffered Systems

PRE-LABORATORY QUESTIONS

1. Using your textbook or a handbook of chemical data, list five strong acids and five strong bases. Write an equation for each acid and each base showing its ionization in water.

2. Using a handbook of chemical data, find the ionization equilibrium constants (K_a or K_b) for the following weak acids and weak bases. Write an equation for each acid or base showing its ionization in water, and write the expression in terms of chemical symbols for the ionization equilibrium constant for the weak acid or base.

Hydrocyanic acid, HCN

Hydrofluoric acid, HF

Formic acid, HCOOH

Methylamine, CH_3NH_2

Diethylamine, $(CH_3CH_2)_2NH$

3. Using a handbook or chemical encyclopedia, find five acid/base indicators other than the indicators listed in the *Procedure* for this experiment, and give their names and pH color change range.

Name ______________________ Section ______________________

Lab Instructor ______________________ Date ______________________

EXPERIMENT 28 Acids, Bases, and Buffered Systems

RESULTS/OBSERVATIONS

A. pH Using Indicators

Identification number of unknown sample ______________________

Approximate pH of unknown ______________________

Which indicators bracketed the pH of your unknown?

Lower pH ______________________ Higher pH ______________________

Best estimate of unknown's pH ______________________

B. Calibration of pH meter

Describe how you calibrated the pH meter for subsequent determinations of pH. Which standard reference buffers did you use?

____

____

____

C. pH of Salt Solutions

Salt	pH	Equation
NaCl		
KBr		
$NaNO_3$		
K_2SO_4		
$NaC_2H_3O_2$		
NH_4Cl		
K_2CO_3		

$(NH_4)_2SO_4$ ____________ ____

$CuSO_4$ ____________ ____

$FeCl_3$ ____________ ____

D. Determination of K_a for a Weak Acid

Identification number of unknown acid ________________________________

pH of half-neutralized solution of unknown ________________________________

$[H^+]$ = ________________________ K_a = ____________________

E. Properties of a Buffered Solution

pH of buffer (half-neutralized
acid unknown) before addition of other reagents ________________________________

pH of buffer with 5 drops HCl ________________________________

pH of buffer with 5 drops NaOH ________________________________

pH of buffer (standard reference)
before addition of other reagents ________________________________

pH of buffer with 5 drops HCl ________________________________

pH of buffer with 5 drops NaOH ________________________________

pH of distilled water ________________________________

pH of water with 5 drops HCl ________________________________

pH of water with 5 drops NaOH ________________________________

Balanced equations demonstrating buffering action:

____

____

QUESTIONS

1. It was mentioned that you should not expect the pH of distilled water to be 7.00 as ordinarily measured in the laboratory. What component(s) of the atmosphere might cause distilled water in the laboratory to have pH < 7.0? Write an equation to show this.

____

____

2. The human body contains several systems that show buffering capabilities. Why would buffering action be vital in these systems? Use your textbook or a chemical encyclopedia to find some of the buffer systems found in the body. Write equations showing how the components of these buffers maintain pH in the body.

3. Calculate the pH of a buffer solution made by adding 10.0 g of anhydrous sodium acetate ($NaC_2H_3O_2$) to 100 mL of 0.100 *M* acetic acid. Assume there is no change in volume on adding the salt to the acid. pK_a for acetic acid is 4.74.

4. In addition to pH, another important aspect of a buffer system is called the *buffer capacity*. Use your textbook or a chemical dictionary to define buffer capacity.

EXPERIMENT

29 Acid/Base Titrations

Choice I. Analysis of an Unknown Acid Sample

Objective

An unknown acid, either a sample of vinegar or an acid salt, will be analyzed by the process of titration, using a standard sodium hydroxide solution. The sodium hydroxide solution to be used for the analysis will be prepared approximately and will then be standardized against a weighed sample of a known acidic salt.

Introduction

The technique of **titration** finds many applications, but is especially useful in the analysis of acidic and basic substances. Titration involves measuring the exact volume of a solution of *known* concentration that is required to react with a measured volume of a solution of *unknown* concentration, or with a *weighed sample* of unknown solid. A solution of accurately known concentration is called a **standard solution**. Typically, to be considered a standard solution, the concentration of the solute in the solution must be known to four significant figures.

In many cases (especially with solid solutes) it is possible to prepare a standard solution by accurate weighing of the solute, followed by precise dilution to an exactly known volume in a volumetric flask. Such a standard is said to have been prepared *determinately.* One of the most common standard solutions used in acid/base titration analyses, however, cannot be prepared in this manner.

Solutions of sodium hydroxide are commonly used in titration analyses of samples containing an acidic solute. Although sodium hydroxide is a solid, it is *not* possible to prepare standard sodium hydroxide solutions by mass. Solid sodium hydroxide is usually of questionable purity. Sodium hydroxide reacts with carbon dioxide from the atmosphere and is also capable of reacting with the glass of the container in which it is provided. For these reasons, sodium hydroxide solutions are generally prepared to be *approximately* a given concentration. They are then standardized by titration of a weighed sample of a primary standard acidic substance. By measuring how many milliliters of the approximately prepared sodium hydroxide are necessary to react completely with a weighed sample of a known primary standard acidic substance, the concentration of the sodium hydroxide solution can be calculated. Once prepared, however, the concentration of a sodium hydroxide solution will change with time (for the same reasons outlined earlier). As a consequence, sodium hydroxide solutions must be used relatively quickly.

In titration analyses, there must be some means of knowing when enough titrant has been added to react exactly and completely with the sample being titrated. In an acid/base titration analysis, there should be an abrupt change in pH when the reaction is complete. For example, if the sample being titrated is an acid, then the titrant to be used will be basic (probably sodium hydroxide). When one excess drop of titrant is added (beyond that needed to react with the acidic sample), the solution being titrated will suddenly become basic. There are various natural and synthetic dyes, called indicators, that exist in different colored forms at different pH values. A suitable indicator can be chosen that will change color at a pH value consistent with the point at which the titration reaction is complete. The indicator to be used in this experiment is phenolphthalein, which is colorless in acidic solutions, but changes to a pink form at basic pH.

Safety Precautions

- Wear safety glasses at all times while in the laboratory.
- The primary standard acidic substance potassium hydrogen phthalate (KHP) will be kept stored in an oven to keep moisture from adhering to the crystals. Use tongs or a towel to remove the KHP from the oven.
- Sodium hydroxide is extremely caustic, and sodium hydroxide dust is very irritating to the respiratory system. Do not handle the pellets with the fingers. Wash hands after weighing the pellets. Work in a ventilated area and avoid breathing NaOH dust.
- Use a rubber safety bulb when pipeting. *Never pipet by mouth.*
- The unknowns to be used are acidic and may be irritating/damaging to the skin. Avoid contact, and wash after using them.

Apparatus/Reagents Required

Two burets and clamp, buret brush, 5-mL pipet and safety bulb, soap, 1-L glass or plastic bottle with stopper, sodium hydroxide pellets, primary standard grade potassium hydrogen phthalate (KHP), phenolphthalein indicator solution, unknown vinegar or acid salt sample

Procedure

Record all data and observations directly in your notebook in ink.

A. Preparation of the Burets and Pipet

For precise quantitative work, volumetric glassware must be scrupulously clean. Water should run down the inside of burets and pipets in *sheets* and should *not* bead up anywhere on the interior of the glassware. Rinse the burets and the pipet with distilled water to see if they are clean.

If not, partially fill with a few milliliters of soap solution, and rotate the buret/pipet so that all surfaces come in contact with the soap.

Rinse with tap water, followed by several portions of distilled water. If the burets are still not clean, they should be scrubbed with a buret brush. If the pipet cannot be cleaned, it should be exchanged.

In the subsequent procedure, it is important that water from rinsing a pipet/buret does not contaminate the solutions to be used in the glassware. This rinse water would change the concentration of the glassware's contents. Before using a pipet/buret in the following procedures, *rinse* the pipet/buret with several small portions of the solution that is to be *used* in the pipet/buret. Discard the rinsings.

B. Preparation of the Sodium Hydroxide Solution

Clean and rinse the 1-L bottle and stopper. Label the bottle "Approx. 0.1 *M* NaOH." Put about 500 mL of distilled water into the bottle.

Weigh out approximately 4 g (0.1 mol) of sodium hydroxide pellets (*Caution*!) and transfer to the 1-L bottle. Stopper and shake the bottle to dissolve the sodium hydroxide.

When the sodium hydroxide pellets have dissolved, add additional distilled water to the bottle until the water level is approximately 1 inch from the top. Stopper and shake thoroughly to mix.

This sodium hydroxide solution is the titrant for the analyses to follow. Keep the bottle tightly stoppered when not actually in use (to avoid exposure of the NaOH to the air).

Set up one of the burets in the buret clamp. See Figure 29-1. Rinse and fill the buret with the sodium hydroxide solution just prepared.

C. Standardization of the Sodium Hydroxide Solution

Clean and dry a small beaker. Take the beaker to the oven that contains the primary standard grade potassium hydrogen phthalate (KHP).

Using tongs or a towel to protect your hands, remove the bottle of KHP from the oven, and pour a few grams into the beaker. If you pour too much, do *not* return the KHP to the bottle. Return the bottle of KHP to the oven, and take the beaker containing KHP back to your lab bench. Cover the beaker of KHP with a watch glass.

Allow the KHP to cool to room temperature. While the KHP is cooling, clean three 250-mL Erlenmeyer flasks with soap and water. Rinse the Erlenmeyer flasks with 5–10-mL portions of distilled water. Label the Erlenmeyer flasks as 1, 2, and 3.

When the KHP is completely cool, weigh three samples of KHP between 0.6 and 0.8 g, one for each of the Erlenmeyer flasks. Record the exact weight of each KHP sample *at least* to the nearest milligram, preferably to the nearest 0.1 mg (if an analytical balance is available). Be certain not to confuse the samples while determining their masses.

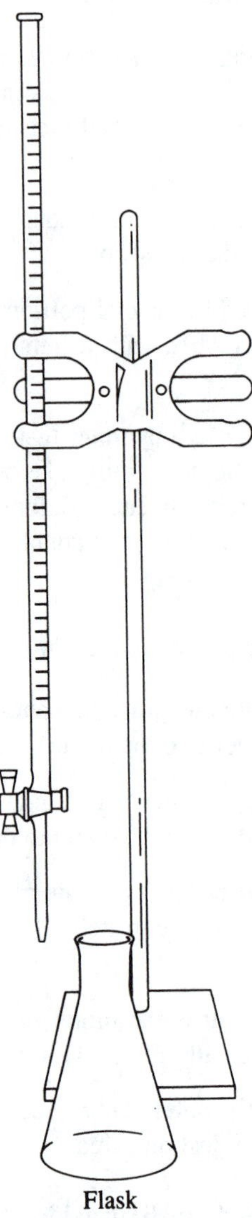

Figure 29-1. Setup for titration. The buret should be below eye-level during filling.

At your lab bench, add 100 mL of water to KHP sample 1. Add 2–3 drops of phenolphthalein indicator solution. Swirl to dissolve the KHP sample completely.

Record the initial reading of the NaOH solution in the buret to the nearest 0.02 mL, remembering to read across the bottom of the curved solution surface (meniscus).

Begin adding NaOH solution from the buret to the sample in the Erlenmeyer flask, swirling the flask constantly during the addition. (See Figure 29-2.) If your solution was prepared correctly, and if your KHP samples are of the correct size, the titration should require at least 20 mL of NaOH solution. As the NaOH solution enters the solution in the Erlenmeyer flask, streaks of red or pink will be visible. They will fade as the flask is swirled.

Eventually the red streaks will persist for a longer and longer period of time. This indicates the approach of the endpoint of the titration.

Begin adding NaOH one drop at a time, with constant swirling, until one single drop of NaOH causes a permanent pale pink color that does not fade on swirling. Record the reading of the buret to the nearest 0.02 mL.

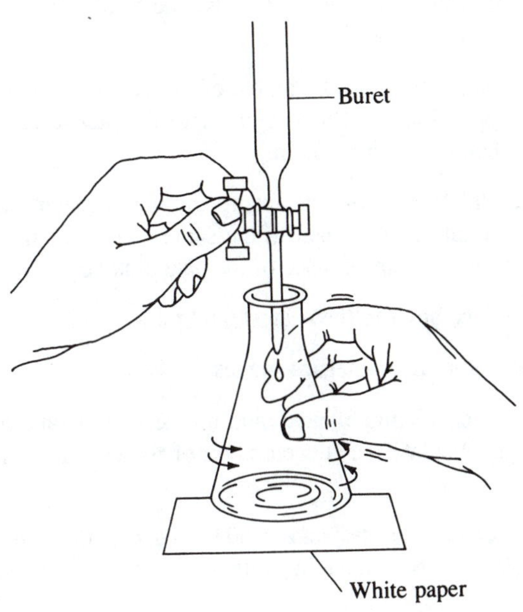

Figure 29-2. Titration technique. A right-handed person should titrate with the left hand, swirling the flask with the right hand. The tip of the buret should be well inside the flask.

Repeat the titration of the remaining KHP samples. Record both initial and final readings of the buret to the nearest 0.02 mL.

Given that the molecular weight of potassium hydrogen phthalate is 204.2, calculate the number of moles of KHP in samples 1, 2, and 3.

From the number of moles of KHP present in each sample, and from the volume of NaOH solution used to titrate the sample, calculate the concentration of NaOH in the titrant solution in moles per liter (molarity of NaOH, *M*). The reaction between NaOH and KHP is of 1:1 stoichiometry.

If your three values for the concentration differ by more than 1%, weigh out an additional sample of KHP and repeat the titration. Use the average concentration of the NaOH solution for subsequent calculations for the unknown.

D. Analysis of the Unknown Acid Sample

Two types of unknown acid samples may be provided. Your instructor may ask you to analyze either or both of these.

1. Analysis of a Vinegar Solution

Vinegar is a dilute solution of acetic acid and can be effectively titrated with NaOH using the phenolphthalein endpoint.

Clean and dry a small beaker, and obtain 25–30 mL of the unknown vinegar solution. Cover the vinegar solution with a watchglass to prevent evaporation. Record the code number of the sample. If the vinegar is a commercial product, record its brand name.

Clean three Erlenmeyer flasks, and label as samples 1, 2, and 3. Rinse the flasks with small portions of distilled water.

Using the rubber safety bulb to provide suction, rinse the 5-mL pipet with small portions of the vinegar solution and discard the rinsings.

Using the rubber safety bulb, pipet a 5-mL sample of the vinegar solution into each of the Erlenmeyer flasks. Add approximately 100 mL of distilled water to each flask, as well as 2–3 drops of phenolphthalein indicator solution.

Refill the buret with the NaOH solution and record the initial reading of the buret to the nearest 0.02 mL. Titrate sample 1 of vinegar in the same manner as in the standardization until one drop of NaOH causes the appearance of the pale pink color.

Record the final reading of the buret to the nearest 0.02 mL.

Repeat the titration for the other two vinegar samples.

Based on the volume of vinegar sample taken, and on the volume and average concentration of NaOH titrant used, calculate the concentration of the vinegar solution in moles per liter.

Given that the formula weight of acetic acid is 60.0, and that the density of the vinegar solution is 1.01 g/mL, calculate the percent by weight of acetic acid in the vinegar solution.

2. Analysis of a Solid Acid

As you saw with the KHP used in the standardization of NaOH, some solid substances are quite acidic. Your instructor will provide you with a solid acidic unknown substance and will tell you approximately what weight of the substance to use in your analysis. Record the code number of the sample.

Clean three Erlenmeyer flasks and label them as samples 1, 2, and 3.

Based on the instructor's directions, weigh out three samples of the solid unknown, one into each Erlenmeyer flask. Make the weight determination at least to the nearest milligram, or preferably, to the nearest 0.1 mg (if an analytical balance is available).

Dissolve the unknown samples in approximately 100 mL of distilled water, and add 2–4 drops of phenolphthalein indicator solution.

Fill the buret with the NaOH titrant and record the initial volume to the nearest 0.02 mL. Titrate sample 1 to the pale pink endpoint as described in the standardization of NaOH above. Record the final volume to the nearest 0.02 mL. Repeat the titration for samples 2 and 3.

Based on the weight of unknown sample taken, and the volume and concentration of the NaOH used to titrate the sample, calculate the molar mass of the solid unknown acid.

Choice II. Analysis of Stomach Antacid Tablets

Objective

Antacid tablets consist of weakly basic substances that are capable of reacting with the hydrochloric acid found in the stomach, converting the stomach acid into neutral or nearly neutral salts. In this experiment, you will determine the amount of stomach acid that two commonly used antacids are capable of neutralizing.

Introduction

In this experiment, you will attempt to evaluate the effectiveness in neutralizing hydrochloric acid of various commercial antacids and will compare these products to the effectiveness of simple baking soda (sodium bicarbonate). The antacids will be dissolved in an excess of 0.1 *M* hydrochloric acid, and then the remaining acid (i.e., the portion of the acid that did not react with the antacid) will be titrated with standard sodium hydroxide solution.

It is not possible to titrate antacid tablets directly for several reasons. First, commercial antacid tablets frequently contain binders, fillers, flavorings, and coloring agents that may interfere with the titration. Second, the bases found in most antacids are weak and become buffered as they are titrated, often leading to an indistinct indicator endpoint.

Safety Precautions

- Wear safety glasses at all times while in the laboratory.
- Only dilute solutions of HCl and NaOH are used in this experiment. But you should wash if these are spilled on the skin, since they may cause minor irritations in sensitive individuals.

Apparatus/Reagents Required

Buret and clamp, 100-mL graduated cylinder, plastic wrap, two brands of commercial antacid tablets, sodium bicarbonate, bromphenol blue indicator solution, standard 0.1 *M* hydrochloric acid solution (record the exact concentration), standard 0.1 *M* sodium hydroxide solution (record the exact concentration)

Procedure

Record all data and observations directly in your notebook in ink.

Clean and rinse a buret with water. Then rinse and fill the buret with the available standard 0.1 *M* sodium hydroxide solution (record the exact concentration). If you performed Choice 1 of this experiment, you may have prepared and standardized your *own* NaOH solution.

Obtain an antacid tablet (record the brand) and wrap it in a piece of plastic wrap. Crush the tablet (use the bottom of a beaker or a heavy object). Weigh a clean, dry Erlenmeyer flask to the nearest 0.01 g. Transfer as much of the crushed antacid tablet as possible to the flask, and reweigh to the nearest 0.01 g.

Obtain about 300 mL of the standard 0.1 *M* hydrochloric acid (record the exact concentration). With a graduated cylinder, measure exactly 100 mL of the hydrochloric acid and add it to the Erlenmeyer flask containing the crushed antacid tablet. Swirl the flask to dissolve the tablet as much as possible. *Note*: As mentioned earlier, commercial antacid tablets contain various other ingredients, which may not completely dissolve.

Add 2–5 drops of bromphenol blue indicator solution to the sample. The indicator should be bright yellow at this point, indicating that the solution is acidic (excess HCl).

If the indicator is blue, this means that not enough hydrochloric acid was added to consume completely the antacid tablet. If the solution is blue, add 0.1 *M* HCl, in 10-mL increments (record the amount used) until the sample is yellow.

Record the initial reading of the buret. Titrate the antacid sample with standard NaOH solution until the solution just barely turns blue. Record the final reading of the buret at the color change.

Repeat the procedure two more times using the *same brand* of antacid tablet.

Repeat the procedure three times with either another brand of commercial antacid tablet, or with samples of baking soda (sodium bicarbonate) on the order of 0.7 g (record the exact weight used). If sodium bicarbonate is used, add the standard 0.1 *M* HCl to it very slowly to prevent excessive frothing as carbon dioxide is liberated.

Calculations

Calculate the number of millimoles of HCl consumed by the antacid tablet in each titration.

> The volume of hydrochloric acid used to dissolve the sample, multiplied by the concentration of the HCl, represents the total number of millimoles of HCl taken.
>
> The volume of sodium hydroxide used in the titration, multiplied by the concentration of the NaOH, represents the number of millimoles of HCl *not* consumed by the tablet.
>
> The difference between these two quantities represents the number of millimoles of HCl that was neutralized by the tablet.

Calculate the volume of the dilute HCl solution that corresponds to the number of millimoles of HCl consumed by the tablet.

> The number of millimoles of HCl consumed by the tablet, divided by the concentration of the HCl solution used, represents the volume of the HCl solution consumed by the tablet.

Assuming that the density of the HCl solution used was 1.00 g/mL, calculate the weight of the hydrochloric acid solution consumed by the tablet.

Calculate the mass of the HCl solution consumed per gram of antacid tablet. Divide the mass of HCl consumed by the mass of the tablet used.

Name ______________________________ Section ______________________

Lab Instructor ______________________________ Date ______________________

EXPERIMENT 29 Acid/Base Titrations

PRE-LABORATORY QUESTIONS

CHOICE I. ANALYSIS OF AN UNKNOWN ACID SAMPLE

1. Using your textbook or a handbook, look up the formula and structure of potassium hydrogen phthalate (KHP) used to standardize the solution of NaOH in this experiment. Calculate the molar mass of KHP.

____

____

____

____

2. Using your textbook or a chemical dictionary, write the definition of *indicator*.

____

____

____

____

3. Suppose a sodium hydroxide solution were to be standardized against pure solid primary standard grade KHP. If 0.4538 g of KHP requires 44.12 mL of the sodium hydroxide to reach a phenolphthalein endpoint, what is the molarity of the NaOH solution?

____

____

____

____

____

____

____

CHOICE II. ANALYSIS OF STOMACH ANTACID TABLETS

1. What is meant by a *standard* solution of acid or base?

____

____

____

____

____

____

2. Some of the common bases used as the active ingredient in commercial antacid tablets are listed. Calculate the number of milliliters of 0.100 M HCl solution that could be neutralized by 1.00 g of each of the substances. Show your calculations.

$CaCO_3$ ____________________ $NaHCO_3$ ____________________

$Mg(OH)_2$ ____________________ $Al(OH)_3$ ____________________

Name ______________________ Section ______________

Lab Instructor ______________________ Date ______________

EXPERIMENT 29 Acid/Base Titrations

CHOICE I. ANALYSIS OF AN UNKNOWN ACID SAMPLE

RESULTS/OBSERVATIONS

Standardization of NaOH Titrant

	Sample 1	Sample 2	Sample 3
Weight of KHP taken	______	______	______
Initial NaOH buret reading	______	______	______
Final NaOH buret reading	______	______	______
Volume NaOH used	______	______	______
Moles KHP present	______	______	______
Molarity of NaOH solution	______	______	______

Mean molarity of NaOH solution and average deviation ______________________

1. Analysis of Vinegar Solution

Identification number of vinegar sample used ______________________

	Sample 1	Sample 2	Sample 3
Quantity of vinegar taken	______	______	______
Initial NaOH buret reading	______	______	______
Final NaOH buret reading	______	______	______
Volume NaOH used	______	______	______
Molarity of vinegar	______	______	______

Mean molarity of vinegar and average deviation ______________________

% by mass acetic acid present ______________________

2. Analysis of a Solid Acid

Identification number of solid acid used ______________________________

	Sample 1	Sample 2	Sample 3
Weight unknown taken	________	________	________
Initial NaOH buret reading	________	________	________
Final NaOH buret reading	________	________	________
Volume of NaOH used	________	________	________
Moles of NaOH used	________	________	________
Molar mass unknown solid	________	________	________

Mean molar mass and average deviation ______________________________

QUESTIONS

1. Commercial vinegar is generally 5.0 ± 0.5% acetic acid by weight. Assuming this to be the true value for your unknown, by how much were you in error in your analysis?

____

____

____

2. The solid acids chosen for the analysis were typically monoprotic acidic salts such as $NaHSO_4$, $KHSO_4$, etc. Explain why such salts behave as strong enough acids to be titratable with NaOH using phenolphthalein as indicator.

____

____

____

____

____

3. Use a chemical dictionary or encyclopedia to explain the difference between an indicator *endpoint* for a titration analysis and the true *equivalence point* for the titration.

____

____

____

Name ______________________________ Section ______________________

Lab Instructor ______________________________ Date ______________________

EXPERIMENT 29 Acid/Base Titrations

CHOICE II. ANALYSIS OF STOMACH ANTACID TABLETS

RESULTS/OBSERVATIONS

Concentrations of standard solutions:

HCl ______________________________ NaOH ______________________________

First antacid: Brand ______________________________

	Sample 1	Sample 2	Sample 3
Mass of tablet used			
Volume of 0.1 *M* HCl used			
Initial reading of buret			
Final reading of buret			
mL of 0.1 *M* NaOH used			
Mass of HCl consumed per gram			

Mean mass of HCl consumed per gram of tablet ______________________________

Second antacid: Brand ______________________________

	Sample 1	Sample 2	Sample 3
Mass of tablet used			
Volume of 0.1 *M* HCl used			
Initial reading of buret			
Final reading of buret			
mL of 0.1 *M* NaOH used			

Mass of HCl consumed per gram ________________ ________________ ________________

Mean mass of HCl consumed per gram of tablet ________________________________

QUESTIONS

1. Which of the two antacids tested consumed more HCl per gram? Which consumed more HCl per tablet? If the prices of the antacids are available, which antacid is a better buy?

____

____

____

____

____

____

2. Generally, the substances used as antacids are either weak bases, or very insoluble bases. Why is a strong soluble base like NaOH not used in antacid tablets?

____

____

____

____

3. Read the label(s) on the commercial antacid tablets you used. List the brand name(s) and the active antacid ingredients.

____

____

____

4. Some people abuse the use of antacids. Use a chemical encylopedia to find out what side effects may occur if antacids are used too frequently.

____

____

____

____

____

EXPERIMENT

30 Determination of Calcium in Calcium Supplements

Objective

The amount of calcium ion in a dietary calcium supplement will be determined by titration with ethylenediaminetetraacetic acid (EDTA).

Introduction

The concentration of calcium ion in the blood is very important for the development and preservation of strong bones and teeth. Absorption of too much calcium may result in the build-up of calcium deposits in the joints. More commonly, a low concentration of calcium in the blood may lead to the leaching of calcium from bones and teeth. The absorption of calcium is controlled, in part, by Vitamin D.

Growing children in particular need an ample supply of calcium for the development of strong bones. Milk, which is a good natural source of calcium and which is consumed primarily by children, is usually fortified with Vitamin D to promote calcium absorption. Similarly, outdoor recess for elementary school children is mandated by law in many states because exposure to sunshine allows the body to synthesize Vitamin D.

In later adulthood, particularly among women, the level of calcium in the blood may decrease to the point where calcium is removed from bones and teeth to replace the blood calcium, making the bones and teeth much weaker and more susceptible to fracture. This condition is called osteoporosis and can become very serious, causing shrinking of the skeleton and severe arthritis. For this reason, calcium supplements are often prescribed in an effort to maintain the proper concentration of calcium ion in the blood. Although the results are not yet conclusive, recent reports indicate 1000 mg of calcium ion, taken orally on a daily basis, may serve to prevent osteoporosis. Typically, such calcium supp :ments consist of calcium carbonate, $CaCO_3$.

A standard analysis for calcium ion involves *titration* with a standard solution of ethylenediaminetetraacetic acid (EDTA). In a titration experiment, a standard reagent of known concentration is added slowly to a measured volume of a sample of unknown concentration until the reaction is complete. From the concentration of the standard reagent, and from the volumes of standard and unknown taken, the concentration of the unknown sample may be calculated. In titration experiments, typically the sample of unknown concentration is measured with a pipet, and the volume of standard solution required is measured with a buret.

EDTA is a type of molecule called a complexing agent, and is able to form stable, stoichiometric (usually 1:1) compounds with many metal ions. Reactions of EDTA with metal ions are especially sensitive to pH, and typically a concentrated buffer solution is added to the sample being titrated to maintain a relatively constant pH. A suitable indicator, which will change color when the reaction is complete, is also necessary for EDTA titrations.

Safety Precautions

- Wear safety glasses at all times while in the laboratory.
- Hydrochloric acid solution is corrosive to eyes, skin, and clothing. Wash immediately if spilled and clean up all spills on the benchtop.
- Ammonia is a respiratory irritant and cardiac stimulant. The ammonia buffer solution is very concentrated and *must be kept in the exhaust hood at all times*. Pour out the ammonia buffer and transfer it to your calcium sample *while still under the exhaust hood.*
- Use a rubber safety bulb when pipeting samples. Do *not* pipet by mouth.
- Eriochrome Black indicator is toxic if ingested and will stain skin and clothing.

Apparatus/Reagents Required

Buret and clamp, 25-mL pipet, 250-mL volumetric flask, 3 *M* HCl, standard 0.0500 *M* EDTA solution, Eriochrome Black T indicator, calcium supplement tablet.

Procedure

Record all data and observations directly in your notebook in ink.

A. Dissolving of the Calcium Supplement

Obtain a calcium supplement tablet. Record the calciumcontent of the tablet as listed on the label by the manufacturer. Place the tablet in a clean 250-mL beaker.

Obtain 25 mL of 3 *M* HCl solution in a graduated cylinder. Over a 5-minute period, add the HCl to the beaker containing the calcium tablet in 5-mL portions, waiting between additions until all frothing of the tablet has subsided.

After the last portion of HCl has been added, allow the calcium mixture to stand for an additional five minutes to complete the dissolving of the tablet. Since such tablets usually contain various binders, flavoring agents, and other inert material, the solution may *not* be entirely homogeneous.

While the tablet is dissolving, clean a 250-mL volumetric flask with soap and tap water. Then rinse the flask with two 10-mL portions of distilled water. Fill your plastic wash bottle with distilled water.

Taking care not to lose any of the mixture, transfer the tablet solution to the volumetric flask using a small funnel. To make sure that the transfer of the calcium solution has been complete, use distilled water from the plastic wash bottle to rinse the beaker that contained the calcium sample into the volumetric flask. Repeat the rinsing of the beaker twice more.

Use a stream of distilled water from the wash bottle to thoroughly rinse any adhering calcium solution from the funnel into the volumetric flask, and then remove the funnel.

Add distilled water to the volumetric flask until the water level is approximately 1-inch below the calibration mark on the neck of the volumetric flask. Then use a medicine dropper to add distilled water until the bottom of the solution meniscus is aligned exactly with the volumetric flask's calibration mark.

Stopper the volumetric flask securely (hold your thumb over the stopper) and then invert and shake the flask 10–12 times to mix the contents.

B. Preparation for Titration

Set up a 50-mL buret and buret clamp. Check the buret for cleanliness. If the buret is clean, water should run down the inside walls in sheets, and should not bead up anywhere. If the buret is not clean enough for use, place approximately 10 mL of soap solution in the buret and scrub with a buret brush for several minutes.

Rinse the buret several times with tap water, and then check again for cleanliness by allowing water to run from the buret. If the buret is still not clean, repeat the scrubbing with soap solution. Once the buret is clean, rinse it with small portions of distilled water, allowing the water to drain through the stopcock.

Obtain a 25-mL volumetric pipet and rubber bulb. Using the bulb to provide suction, fill the pipet with tap water and allow the water to drain out. During the draining, check the pipet for cleanliness. If the pipet is clean, water will not bead up anywhere on the interior during draining.

If the pipet is not clean, use the rubber bulb to pipet 10–15 mL of soap solution. Holding your fingers over both ends of the pipet, rotate and tilt the pipet to rinse the interior with soap for 2–3 minutes. Allow the soap to drain from the pipet, rinse several times with tap water, and check again for cleanliness. Repeat the cleaning with soap if needed. Finally, rinse the pipet with several 5–10 mL portions of distilled water.

Obtain 200 mL of standard 0.0500 *M* EDTA solution in a 400-mL beaker. Keep the EDTA solution covered with a watchglass when not in use.

Transfer 5–10 mL of the EDTA solution to the buret. Rotate and tilt the buret to rinse and coat the walls of the buret with the EDTA solution. Allow the EDTA solution to drain through the stopcock to rinse the tip of the buret.

Rinse the buret twice more with 5–10-mL portions of the standard EDTA solution.

After the buret has been thoroughly rinsed with EDTA, fill the buret to slightly above the zero mark with the standard EDTA.

Allow the buret to drain until the solution level is slightly below the zero mark. Read the volume of the buret (to two decimal places), estimating between the smallest scale divisions. Record this volume as the initial volume for the first titration to be performed.

Rinse four clean 250-mL Erlenmeyer flasks with distilled water. Label the flasks as samples 1, 2, 3, and 4.

Transfer approximately 50 mL of your calcium solution from the volumetric flask to a clean, *dry* 150-mL beaker. Using the rubber bulb for suction, rinse the pipet with several 5–10-mL portions of the calcium solution to remove any distilled water still in the pipet.

Pipet exactly 25 mL of the calcium solution into each of the four Erlenmeyer flasks.

C. Titration of the Calcium Samples

Each sample in this titration must be treated one at atime. Do not add the necessary reagents to a particular sample until you are ready to titrate. Because the buffer solution used to control the pH of the calcium samples during the titration contains concentrated ammonia, it will be kept stored in the exhaust hood. Take calcium sample 1 to the exhaust hood and add 10 mL of the ammonia buffer (*Caution*!).

At your bench, add 5 drops of Eriochrome T indicator solution to calcium sample 1. The sample should be wine-red at this point; if the sample is blue, consult with the instructor.

Add EDTA from the buret to the sample a few milliliters at a time, with swirling after each addition, and carefully watch the color of the sample. The color change of Eriochrome T is from red to blue, but the sample will pass through a gray transitional color as the endpoint is neared.

As the sample becomes gray in color, begin adding EDTA *dropwise* until the pure blue color of the endpoint is reached. Record the final volume used to titrate sample 1 to two decimal places, estimating between the smallest scale divisions on the buret.

Refill the buret, and repeat the titration procedure for the three remaining calcium samples. Remember not to add ammonia buffer or indicator until you are actually ready to titrate a particular sample.

D. Calculations

For each of the four titrations, use the volume of EDTA required to reach the endpoint and the concentration of the standard EDTA solution to calculate how many moles of EDTA were used. Record.

$$\text{moles EDTA} = (\text{volume used to titrate in liters}) \times (\text{molarity})$$

Based on the stoichiometry of the EDTA/calcium reaction, calculate how many moles of calcium ion were present in each sample titrated. Record.

Using the atomic mass of calcium, calculate the mass of calcium ion present in each of the four samples titrated, and the average mass of calcium present.

Based on the fact that the average mass of calcium calculated above *represents a 25-mL sample*, taken from a *total volume of 250 mL used to dissolve the tablet, calculate the mass of calcium present in the original tablet.*

Compare the mass of calcium present in the tablet calculated from the titration results with the nominal mass reported on the label by the manufacturer. Calculate the percent difference between these values.

Name ______________________ Section ______________________

Lab Instructor ______________________ Date ______________________

EXPERIMENT 30 Determination of Calcium in Calcium Supplements

PRE-LABORATORY QUESTIONS

1. Using your textbook or an encyclopedia of chemistry, discuss some uses of calcium in the human body.

2. The titrant reagent used in this experiment, EDTA, is a **complexing agent** for calcium and for many other metal ions. Use your textbook or an encyclopedia of chemistry to write a definition of this term.

3. Use an encyclopedia of chemistry to find and sketch the structural formula for EDTA.

Name ______________________________ Section ______________________

Lab Instructor ______________________________ Date ______________________

EXPERIMENT 30 Determination of Calcium in Calcium Supplements

RESULTS/OBSERVATIONS

Dissolving of the Calcium Supplement

Observation

____

____

Did the tablet dissolve *completely*? ______________________________

Was the solution of the tablet *homogeneous*? ______________________________

Titrations

Concentration of *standard* EDTA solution used, *M* ______________________________

Volume of calcium solution taken for titration, mL ______________________________

Sample	1	2	3	4
Final volume EDTA, mL	______	______	______	______
Initial volume EDTA, mL	______	______	______	______
Volume EDTA used, mL	______	______	______	______
Moles EDTA used	______	______	______	______
Moles Ca^{2+} ion present	______	______	______	______
Mass of Ca^{2+} present, g	______	______	______	______

Average mass of Ca^{2+} present, g ______________________

Average mass of Ca^{2+} present in original tablet, g ______________________

Listed mass of Ca^{2+} present in tablet (from label) ______________________

Percent difference between experimental mass and listed mass ______________________

QUESTIONS

1. Most calcium supplements consist of *calcium carbonate*, $CaCO_3$, since this substance is readily available and relatively cheap. Based on your experimentally determined average mass of calcium ion, calculate the mass of calcium carbonate that would be equivalent to this amount of calcium ion.

2. Suggest at least two reasons why your experimentally determined amount of calcium might differ from that listed by the manufacturer of the calcium supplement you used.

3. Use your textbook or an encyclopedia of chemistry to write at least three *natural* sources of calcium ion that should be included in the diet.

EXPERIMENT

31 Determination of Iron by Redox Titration

Objective

In this experiment, one of the main types of chemical reactions, oxidation/reduction (or redox) reactions, will be used in the titration analysis of an iron compound.

Introduction

Oxidation/reduction processes form one of the major classes of chemical reactions. In **redox reactions**, electrons are transferred from one species to another. For example, in the following simple reaction

$$Zn(s) + Cu^{2+}(aq) \rightarrow Zn^{2+}(aq) + Cu(s)$$

electrons are transferred from elemental metallic zinc to aqueous copper(II) ions. This is most easily seen if the overall reaction is written instead as two *half-reactions*, one for the oxidation and one for the reduction:

$$Zn(s) \rightarrow Zn^{2+}(aq) + 2e^- \qquad \textit{oxidation}$$

$$Cu^{2+}(aq) + 2e^- \rightarrow Cu(s) \qquad \textit{reduction}$$

In this experiment you will use potassium permanganate, $KMnO_4$, as the titrant in the analysis of an unknown sample containing **iron.** Although this method can easily be applied to the analysis of realistic iron *ores*—the native state in which an element is found in nature—for simplicity your unknown sample will contain only iron in the +2 oxidation state. In acidic solution, potassium permanganate rapidly and quantitatively *oxidizes* iron(II) to iron(III), while itself being *reduced* to manganese(II). The half reactions for the process are:

$$MnO_4^- + 8H^+ + 5e^- \rightarrow Mn^{2+} + 4H_2O \qquad \textit{reduction}$$

$$Fe^{2+} \rightarrow Fe^{3+} + e^- \qquad \textit{oxidation}$$

When these half-reactions are combined to give the overall balanced chemical reaction equation, a factor of *five* has to be used with the iron half-reaction so that the number of electrons lost in the overall oxidation will equal the number of electrons gained in the reduction:

$$MnO_4^- + 8H^+ + 5Fe^{2+} \rightarrow Mn^{2+} + 4H_2O + 5Fe^{3+}$$

Potassium permanganate has been one of the most commonly used oxidizing agents because it is extremely powerful, as well as inexpensive and readily available. It does have some drawbacks, however. Because $KMnO_4$ is such a strong oxidizing agent, it reacts with practically *anything* that can be oxidized: this tends to make solutions of $KMnO_4$ difficult to store without decomposition or a change in concentration. Because of this limitation, it is common to prepare, standardize, and then use $KMnO_4$ solutions for an analysis all on the same day. It is *not* possible to prepare directly $KMnO_4$ standard solutions determinately by mass: solid potassium permanganate cannot be obtained in a completely pure state due to the high reactivity mentioned above. Rather, potassium permanganate solutions are prepared to be an approximate concentration, and are then standardized against a known primary standard sample of the same substance which is to be analyzed in the unknown sample.

Potassium permanganate is especially useful among titrants since it requires no indicator to signal the endpoint of a titration. Potassium permanganate solutions—even at fairly dilute concentrations—are intensely colored purple. The product of the permanganate reduction half-reaction, manganese(II), in dilute solution shows almost *no* color. Therefore, during a titration using $KMnO_4$, when one drop *excess* of potassium permanganate has been added to the sample, the sample will take on a pale red/pink color (since there are no more sample molecules left to convert the purple MnO_4^- ions to the colorless Mn^{2+} ions).

Safety Precautions

- Wear safety glasses at all times while in the laboratory.
- Potassium permanganate is a strong oxidizing agent and can be damaging to skin, eyes, and clothing. Wash after handling.
- Potassium permanganate solutions will stain skin and clothing if spilled.
- Sulfuric acid solutions are damaging to the skin, eyes, and clothing—especially if allowed to concentrate through evaporation of water. If the sulfuric acid solution is spilled on the skin, wash immediately and inform the instructor.
- Iron salts may be irritating to the skin. Wash after handling.

Apparatus/Reagents Required

Potassium permanganate, ferrous ammonium sulfate, 3 *M* sulfuric acid, buret and clamp, 250-mL Erlenmeyer flasks, 1-L glass bottle with cap, iron unknown sample

Procedure

A. Preparation of Potassium Permanganate Solution

Clean out a 1-L glass bottle and its cap. Wash first with soap solution, then rinse with tap water to remove all the soap. Finally rinse with several small portions of distilled water. Then place approximately 600 mL of distilled water in the bottle.

In a small beaker, weigh out 2.0 ± 0.1 g of potassium permanganate crystals (*Caution!*). Using a funnel, transfer the permanganate crystals to the 1-L bottle. Cap the bottle *securely* and shake the mixture to dissolve the permanganate. Potassium permanganate is often very *slow* in dissolving: continue shaking the solution until you are absolutely certain that all the permanganate crystals have dissolved completely.

B. Preparation of the Iron Standards

Clean out three 250-mL Erlenmeyer flasks with soap and water, giving a final rinse with several small portions of distilled water. Label the flasks as Samples 1, 2, and 3.

The primary standard iron(II) compound to be used for the standardization of the potassium permanganate solution is the salt ferrous ammonium sulfate hexahydrate (Mohr's salt), $FeSO_4 \cdot (NH_4)_2SO_4 \cdot 6H_2O$, which is almost universally abbreviated as FAS.

Place approximately 3 g of primary standard FAS into a small, clean dry beaker. Determine the mass of the beaker and FAS to at least the nearest milligram (0.001 g), or better, if an analytical balance is available, make the mass determination to the nearest 0.1 milligram (0.0001 g).

Using a funnel, carefully transfer approximately *one-third* of the FAS in the beaker (approximately 1 g) to the Erlenmeyer flask labeled Sample 1. Distilled water from a wash bottle can be used to rinse the salt into the flask if the funnel should clog.

Weigh the beaker containing the remainder of the FAS. The difference in mass from the previous weighing represents the mass of FAS transferred to the Sample 1 Erlenmeyer flask.

Using a funnel, carefully transfer approximately *half* the FAS remaining in the beaker (one-third of the original FAS sample) to the Erlenmeyer flask labeled Sample 2. Again, rinse the FAS from the funnel into the flask with distilled water from a wash bottle.

Weigh the beaker containing the remaining FAS. The difference in mass from the previous weighing represents the mass of Sample 2 that was transferred to the Erlenmeyer flask.

Using a funnel as above, transfer as much as possible of the FAS remaining in the beaker to the Sample 3 Erlenmeyer flask. Finally weigh the empty beaker and any residual FAS that may not have been transferred completely. The difference in mass from the previous weighing represents the mass of Sample 3.

Add 25 mL of distilled water to each sample and swirl the flasks to dissolve the FAS.

Add 15 mL of 3 *M* sulfuric acid, H_2SO_4, to each sample (*Caution!*). Sulfuric acid is added to the samples to provide the hydrogen ions, H^+, required for the reduction of the permanganate ion.

C. Standardization of the Potassium Permanganate Solution

Clean out a buret with soap and water, and rinse with several portions of tap water, followed by several small rinsings with distilled water.

Rinse the buret with several 4–6-mL portions of the potassium permanganate solution. Tilt and rotate the buret so that the inside walls are completely rinsed with the permanganate, and allow the solution to run out of the tip of the buret to rinse the tip also. Discard the rinsings.

Finally, using a funnel *and with the buret below eye-level,* fill the buret to slightly *above* the zero mark with the permanganate solution. Allow a few mL of permanganate to run out of the tip of the buret to remove any air bubbles from the tip.

Since potassium permanganate solutions are so intensely colored, it is generally impossible to see the curved meniscus that the solution surface forms in the buret. In this case, it is acceptable to make your liquid level readings at the point where the *top* surface of the permanganate solution comes in contact with the wall of the buret. Remove the funnel from the buret and take the initial volume reading, to the nearest 0.02 mL. Record.

Place the Erlenmeyer flask containing Sample 1 under the tip of the buret, and begin adding potassium permanganate to the sample a few mL at a time, swirling the flask after each addition of permanganate. As the permanganate solution is added to the sample, red streaks may be visible in the sample until the permanganate has a chance to mix with and react with the iron(II) present.

Continue adding permanganate a few mL at a time, with swirling, until the red streaks begin to become more persistent. At this point, begin adding the permanganate *one drop* at a time, swirling the flask constantly to mix. The endpoint is the first appearance of a *permanent*, pale pink color.

After the endpoint has been reached, record the final buret reading (to the nearest 0.02 mL). Calculate the volume of potassium permanganate solution used to titrate FAS Sample 1.

Titrate FAS Samples 2 and 3 in a similar manner, recording initial and final volumes of potassium permanganate solution used to the nearest 0.02 mL.

From the mass of each FAS sample, and from the volume of $KMnO_4$ used to titrate the respective samples, calculate three values for the molarity of your potassium permanganate solution, as well as the average molarity. If any one of your invidual molarities seems out-of-line from the other two, you should consult with the instructor about performing a fourth standardization titration (using an additional 1-g sample of FAS).

D. Analysis of the Iron Unknown

Your instructor will provide you with an iron unknown, and will inform you of the approximate sample size to weigh out. Prepare three samples of the iron unknown as you performed in Part B for the primary standard FAS, being sure to add the sulfuric acid required for the reaction, and being sure to keep the masses of the samples to the amount indicated by the instructor.

Titrate the iron unknown samples as described in Part C above, recording initial and final volumes to the nearest 0.02 mL. Calculate the volume of permanganate solution required to reach the pale pink endpoint for each sample.

From the volume required to titrate each unknown sample, and the average molarity of the $KMnO_4$ solution, calculate the number of moles of potassium permanganate required for each iron sample.

From the number of moles of potassium permanganate required for titration of each iron sample, calculate the number of moles of iron present in each unknown sample.

From the number of moles of iron present in each sample, and the molar mass of iron, calculate the mass of iron present in each unknown sample.

From the mass of iron present in each sample, and from each sample's mass, calculate the percent iron in each sample.

Calculate the average percent iron by mass present in your unknown sample.

Name ______________________________ Section ______________________________

Lab Instructor ______________________________ Date ______________________________

EXPERIMENT 31 Determination of Iron by Redox Titration

PRE-LABORATORY QUESTIONS

1. Another common analysis for iron is titration of iron(II) with potassium dichromate, $K_2Cr_2O_7$. Divide the following reaction into half-reactions, balance the half-reactions, and combine the balanced half-reactions into the balanced overall equation for the reaction.

$$Fe^{2+} + Cr_2O_7^{2-} + H^+ \rightarrow Fe^{3+} + Cr^{3+} + H_2O$$

2. Use a chemical dictionary, handbook, or encyclopedia to write several *other* uses of potassium permanganate (other than in the analysis of iron samples as in this experiment).

3. If 26.23 mL of potassium permanganate solution is required to titrate 1.041 g of ferrous ammonium sulfate hexahydrate, $FeSO_4(NH_4)_2SO_4 \cdot 6H_2O$, calculate the molarity of the $KMnO_4$ solution.

4. If a 2.893 g sample of an unknown containing iron requires 28.45 mL of the permanganate solution described in Pre-Laboratory Question 3 to reach the endpoint, calculate the % Fe in the unknown.

Name ______________________ Section ______________________

Lab Instructor ______________________ Date ______________________

EXPERIMENT 31 Determination of Iron by Redox Titration

RESULTS/OBSERVATIONS

Standardization of $KMnO_4$ Solution

	Sample 1	Sample 2	Sample 3
Mass of FAS taken			
Initial $KMnO_4$ volume			
Final $KMnO_4$ volume			
Volume $KMnO_4$ used			
Moles iron present			
Moles $KMnO_4$ present			
Molarity of $KMnO_4$ solution			

Mean molarity and average deviation ______________________

Analysis of Unknown

	Sample 1	Sample 2	Sample 3
Mass of unknown taken			
Initial $KMnO_4$ volume			
Final $KMnO_4$ volume			
Volume $KMnO_4$ used			
Moles $KMnO_4$ present			
Moles iron present			
Mass of iron present			
% of iron present			

Mean % iron present and average deviation ______________________________

QUESTIONS

1. Typically, a solid iron unknown for titration is dried in an oven to remove adsorbed water before analysis (the unknowns used in this experiment were dried before dispensing). How would the % Fe determined be affected if the unknowns had not been dried?

2. For what purpose was sulfuric acid added to the iron samples before titrating?

3. Potassium permanganate solutions are used directly for the analysis of samples containing iron in the +2 oxidation state. Consult a handbook or textbook of chemical analysis to write how samples containing elemental iron and iron in the +3 oxidation state might be analyzed for their iron content.

EXPERIMENT

32 Electrochemistry I: Chemical Cells

Objective

Oxidation/reduction reactions find their most important use in the construction of voltaic cells (chemical batteries). In this experiment, several such cells will be constructed and their properties studied.

Introduction

Electrochemistry is the detailed study of *electron-transfer* (or oxidation/reduction) reactions. Electrochemistry is a very wide field of endeavor, covering such subjects as batteries, corrosion and reactivities of metals, and electroplating. This experiment will briefly examine some of these topics.

Consider the following reaction equation:

$$Zn(s) + Cu^{2+}(aq) \rightarrow Zn^{2+}(aq) + Cu(s)$$

In this process, metallic elemental zinc has been added to a solution of dissolved copper(II) ion. Reaction occurs, and the metallic zinc dissolves, producing a solution of zinc ion. Concurrently, metallic elemental copper forms from the copper(II) ion that had been present in solution. To see what really is happening in this reaction equation, it is helpful to split the given equation into two *half-reactions:*

$$Zn(s) \rightarrow Zn^{2+}(aq) + 2e^- \qquad \textit{oxidation}$$

$$Cu^{2+}(aq) + 2e^- \rightarrow Cu(s) \qquad \textit{reduction}$$

The zinc half-reaction is called an *oxidation* half-reaction. **Oxidation** is a process in which a species *loses electrons* to some other species. In the zinc half-reaction, metallic zinc loses electrons in becoming zinc(II) ions. The copper half-reaction is called a *reduction* half-reaction. **Reduction** is a process in which a species *gains electrons* from some other species. In the copper half-reaction, copper(II) ions gain electrons (i.e., the electrons that had been lost by the zinc atoms) and become metallic copper.

In the zinc/copper reaction, metallic zinc has *replaced* copper(II) ion from a solution. This has happened because metallic zinc is *more reactive* than metallic copper, and zinc is more likely to be found *combined* in a compound rather than as the free elemental metal.

The common metallic elements have been investigated for their relative reactivities and have been arranged into what is called the **electromotive series**. A portion of this series follows:

K, Na, Ba, Ca, Mg, Al, Mn, Zn, Cr, Cd, Fe, Co, Ni, Sn, Pb, H, Sb, Bi, As, Cu, Hg, Ag, Pt, Au

The more reactive elements are at the *left* of this series, and the elements become progressively less reactive moving toward the right of the series. For example, you will notice that zinc comes considerably *before* copper in the series, showing that zinc is more reactive than copper.

Notice that H appears in the series. Metals to the left of H are capable of replacing hydrogen ion (H^+) from acids, with evolution of gaseous elemental hydrogen (H_2). In fact, the first four elements of the series are so reactive that they will even replace hydrogen from pure cold water. Elements in the series that come to the *right* of H will *not* replace hydrogen from acids and will consequently generally *not* dissolve in acids. You will note that the elements at the far right side of the series include the so-called noble metals: silver, platinum, and gold. These metals are used in jewelry because they have such low reactivities and can maintain a shiny, attractive appearance. In addition to hydrogen, an element at the left of the electromotive series can replace any element to its right.

While the fact that a reactive metal can replace a less reactive metal from its compounds might be interesting in itself, there is much more implied by the electromotive series. In the zinc/copper reaction discussed at the beginning of this introduction, we considered putting a piece of metallic zinc into *direct contact* with a solution of copper(II) ion so that electrons could flow directly from zinc atoms to copper ions. A far more useful version of this same experiment would be to set up the reaction so that the zinc metal and copper(II) ion solution are *physically separate* from one another (in separate beakers, for example) but are connected *electrically* by a conducting wire. (See Figure 32-1.) Since the reaction that occurs is a transfer of electrons, this can now occur through the wire, thereby producing an electrical current. We could place a motor or light bulb along the wire joining the zinc/copper beakers and make use of the electrical current produced by the reaction. We have constructed a **battery** (or voltaic cell) consisting of a zinc half-cell and a copper half-cell. A second connection will have to be made between the two beakers to complete the electrical circuit, however.

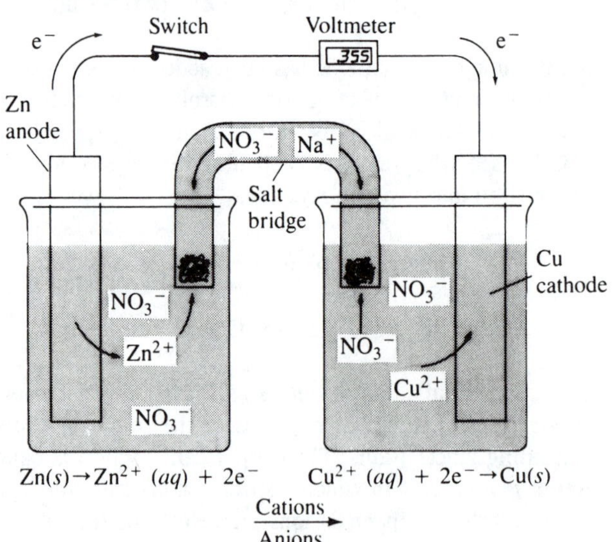

Figure 32-1. A zinc/copper voltaic cell. Electrons flow spontaneously through the wire from the zinc half-cell to the copper half-cell when the switch is closed.

In common practice, a glass tube containing a nonreactive salt solution is used to do this (a **salt bridge**), or alternatively a porous porcelain cup is used to contain one half-reaction, with the porcelain cup then being placed in a beaker containing the second half-reaction.

Reactions of voltaic cells are exergonic. They take place with the *release* of energy. This energy can be put to use if the cell is set up correctly.

It was indicated earlier that gold is the least reactive metal in the electromotive series. This would mean that metallic gold could not be produced by some more active metal replacing gold ions from solution. Yet gold is quite commonly electroplated from solutions of Au(III) ions onto more common, cheaper metals. The reduction of Au(III) ions to metallic gold is an endergonic process, requiring the *input* of energy from an external source to overcome the reluctance of Au(III) ions to undergo reduction. If an electrical current of sufficient voltage from an outside source is passed through a solution of Au(III) ions, it will provide the required energy, and metallic gold will be produced. When an electrical current is used to force a reaction to occur that would ordinarily not be capable of spontaneously occurring, **electrolysis** is said to be taking place. You will perform several electrolysis reactions in Experiment 33 in this manual.

In this experiment, you will examine the relative reactivity of some metals and determine a small portion of the electromotive series. You will also set up several batteries and measure the voltage delivered by the cells.

Safety Precautions

- Wear safety glasses at all times while in the laboratory.
- Salts of metal ions may be toxic. Wash hands after use.
- Dispose of all solutions as directed by the instructor.
- Wash hands after using sulfuric acid. Although the acid used in this experiment is dilute, it will become more concentrated as water evaporates from it.

Apparatus/Reagents Required

Equipment for voltaic cells (beakers, porous porcelain cup, voltmeter), 1 *M* sulfuric acid, 1 *M* magnesium sulfate, 1 *M* copper(II) sulfate, 1 *M* sodium sulfate, 1 *M* zinc sulfate, 0.1 *M* copper sulfate, 0.1 *M* zinc sulfate, small strips of metallic zinc and copper, magnesium turnings, 24-well plate, 4-inch strips of magnesium, copper, and zinc metals

Procedure

Record all data and observations directly in your notebook in ink.

A. The Electromotive Series

In separate wells of the 24-well test plate, add 10 drops of solutions of one of the following: 1 *M* sulfuric acid, 1 *M* magnesium sulfate, 1 *M* copper(II) sulfate, 1 *M* sodium sulfate, and 1 *M* zinc sulfate. Place a small strip of metallic zinc in each well so that the metal is partially covered by the solution in the test plate.

Allow the solutions to stand for about 15 minutes. Examine the zinc strips for evidence of reaction, both during the 15-minute waiting period and after removing from the test tubes. Determine which ionic species zinc is capable of *replacing* from solution and write equations for the reactions that take place.

Repeat the process using new 10-drop samples of the same solutions, but substituting first copper and then magnesium in place of the zinc metal.

On the basis of your results, arrange the following elements in order of their activity: H, Mg, Cu, Zn, Na.

B. Voltaic Cells

Using strips of copper, zinc, and magnesium metals as electrodes, and solutions of the sulfates of these metals, you will set up three voltaic cells, and will measure the cell potentials (voltages). The following procedure is described in terms of a copper/zinc voltaic cells. You will also set up copper/magnesium and zinc/magnesium voltaic cells.

Obtain a porous porcelain cup from your instructor and place it in a 400-mL beaker of distilled water for 5 minutes to wet the cup.

Add 15–20 mL of 1 *M* $CuSO_4$ to a 100-mL beaker. Obtain a 4-inch strip of copper metal and clean it with sandpaper. Place the copper metal strip into the beaker containing the copper sulfate solution to serve as an electrode.

Add 10–15 mL of 1 *M* ZnSO4 to the porous porcelain cup you have soaked in water. The porcelain cup is very fragile and expensive: be careful with it. Obtain a 4-inch strip of zinc metal and clean it with sandpaper. Place the zinc metal strip into the porous cup containing the zinc sulfate solution to serve as an electrode.

Connect one lead of the voltmeter to the copper strip, and the other lead of the voltmeter to the zinc strip.

Place the porous cup containing the $Zn|Zn^{2+}$ half cell into the beaker containing the $Cu|Cu^{2+}$ half cell. Allow the cell to stand until the voltage reading on the voltmeter has stabilized, then record the highest voltage obtained.

Using the table of standard reduction potentials in your textbook, calculate the *standard potential* for the copper/zinc voltaic cell. Calculate the % *difference* between your experimetally determined voltage and the standard voltage.

Using the same method as discussed for the copper/zinc cell, construct copper/magnesium and zinc/magnesium cells and measure their potentials. Calculate the standard cell potential for both of these cells, and the % *difference* between your experimental voltage and the standard voltage.

C. Effect of Concentration on Cell Potential

Prepare a copper/zinc voltaic cell as in Part B, using 1 *M* $ZnSO_4$ solution as before, but replace the 1 *M* $CuSO_4$ solution with 0.1 *M* $CuSO_4$ solution. Measure the voltage of the cell. How does the decrease in concentration affect the voltage of the cell?

Prepare a copper/zinc voltaic cell as in Part B, using 1 *M* $CuSO_4$ solution as before, but replace the 1 *M* $ZnSO_4$ solution with 0.1 *M* $ZnSO_4$ solution. Measure the voltage of the cell. Does the decrease in concentration of Zn^{2+} ion affect the voltage measured? Why?

Name ______________________ Section ______________________

Lab Instructor ______________________ Date ______________________

EXPERIMENT 32 Electrochemistry I: Chemical Cells

PRE-LABORATORY QUESTIONS

1. Textbooks often describe oxidation as a process in which the oxidation number of a species increases. Demonstrate that this definition is consistent with the definition given for oxidation in the introduction to this experiment.

2. Find specific definitions in your textbook or a chemical dictionary that will distinguish between a voltaic cell and an electrolysis cell. Describe that distinction.

3. How is the voltage (potential) developed by a voltaic cell dependent on the concentrations of the ionic species involved in the cell reaction?

4. Given the following two standard half-cells and their associated standard reduction potentials, what cell reaction occurs when the half-cells are connected to form a battery? What is the standard potential of the battery?

$Pb^{2+} + 2\ e^- \rightarrow Pb$ $E^\circ = -0.13$ V

$Ni^{2+} + 2\ e^- \rightarrow Ni$ $E^\circ = -0.23$ V

Name ______________________ Section ______________________

Lab Instructor ______________________ Date ______________________

EXPERIMENT 32 Electrochemistry I: Chemical Cells

RESULTS/OBSERVATIONS

A. Electromotive Series

Observations for reactions of metallic zinc

with 0.1 *M* sulfuric acid ______________________

with 1 *M* $MgSO_4$ ______________________

with 1 *M* $CuSO_4$ ______________________

with 1 *M* Na_2SO_4 ______________________

Observations for reactions of metallic copper

with 1 *M* sulfuric acid ______________________

with 1 *M* $MgSO_4$ ______________________

with 1 *M* $ZnSO_4$ ______________________

with 1 *M* Na_2SO_4 ______________________

Observations for reactions of metallic magnesium

with 1 *M* sulfuric acid ______________________

with 1 *M* $CuSO_4$ ______________________

with 1 *M* $ZnSO_4$ ______________________

with 1 *M* Na_2SO_4 ______________________

Write balanced equations for any reactions that occurred.

____

____

____

Order of activity ______________________

B. Voltaic Cells

Copper/zinc cell

Observation of zinc electrode ______________________________

Observation of copper electrode ______________________________

Voltage measured for the cell ______________ % difference from E° ______________

Balanced chemical equation for the cell reaction:

____

Copper/magnesium cell

Observation of magnesium electrode ______________________________

Observation of copper electrode ______________________________

Voltage measured for the cell ______________ % difference from E° ______________

Balanced chemical equation for the cell reaction:

____

Zinc/magnesium cell

Observation of magnesium electrode ______________________________

Observation of zinc electrode ______________________________

Voltage measured for the cell ______________ % difference from E° ______________

Balanced chemical equation for the cell reaction:

____

C. Effect of Concentration on Cell Potential

Voltage measured with 0.1 *M* Cu(II) ______________________________

Why is the voltage measured lower when [Cu(II)] is decreased?

____

____

Voltage measured with 0.1 *M* Zn(II) ______________________________

How was the measured voltage affected by the decrease in Zn(II) concentration? Why?

____

QUESTIONS

1. Voltages listed in references for voltaic cells are given in terms of standard cell potentials (voltages). What is meant by a standard cell? Was your initial voltaic cell a standard cell?

2. As a standard voltaic cell runs, the voltage delivered by the cell drops with time. Why does this happen?

3. Use a table of standard reduction potentials to calculate E° and E for each of the following voltaic cells at 25°C.

a. $Mg(s)|Mg^{2+}(1\ M)||Al^{3+}(0.01\ M)|Al(s)$

b. $Al(s)|Al^{3+}(0.1\ M)||Ag^{+}(0.01\ M)|Ag(s)$

c. $Zn(s)|Zn^{2+}(0.1\ M)||Ag^{+}(1\ M)|Ag(s)$

EXPERIMENT

33 Electrochemistry II: Electrolysis

Objective

Electrolysis is the use of an electrical current to force a chemical reaction to occur that would ordinarily not proceed spontaneously. When an electrical current is passed through a molten or dissolved electrolyte, between two physically separated electrodes, two chemical changes take place. At the positive electrode (also called the **anode**), an oxidation half-reaction takes place. At the negative electrode (referred to as the **cathode**) a reduction half-reaction takes place. Exactly what half-reaction occurs at each electrode is determined by the relative ease of oxidation/reduction of all the species present in the electrolysis cell. In this experiment you will study the electrolysis of water (Choice I), and the electrolysis of a salt solution (Choice II).

Choice I. Electrolysis of Water

Introduction

When an electrical current is passed through water, two electrochemical processes take place. At the anode, water molecules are *oxidized:*

$$2H_2O \rightarrow O_2 + 4H^+ + 4e^-$$

Gaseous elemental oxygen is produced at the anode and can be collected and tested. The solution in the immediate vicinity of the electrode becomes acidic as H^+ ions are released. At the cathode during the electrolysis of water, water molecules are *reduced*:

$$4e^- + 4H_2O \rightarrow 2H_2 + 4OH^-$$

Gaseous elemental hydrogen is produced at the cathode and may be collected and tested. The solution in the immediate vicinity of the electrode becomes basic as OH^- ions are released.

The two processes above are called **half-reactions**. It is the combination of the two half-reactions, which are taking place at the same time but in different locations, that constitutes the overall reaction in the electrolysis cell. The overall cell reaction that takes place is obtained by adding together the two half-reactions and canceling species common to either side:

$$6H_2O \rightarrow O_2 + 2H_2 + 4H^+ + 4OH^-$$

However, when the hydrogen ions and hydroxide ions migrate toward each other in the cell, they will react with each other,

$$H^+ + OH^- \rightarrow H_2O$$

resulting in the production of four water molecules. This leaves the final overall equation for what occurs in the cell as simply:

$$2H_2O \rightarrow O_2 + 2H_2$$

Note the *coefficients* in this final overall equation. Twice as many moles of elemental hydrogen gas are produced as elemental oxygen gas. If the gases produced are collected, then according to Avogadro's law, the volume of hydrogen collected should be twice the volume of oxygen collected.

Safety Precautions

- Wear safety glasses at all times in the laboratory.
- A 9-volt transistor battery will be used as the source of electrical current. Be aware that even a small battery can cause an electrical shock if care is not exercised.
- Hydrogen gas is produced in the reaction. Hydrogen is flammable. Use caution during its generation and testing.

Apparatus/Reagents Required

Electrolysis apparatus (9-volt battery and leads, graphite electrodes, two test tubes for collecting gases evolved), two rubber stoppers to fit the test tubes tightly, wood splints, ruler, 1 *M* sodium sulfate solution

Procedure

Record all data and observations directly in your notebook in ink.

A. Electrolysis of Water

Place approximately 200 mL of distilled water in a 400-mL beaker. Add 2–3 mL of 1 *M* sodium sulfate to the water and stir. The sodium sulfate is added to help the electrical current pass more easily through the cell.

Arrange the dc power supply (9-volt battery) and graphite electrodes as indicated in Figure 33-1, using connecting wires terminating in alligator clips to make the connections. *Do not connect the battery at this point, however.* Make sure that the electrodes do not touch each other, and be certain that the electrodes are arranged in such a way that the test tubes can be inverted over them easily.

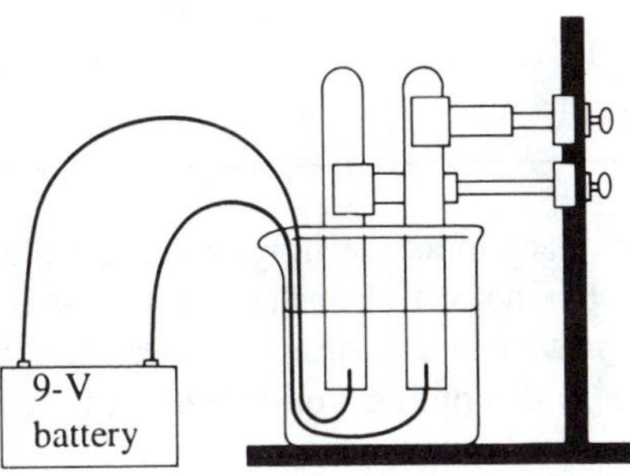

Figure 33-1. Apparatus for electrolysis of water with collection of the evolved gases. Beware of electrical shock hazard.

Fill each of the test tubes to be used for collecting gas with some of the water to be electrolyzed. Take one of the test tubes and place your finger over the mouth of the test tube to prevent loss of water.

Invert the test tube and lower the test tube into the water in the beaker. Remove your finger, and place the test tube over one of the electrodes so that the gas evolved at the electrode surface will be directed into the test tube.

If the liquid in the test tube is lost during this procedure, remove the test tube, refill with water, and repeat the transfer. Repeat the procedure with the other test tube and the remaining electrode.

Have the instructor check your set-up before continuing.

If the instructor approves, connect the battery to begin the electrolysis. Allow the electrolysis to continue until one of the test tubes is just filled with gas (hydrogen).

Disconnect the battery.

Stopper the test tubes while they are still *under the surface* of the water in the beaker and remove. One test tube should be filled with gas (hydrogen), while the other test tube should be only about half-filled with gas (oxygen), with the remainder of the test tube filled with water.

B. Testing of the Gases Evolved

With a ruler, measure the approximate height of gas contained in each test tube as an index of the volume of gas that was generated. Do the relative amounts of hydrogen and oxygen generated seem to correspond to the stoichiometry of the reaction?

Ignite a wooden splint. Using a clamp or test tube holder to protect your hands, hold the test tube containing the hydrogen gas upside down (hydrogen is lighter than air) and remove the stopper. Bring the flame near the open mouth of the test tube. Describe what happens to the hydrogen when ignited.

Obtain a wooden splint. Ignite the splint in a burner flame or match; then *blow out* the splint quickly so that the wood is still glowing. Remove the stopper from the oxygen test tube and insert the splint. Describe what happens to the splint.

Choice II. Electrolysis of Potassium Iodide Solution

Introduction

The reactions that take place in an electrolysis cell are always those that require the least expenditure of energy. In Choice I of this experiment, you electrolyzed water. Since water was the only reagent present in any quantity, water was both oxidized at the anode and reduced at the cathode. In this option, you will electrolyze a solution of the salt potassium iodide, KI.

Two possible oxidation half-reactions must be considered. Depending on which species present in the solution is more easily oxidized, one of these half-reactions will represent what actually occurs in the cell

$$2H_2O \rightarrow O_2 + 4H^+ + 4e^-$$

$$2I^- \rightarrow I_2(s) + 2e^-$$

In the first half-reaction, *water* is being oxidized. This half-reaction would generate elemental oxygen gas, whose presence can be detected with a glowing splint as in Choice I. In addition, the pH of the solution in the region of the anode would be expected to decrease as hydrogen ion is generated by the electrode process. An indicator might be added to determine if the pH changes in the region of the anode. In the second possible half-reaction, elemental *iodine* is generated. Elemental iodine is slightly soluble in water, producing a brown solution. If this is the correct oxidation half-reaction, you would notice the solution surrounding the anode in the cell becoming progressively more brown as the electrolysis occurs. You might also notice that the anode becomes coated with dark crystals of solid iodine as the electrolysis continues.

The reduction half-reaction that takes place at the cathode in this experiment could be either of two processes, again depending on which reduction requires a lower expenditure of energy:

$$2H_2O + 2e^- \rightarrow H_2 + 2OH^-$$

$$K^+ + e^- \rightarrow K(s)$$

If the reduction of *water* is the actual half-reaction, as in the first example, hydrogen gas will be generated at the cathode and can be collected and tested for its flammability. Notice that hydroxide ion is also produced. Hydroxide ion will make the solution basic in the region of the cathode. As before, an indicator might be added to detect this change in pH in the region of the cathode. If the actual reduction, on the other hand, is that of *potassium ion*, the cathode would be expected to increase in size (and mass) as potassium metal is plated out on the surface of the cathode.

By careful observation in this experiment, you should be able to determine which oxidation and which reduction of those suggested actually take place in the cell.

Safety Precautions

- Wear safety glasses at all times while in the laboratory.
- A 9-volt transistor battery will be used as the source of electrical current for the electrolysis. Be aware that even a small battery can cause an electrical shock if care is not exercised.
- Potassium iodide may be irritating to the skin. Avoid contact.
- Elemental iodine will stain the skin. The staining is generally not harmful unless a person is hypersensitive to iodine, but the stain will take 2–3 days to wear off.
- Hydrogen gas is flammable. Exercise caution during its generation and testing.

Apparatus/Reagents Required

Electrolysis apparatus (9-volt battery and leads, graphite electrodes, and connecting wires), two test tubes and tightly fitting stoppers for isolation of the products, potassium iodide, sodium thiosulfate, Universal indicator solution and color chart, starch solution

Procedure

Record all data and observations directly in your notebook in ink.

Weigh out approximately 2.5 g of potassium iodide and dissolve in 150 mL of distilled water. This results in an approximately 0.1 *M* KI solution.

The KI solution should be *colorless.* If the solution is brown at this point, some of the iodide ion present has been oxidized. If this has happened, add a *single* crystal of sodium thiosulfate and stir. If the brown color does not fade, add more single crystals of sodium thiosulfate until the potassium iodide solution is colorless.

Add 4–5 drops of Universal indicator solution to the beaker and stir. Record the color and pH of the solution. Keep handy the color chart provided with the indicator.

Arrange the dc power supply (9-volt battery) and graphite electrodes as indicated in Figure 33-1 (Choice I), but *do not connect the battery yet.* Make sure that the electrodes do not touch each other and that the electrodes are arranged in such a way that the test tubes can be inverted over them easily.

Fill each of the test tubes with some of the KI solution to be electrolyzed. Take one of the test tubes and place your finger over its mouth to prevent loss of solution.

Invert the test tube and lower the test tube into the solution in the beaker. Remove your finger, and place the test tube over one of the electrodes so that the substances evolved at the electrode surface will be directed into the test tube.

If the liquid in the test tube is lost during this procedure, remove the test tube, refill with solution, and repeat the transfer. Repeat the procedure with the other test tube and the remaining electrode.

Wash your hands at this point to remove potassium iodide.

Have the instructor check your set-up before continuing.

If the instructor approves, connect the 9-volt battery to begin the electrolysis. Examine the electrodes for evolution of gas, or deposition of a solid. Allow the electrolysis to continue for several minutes. (If a gas is generated in the cell reaction, stop the electrolysis when the test tube above the electrode is filled with the gas.)

Observe and record the color changes that take place in the solution in the region of the electrodes. Be careful to distinguish between color changes associated with the indicator and the possible production of elemental iodine (brown color). By reference to the color chart provided with the indicator, determine what pH changes (if any) have occurred near the electrodes.

While still under the surface of the solution in the beaker, stopper the test tubes, and remove them from the solution in the beaker.

The possible oxidation and reduction half-reactions for this system were listed in the introduction to this choice. By testing the contents of the two test tubes, determine which half-reactions actually occurred.

If a gas is present in either test tube, use a clamp to hold the test tube and test the gas with a glowing wood splint. If you suspect the gas is hydrogen, invert the test tube (hydrogen is lighter than air), remove the stopper, and bring the wood splint near the mouth of the test tube. Hydrogen will explode with a loud pop. If you suspect the gas is oxygen, hold the test tube upright with the clamp, remove the stopper, and insert the glowing splint. Oxygen will cause the splint to burst into full flame.

If elemental iodine were produced, one of the test tubes would contain a brown solution. Confirm the presence of iodine by addition of a few drops of starch (blue/black color). If iodine were produced, the electrode at which the oxidation occurred would probably be coated with a thin layer of gray-black iodine crystals.

If metallic potassium were produced, it would have plated out as a thin gray-white coating on one of the electrodes.

After determining what oxidation and what reduction have actually occurred in the electrolysis cell, combine the appropriate half-reactions into the overall cell reaction for the electrolysis.

Name ______________________________ Section______________________

Lab Instructor ______________________________ Date______________________

EXPERIMENT 33 Electrochemistry II: Electrolysis

PRE-LABORATORY QUESTIONS

CHOICE I. ELECTROLYSIS OF WATER

1. In this experiment we collect the gases produced by the electrolysis of water and measure their volumes to see if the volumes correspond to the stoichiometry of the reaction. Look up the solubilities of gaseous hydrogen and gaseous oxygen in a handbook to see if there will be any problem with one gas dissolving more than the other as it is generated, thereby affecting the volumes of gases measured. Write what you find.

____

____

____

____

____

____

2. Suppose 25 mL of gaseous hydrogen is collected through the electrolysis of water. What volume of gaseous oxygen should also be collected? Why?

____

____

____

____

3. Describe the qualitative tests for oxygen gas and for hydrogen gas.

____

____

____

____

CHOICE II. ELECTROLYSIS OF POTASSIUM IODIDE SOLUTION

1. When the electrolysis of an aqueous solution of an ionic salt is undertaken, very often the half-reactions that actually occur involve *water* molecules rather than the ions of the salt. Consult your textbook for information as to how you can predict what half-reactions are most likely to occur in an electrolysis experiment. Summarize your findings.

2. Suppose an aqueous solution of sodium fluoride were to be electrolyzed. What are the *possible* half-reactions that might take place in the system? Which are the *most likely* half-reactions? Why?

Name ______________________ Section ______________

Lab Instructor ______________________ Date ______________

EXPERIMENT 33 Electrochemistry II: Electrolysis

CHOICE I. ELECTROLYSIS OF WATER

RESULTS/OBSERVATIONS

Observations on electrolysis

____

____

____

____

Approximately how long did it take to fill the test tube with H_2? ______________

Height of gas in O_2 tube ______________ in H_2 tube ______________

Ratio of H_2/O_2 heights ______________ error ______________

Observation on testing H_2 with flame

____

____

Observation on testing O_2 with wood splint

____

____

QUESTIONS

1. How could the speed of an electrolysis such as you performed be increased?

____

____

____

2. Why was sodium sulfate added to the water to be electrolyzed? Was there any particular reason sodium sulfate, rather than some other salt, was used?

3. In order for an electrolysis to take place, the current used must be of sufficient voltage. Use your textbook to determine the minimum voltage current needed to electrolyze water.

Name ______________________________ Section ______________________

Lab Instructor ______________________________ Date ______________________

EXPERIMENT 33 Electrochemistry II: Electrolysis

CHOICE II. ELECTROLYSIS OF POTASSIUM IODIDE SOLUTION

RESULTS/OBSERVATIONS

Was it necessary to add crystals of sodium thiosulfate to the potassium iodide solution? If so, how many crystals did you add?

____

Color and pH of KI solution before electrolysis ______________________________

Color and pH of KI solution in region of the anode ______________________________

Color and pH of KI solution in region of the cathode ______________________________

What gas(es) were evolved during the electrolysis? At which electrode? How did the gas(es) respond when tested with the glowing wood splint?

____

____

____

____

Was elemental iodine produced? How was this confirmed?

____

____

____

Was elemental potassium produced? How was this confirmed?

____

____

____

Appearance of cathode after electrolysis

____

____

Appearance of anode after electrolysis

____

____

What half-reactions occurred in the electrolysis of aqueous KI?

____

____

What is the overall cell reaction?

____

QUESTIONS

1. Why were you advised to add a crystal of sodium thiosulfate to your potassium iodide solution if it had been brown when prepared? Write a balanced equation for the iodine/thiosulfate reaction.

____

____

____

____

____

2. If you had electrolyzed a potassium bromide solution, instead of potassium iodide, what would the likely half-reactions have been?

____

____

____

____

EXPERIMENT

34 Gravimetric Analysis

Objective

An unknown compound will be analyzed for one of its constituents by precipitation and weighing of the constituent of interest. In Choice I, the level of chloride ion is determined by precipitation with silver ion. In Choice II, sulfate ion will be determined by precipitation with barium chloride.

Introduction

Gravimetric analysis is a standard classical method for determining the amount of a given component present in many solution- or solid-unknown samples. The method involves *precipitating* the component of interest from the unknown by means of some added reagent. From the *mass* of the precipitate, the percentage of the unknown component in the original sample may be calculated.

Generally, the reagent causing precipitation is chosen to be as specific as possible for the component of interest. The process is intended to remove and weigh *only* the particular component of interest from the unknown sample. The precipitating agent must also be chosen carefully so that it *completely* precipitates the component under study. The resulting precipitate must have an extremely low solubility, must be of known composition, and must be chemically stable. Handbooks of chemical data list suggested precipitating agents for routine gravimetric analysis of many unknowns.

A complete gravimetric analysis includes a series of distinct steps. First, a precise, known amount of original unknown sample must be taken for the analysis. If the unknown is a solid, an appropriately sized portion is taken and weighed as precisely as permitted by the balances available (typically to the nearest 0.001 or 0.0001 g). If the unknown is a solution, an appropriately sized aliquot (a known fraction) is taken with a pipet. The sample of unknown taken for the analysis must be large enough that precision may be maintained at a high level during the analysis but not so large that the amount of precipitate generated cannot be handled easily.

Next, the unknown sample must be brought into solution (if it is not already dissolved). For some solid samples, such as metal ores, this may involve heating with acid to effect the dissolving. Frequently, the pH of the solution of unknown must be adjusted before precipitation can take place; this is usually done with concentrated buffer systems to maintain the pH constant throughout the analysis.

The reagent that causes the precipitation is then added to the sample. The precipitating reagent is added slowly, and in fairly dilute concentration, to allow large, easily filterable crystals of precipitate to form. The precipitate formed is allowed to stand for an extended period, perhaps at an elevated temperature, to allow the crystals of precipitate to grow as large as possible. This waiting period is called **digestion** of the precipitate.

The precipitate must then be filtered to remove it from the liquid. Although filtration could be accomplished with an ordinary gravity funnel and filter paper, this would probably be very slow. The precipitates produced in gravimetric analysis are often very finely divided and would tend to clog the pores of the filter paper. Specialized *sintered glass* filtering funnels have been prepared for routine gravimetric analyses. Rather than filter paper, such funnels have a *frit plate* constructed of several layers of very fine compressed glass fibers that act to hold back the particles of precipitate. Such glass funnels can use *suction* to speed up the removal of liquid from a precipitate, can be cleaned easily before and after use, and are not affected by reagents in the solution. A typical sintered glass funnel is shown in Figure 34-1.

Figure 34-1. A sintered glass funnel (crucible)

Once the precipitate has been transferred to the sintered glass funnel, it must be washed to remove any adhering ions. As the precipitate forms, certain of the other species present in the mixture may have been *adsorbed* on the surface of the crystals. The liquid chosen for washing the crystals of precipitate is usually chosen so as to remove chemically such adhering ions. The washing of the precipitate must be performed carefully to prevent *peptization* of the precipitate. The wash liquid must not redissolve the precipitate or break up the crystals to the point where they might be lost through the pores of the filter.

After the precipitate has been filtered and washed, it must be dried. This is usually accomplished in an oven whose temperature is rigorously controlled at 110°C. The oven must be hot enough to boil off water adhering to the crystals, but cannot be so hot that it might decompose the crystals. Some precipitating reagents are organic in nature and cannot stand very strong heating.

Finally, the dried precipitate is weighed. From the mass and composition of the precipitate, the mass of the component of interest in the original sample is determined. Generally, gravimetric analyses are done in triplicate (or even quadruplicate) as a check on the determination. The precision expected in a good gravimetric analysis is very high, and if there is any major deviation in the analyses, it is likely due to some source of error either in the procedure or by the operator.

Preparation of the Sintered Glass Filters

Rather than using filter paper, gravimetric determinations are more conveniently performed using sintered glass filtering crucibles or funnels. Each of the following choices requires the cleaning and drying (to constant weight) of two such funnels. The procedure indicated here should be followed in connection with each individual analysis.

Procedure

Obtain two medium porosity sintered glass funnels or crucibles. If the funnels appear to be dirty, wash them with soap and water, but *do not scrub the fritted plate*.

Rinse thoroughly with tap water and then with small portions of distilled water. Set up a suction filtration apparatus, and place one of the funnels into position for filtration.

Add 20 to 30 mL of distilled water to the funnel and apply suction to pull the water through the fritted plate of the funnel. Add another portion of distilled water and repeat. Repeat the process with the other funnel.

If the funnels do not appear clean at this point, consult with the instructor.

Obtain two small beakers that can accommodate the funnels and label the beakers as 1 and 2; also label the beakers with your name. Transfer one funnel to each beaker, then place the beakers into the oven for at least one hour to dry. The oven should be kept at a temperature of at least 110°C.

While the funnels are drying, go on to the next part of the experiment.

Choice I. Gravimetric Determination of Chloride Ion

Introduction

One of the most common gravimetric analyses is that of the chloride ion, Cl^-. Chloride samples determined by this method typically include simple inorganic salts, samples of brine, or even body fluids.

In the analysis, a weighed sample containing chloride ion is first dissolved in dilute nitric acid. Then a dilute solution of silver nitrate is added, which causes the formation of a precipitate of silver chloride. The silver chloride is then digested, filtered, dried, and weighed. The net ionic reaction for the process is simply

$$Ag^+ + Cl^- \rightarrow AgCl(s)$$

The mass of chloride ion present in the original sample is given by

$$(\text{mass of AgCl precipitated}) \times (\text{molar mass of Cl/molar mass of AgCl})$$

and the percent of chloride ion in the original sample is given by

$$[(\text{mass of Cl})/(\text{mass of sample})] \times 100$$

The sample is dissolved originally in dilute nitric acid (rather than distilled water) to prevent precipitation of any *other* negative ions that might be present in the sample. For example, if carbonate ion (or bicarbonate ion) were present in a sample, they too would be precipitated by silver ion. Nitric acid destroys or masks such interferences.

As initially formed, the particles of AgCl are too small to be filtered easily. Digestion of the AgCl allows smaller particles of AgCl to combine to form larger crystals that can be filtered more quantitatively. Digestion of the precipitate may be effected by either of two methods. If it is desired to complete the analysis during a single laboratory period, then the precipitate and supernatant solution can be heated almost to boiling for approximately an hour. If the analysis is to be finished at a later time, the precipitate and supernatant liquid can be allowed to stand in a dark place for a few days. Silver ion can be reduced by light to metallic silver, with loss of chloride ion to the atmosphere. Loss of chloride ion by this mechanism would affect the weight of precipitate obtained.

Filtration of a precipitate of the type formed by AgCl is most easily performed using a sintered glass filtering funnel. Such a funnel contains a glass plate (or frit) that has small pores to retain the particles of precipitate. The filtration can be accomplished with suction, and the glass crucible can be transferred easily to an oven for drying.

Precipitation with silver ion can also be applied to the gravimetric determination of the other halides (bromide and iodide) and to the determination of several other anions.

Safety Precautions

- Wear safety glasses at all times when in the laboratory.
- Silver nitrate solutions will stain the skin. The stain will wear off in 2–3 days. Nitrates are strong oxidizing agents and are toxic.
- Nitric acid solutions are corrosive. Exercise caution in their use.
- Handle hot glassware with tongs or paper towels.

Apparatus/Reagents Required

Medium porosity sintered glass funnels; suction filtration set-up; oven set reliably to 110°C; several clean, dry beakers for storage of funnels/samples; stirring rods with "rubber policeman"; polyethylene squeeze bottle; chloride unknown; 6 *M* nitric acid; 5% silver nitrate solution

Procedure

All observations and data should be entered directly into your notebook in ink.

A. Preparation of Filters

Prepare two sintered glass funnels or crucibles as directed above.

B. Preparation and Precipitation of the Chloride Samples

Obtain an unknown chloride ion sample, and record the identification number in your notebook.

Clean two 400- or 600-mL beakers and label them as 1 and 2.

Weigh out two samples of approximately 0.4 g of the chloride unknown (one into each beaker), making the weight determinations to the precision of the balance (at least to the nearest 0.001 g).

Add 150–200 mL of distilled water to each beaker, followed by 5 mL of 6 *M* nitric acid. Stir to dissolve the samples, using separate stirring rods for each beaker (do not remove the stirring rods from the beaker from this point on, to prevent loss of the sample).

Obtain approximately 50 mL of 5% silver nitrate solution. Each of your samples should require no more than 20–25 mL of silver nitrate to completely precipitate the chloride ion (assuming your chloride samples are close to 0.4 g in mass).

The two chloride samples may be treated one at a time; or, if two burners are available, they may be treated in parallel.

Heat the dissolved chloride sample almost to the boiling point (but do not let it actually boil) and add 5 mL of 5% silver nitrate solution. A white precipitate of silver chloride will form immediately.

While stirring the solution, continue to add silver nitrate solution slowly until a total of 15–18 mL has been added. Continue to heat the solution without boiling to allow the silver chloride precipitate to digest.

When it appears that the silver chloride has settled completely to the bottom of the beaker, and the solution above the precipitate has become clear, slowly add another 1 mL of silver nitrate. Observe the silver nitrate as it enters the chloride sample beaker to see if additional precipitate forms.

If no additional precipitate forms at this point, continue to heat the solution for 10 minutes. Then transfer to a cool, dark place for approximately one hour.

If additional precipitate has formed during this period, add another 3–4 mL of silver nitrate, stir, and allow the sample to settle again. After the precipitate has settled completely, test the solution above the sample with 1 mL of silver nitrate to see if the precipitation is complete yet. Continue in this manner until it is certain that all the chloride ion has been precipitated. Finally, transfer the sample to a cool, dark place for approximately one hour.

C. Preparation for Filtration

While the silver chloride sample is settling, remove the funnels or crucibles from the oven, and allow them to cool for 10 minutes. Cover the beakers containing the funnels to avoid any possible contamination.

When the funnels are cool, weigh them to the precision of the balance (at least to the nearest 0.001 g). Be sure not to mix up the funnels during the weighing process. Handle the funnels with tongs or strips of paper during the weighing to avoid getting finger marks on the funnels.

Clean out the suction filtration apparatus in preparation for the filtration of the silver chloride.

Prepare a washing solution consisting of approximately 2 mL of 6 *M* nitric acid in 200 mL of distilled water. Place this solution into a squeeze bottle.

D.. Filtration of the Silver Chloride

Set up funnel 1 on the suction filtration apparatus.

Remove chloride sample 1 from storage, and slowly pour some of the supernatant liquid into the funnel under suction. Continue to pour off the supernatant liquid into the funnel until only a small amount of liquid remains in the chloride sample.

Carefully begin the transfer of the precipitated silver chloride, a little at a time, waiting for the liquid to be drawn off by the suction before transferring more solid. Use the stirring rod to help transfer the precipitate. Continue this process until most of the AgCl has been transferred to the funnel.

Using the rubber policeman on your stirring rod, scrape any remaining silver chloride from the beaker into the funnel. Using 5–10-mL portions of the dilute nitric acid solution in the squeeze bottle, rinse the beaker two or three times into the funnel. When it is certain that all the silver chloride has been transferred from the beaker to the funnel, scrape any silver chloride adhering to the stirring rod or rubber policeman into the funnel, and put the funnel into its numbered beaker. Transfer the beaker to the oven and dry the sample for one hour.

Repeat the procedure for sample 2, using funnel 2.

When the funnels have dried for one hour, remove them from the oven and allow them to cool in the covered beaker for 10 minutes.

When the funnels have cooled, weigh them to the precision of the balance (to at least the nearest 0.001 g).

From the difference in mass for each funnel, calculate the quantity of silver chloride precipitated from sample 1 and sample 2.

Calculate the quantity of chloride present in each sample.

Calculate the percent chloride present in each sample.

Calculate the mean percent chloride for your two determinations.

Choice II. Gravimetric Determination of Sulfate Ion

Introduction

Sulfate ion is removed from acidic solution by barium ion, producing a very finely divided crystalline precipitate of barium sulfate:

$$Ba^{2+} + SO_4^{2-} \rightarrow BaSO_4(s)$$

Although the analysis of sulfate ion itself is important, this analysis may be extended to nearly any compound containing the element sulfur, by first oxidizing the sulfur present to sulfate ion. For example, sulfur in metallic sulfides or sulfite compounds can be oxidized to sulfate ion by reaction with potassium permanganate, or the sulfur compounds may be heated to a high temperature in the presence of sodium peroxide to effect the oxidation. Similarly, organic compounds containing sulfur (such as some of the amino acids) can also be oxidized with sodium peroxide.

Sulfates are precipitated from acidic solution for two reasons. First, barium sulfate ordinarily crystallizes in very tiny particles that often occlude impurities as they form: the presence of acid helps larger crystals of $BaSO_4$ to form. Secondly, in a real sample, other anions (such as carbonate) would also be precipitated by barium ion: Adding acid minimizes the precipitation of other anions that might be present.

Safety Precautions

- Wear safety glasses at all times when in the laboratory.
- Barium compounds are toxic. Wear disposable gloves during their use.
- Hydrochloric acid can burn clothing and skin.
- Use tongs to handle the hot crucibles.

Apparatus/Reagents Required

Two sintered glass funnels or crucibles, suction filtration setup, unknown sulfate ion sample, 5% barium chloride solution, 6 *M* hydrochloric acid, 0.1 *M* silver nitrate solution, stirring rods with rubber policeman, hotplate, distilled water, disposable gloves

Procedure

Record all observations and data directly in your notebook in ink.

A. Preparation of Filters

Prepare two sintered glass filtering funnels or crucibles as directed earlier.

B. Preparation and Precipitation of the Sulfate Samples

Obtain an unknown sulfate sample and record the identification number in your notebook.

Clean two 600-mL beakers, rinse with distilled water, and label them as 1 and 2. Weigh out two 0.6-g samples of the unknown sulfate sample, one into each beaker. Make the weight determinations to at least the nearest milligram (0.001 g), and do not go much above 0.65 g total weight of sample.

Add approximately 150 mL of distilled water to each beaker, followed by 2 mL of 6 *M* hydrochloric acid. Wearing disposable gloves, obtain about 100 mL of 5% barium chloride solution, and divide into two 50-mL portions in separate clean beakers.

The precipitation of barium sulfate is carried out at an elevated temperature to assist in the formation of larger crystals. On a hotplate, heat sulfate ion sample 1 and one of the beakers of barium chloride solution nearly to boiling.

Using paper towels to protect your hands, slowly pour the hot barium chloride solution into the sulfate ion sample. Stir vigorously for several minutes while continuing to heat the mixture on the hotplate.

Allow the precipitate of barium sulfate to settle to the bottom of the beaker, leaving a clear layer of supernatant liquid. Test for complete precipitation of sulfate ion by adding 1–2 drops of barium chloride solution to the clear layer.

If additional $BaSO_4$ precipitate forms in the clear layer, add an additional 5 mL of barium chloride, stir, and allow the precipitate to settle again. Repeat the testing process until it is certain that all sulfate ion has been precipitated.

After all barium ion has been precipitated, cover the beaker with a watchglass, and continue to heat the mixture to 75–90°C on the hotplate for at least one hour to digest the precipitate. While the first sample is heating, precipitate the second sulfate ion sample by the same process.

C. Preparation for Filtration

Remove the two filtering funnels from the oven and allow them to cool for 10 minutes. When the funnels have reached room temperature, weigh them to the nearest milligram (0.001 g). Be careful not to confuse funnels 1 and 2 during the weighing. Return the funnels to their storage beaker for later use. Handle the funnels with tongs or paper strips during the weighings to avoid getting finger marks on them.

Clean out the suction filtration apparatus in preparation for the filtration of the barium sulfate.

D. Filtration of the Barium Sulfate

The barium sulfate samples must be filtered while still hot, and must be washed (to remove contaminating ions) with hot water. Heat 200–250 mL of water to 80–90°C before beginning the filtration.

Set up funnel 1 on the suction filtration apparatus. Using paper towels to protect your hands from the heat, remove sulfate sample 1 from the hotplate, and slowly pour some of the supernatant liquid into the funnel under suction. Continue to pour off the supernatant liquid into the funnel until only a small amount of liquid remains in the sulfate sample.

Carefully begin the transfer of the precipitated barium sulfate, a little at a time, waiting for the liquid to be drawn off by the suction before transferring more solid. Use the stirring rod to help transfer the precipitate. Continue this process until most of the $BaSO_4$ has been transferred to the funnel.

Using the rubber policeman on your stirring rod, scrape any remaining barium sulfate from the beaker into the funnel. Using 5–10-mL portions of the hot distilled water, rinse the beaker two or three times into the funnel. When it is certain that all the barium sulfate has been transferred from the beaker to the funnel, scrape any barium sulfate adhering to the stirring rod or rubber policeman into the funnel.

Wash the precipitate with three 10-mL portions of hot distilled water. Wash the precipitate with a fourth 10-mL portion of hot water, but catch the wash liquid (after it passes through the funnel) in a test tube. To the wash liquid in the test tube, add 2–3 drops of 0.1 *M* silver nitrate solution to check for the presence of chloride ion (from the precipitating reagent). If a precipitate (of AgCl) forms in the wash liquid, wash the $BaSO_4$ precipitate in the funnel three more times with small portions of hot water. After the repeat washings, test again for chloride ion and wash the precipitate again if necessary.

After it is certain that all chloride ion has been washed from the $BaSO_4$ precipitate, continue suction for 2–3 minutes. Finally, disconnect the suction and transfer the funnel into its numbered beaker. Transfer the beaker to the oven and dry the sample for one hour.

Repeat the procedure for sample 2, using funnel 2.

When the funnels have dried for one hour, remove them from the oven and allow them to cool in the covered beaker for 10 minutes. When the funnels have cooled, weigh them to the precision of the balance (to at least the nearest 0.001 g).

Calculations

From the difference in mass between the empty filtering funnels and the mass of the funnel containing precipitate, calculate the mass of $BaSO_4$ precipitated from each sample. From the molar masses of the sulfate ion and barium sulfate, calculate the mass of sulfate ion contained in each sample.

Based on the mass of sulfate ion present and the mass of the original sample, calculate the percentage by mass of sulfate in the original unknown.

Calculate the mean percentage of sulfate for your two determinations.

Name ____________________ Section ____________________

Lab Instructor ____________________ Date ____________________

EXPERIMENT 34 Gravimetric Analysis

PRE-LABORATORY QUESTIONS

CHOICE I. GRAVIMETRIC DETERMINATION OF CHLORIDE ION

1. A chloride unknown weighing 0.4221 g is dissolved in acidic solution and is treated with silver nitrate. The silver chloride precipitate that forms is filtered, dried, and weighed. The weight of silver chloride obtained is 0.7632 g. Calculate the percentage of Cl in the unknown chloride sample.

2. Given that the solubility product, K_{sp}, for AgCl is 1.8×10^{-10}, calculate the solubility of AgCl in moles per liter and in grams per liter.

CHOICE II. GRAVIMETRIC DETERMINATION OF SULFATE ION

1. A 1.543-g sample containing sulfate ion was treated with barium chloride reagent, and 0.2243 g of barium sulfate was isolated. Calc ılate the percentage of sulfate ion in the sample.

2. Given that the solubility product, K_{sp}, for $BaSO_4$ is 1.1×10^{-10}, calculate the solubility of $BaSO_4$ in moles per liter and in grams per liter.

Name ______________________ Section ______________________

Lab Instructor ______________________ Date ______________________

EXPERIMENT 34 Gravimetric Analysis

CHOICE I. GRAVIMETRIC DETERMINATION OF CHLORIDE ION

RESULTS/OBSERVATIONS

Identification number of sample ______________________

	Sample 1	*Sample 2*
Mass sample taken, g	________	________
Mass of empty funnel, g	________	________
Mass of funnel with AgCl, g	________	________
Mass of AgCl collected, g	________	________
Mass of Cl in the AgCl collected, g	________	________
% chloride in the sample	________	________

Mean % chloride ______________________

QUESTIONS

1. Why is it normal to add a slight *excess* of silver nitrate solution when precipitating chloride ion?

____

____

____

____

____

____

____

2. What effect on the % Cl determined for a sample would be observed if the precipitated silver chloride were left exposed to bright *light*? Would the calculated % Cl be too high or too low?

3. A small amount of nitric acid is present during the precipitation of the chloride ion. What effect might be observed if too much nitric acid were present?

Name ______________________ Section ______________________

Lab Instructor ______________________ Date ______________________

EXPERIMENT 34 Gravimetric Analysis

CHOICE II. GRAVIMETRIC DETERMINATION OF SULFATE ION

RESULTS/OBSERVATIONS

Identification number of sample ______________________

	Sample 1	*Sample 2*
Mass sample taken, g	________	________
Mass of empty funnel, g	________	________
Mass of funnel with $BaSO_4$, g	________	________
Mass of $BaSO_4$ collected, g	________	________
Mass of sulfate in the $BaSO_4$, g	________	________
% sulfate in the sample	________	________

Mean % sulfate ______________________

Appearance of sulfate solution ______________________

Appearance of precipitate ______________________

QUESTIONS

1. This analysis, performed on a real sample, is subject to error because of the nature of the precipitate formed: the crystals of barium sulfate formed initially tend to occlude foreign ions (which add to the mass of the precipitate). How might this effect be prevented?

____

____

____

____

____

2. Barium sulfate is used in medicine as the active ingredient in the "barium cocktails" given before a series of X-rays is taken of the upper gastrointestinal tract. Use your textbook or a chemical encyclopedia to find out the particular property of barium sulfate that makes it useful in this connection.

3. Historically, the results of analyses of samples containing sulfate ion were often reported in terms of the equivalent amount of sulfur trioxide, SO_3 (even if the sample did not really contain SO_3 as such). For your unknown, calculate the equivalent % SO_3. Use the ratio of the molar masses of SO_3 and $BaSO_4$ as the conversion factor for your calculation.

EXPERIMENT

35 Coordination Compounds

Objective

The class of substances referred to as **coordination compounds** generally contains a central metal atom, to which a fixed number of molecules or ions (called *ligands*) are coordinately covalently bonded in a characteristic geometry. Coordination complexes are very important, both in inorganic and in biological systems. In Choice I, you will prepare a particularly striking coordination compound of copper(II). You also will perform some simple tests to characterize the compound. In Choice II, you will prepare an ammonia complex of cobalt(III), by oxidation of a cobalt(II) salt in the presence of the ligand.

Choice I. Preparation of a Coordination Complex of Copper

Introduction

Coordination complex compounds play a vital role in our everyday lives. For example, the molecule *heme* in the oxygen-carrying protein hemoglobin contains coordinated iron atoms. *Chlorophyll,* the molecule that enables plants to photosynthesize, is a coordination compound of magnesium.

In general, coordination compounds contain a central metal ion that is bound to several other molecules or ions by coordinate covalent bonds. For example, the compound to be synthesized in this experiment, tetramminecopper(II) sulfate, consists of a copper(II) ion, surrounded by four coordinated ammonia molecules. The unshared pair of electrons on the nitrogen atom of the ammonia molecule is used in forming the coordinate bond to the copper(II) ion.

One property of most transition metal coordination compounds that is especially striking is their color. Generally the coordinate bond between the metal ion and the ligand is formed using empty low-lying *d*-orbitals of the metal ion. Transitions of electrons within the *d*-orbitals correspond to wavelengths of visible light, and generally these transitions are very intense. Coordination complexes of metal ions are some of the most beautifully colored chemical substances known; frequently they are used as pigments for paints. The complex you will prepare today has an intensely rich blue color.

Safety Precautions

- Wear safety glasses at all times while in the laboratory.
- Copper compounds are toxic and may be irritating to the skin. Wash after handling them.
- Ammonia solutions are caustic and irritating to the skin. Their vapor is a respiratory irritant and cardiac stimulant. Work with ammonia only in the exhaust hood.
- Ethyl alcohol is flammable. Avoid flames during its use.
- Concentrated hydrochloric acid is damaging to skin, and its vapor is a respiratory irritant. Wash after use, and rinse with a large amount of water if spilled on the skin. Use concentrated HCl in the exhaust hood.

Apparatus/Reagents Required

Suction filtration apparatus, pH paper, copper(II) sulfate pentahydrate, concentrated ammonia solution, concentrated hydrochloric acid, ethyl alcohol (ethanol)

Procedure

Record all data and observations directly in your notebook in ink.

A. Preparation of the Complex

Weigh out approximately 1 g of copper(II) sulfate pentahydrate to the nearest 0.01 g (record).

Dissolve the copper salt in approximately 10 mL of distilled water in a beaker or flask. Stir thoroughly to make certain that all the copper salt has dissolved before proceeding. Record the color of the solution at this point [that is, the color of the tetraaquo copper(II) complex].

Transfer the copper solution to the *exhaust hood*, and with constant stirring, slowly add 5 mL of concentrated ammonia solution (*caution*). The first portion of ammonia added will cause a light blue precipitate of copper(II) hydroxide to form. But on adding more ammonia, this precipitate will dissolve as the ammonia complex forms. Record the color of the mixture after all the ammonia has been added.

To decrease the solubility of the tetramminecopper(II) complex, add approximately 10 mL of ethyl alcohol with stirring. A deep blue solid should precipitate.

Allow the solid precipitate to stand for 5–10 minutes, and then filter the precipitate under suction.

While it is on the filtering funnel, wash the precipitate with two 5-mL portions of alcohol, stir, and apply suction to the precipitate until it appears dry.

Weigh the dried precipitate and record the yield. Based on the weight of copper(II) sulfate pentahydrate taken, calculate the theoretical and percentage yields of the product, tetramminecopper(II) sulfate monohydrate, $[Cu(NH_3)_4]SO_4 \cdot H_2O$.

B. Tests on the Product

Dissolve a small amount of the product in a few milliliters of water. In the exhaust hood, add concentrated HCl dropwise (*Caution!*) until a color change is evident. Ammonia molecules have a stronger affinity for protons than they do for copper ions. When HCl is added to the copper complex, the ammonia molecules are converted to ammonium ions, which no longer have an unshared pair of electrons, and which cannot bond to the copper(II) ion.

Heat a very small amount of the copper complex in a test tube in a burner flame. Hold a piece of moist pH test paper near the mouth of the test tube while it is being heated. When the copper complex is heated, the ammonia molecules are driven out of it. Ammonia is a base and should change the color of the pH test paper. Cautiously waft some of the vapor produced from the complex as it is heated toward your nose. The odor of ammonia should be detected.

Choice II. Preparation of a Coordination Complex of Cobalt(III)

Introduction

Although compounds of both cobalt(II) and cobalt(III) are known, in aqueous solution, cobalt usually exists as the hexaquo cobalt(II) ion, $Co(H_2O)_6^{2+}$. The potential for the oxidation of cobalt(II) to cobalt(III) is ordinarily very unfavorable. In acidic solution, for example, cobalt(III) will rapidly oxidize water [and will itself be reduced to cobalt(II)]. If a complexing agent is added, however, the 3+ oxidation state of cobalt is significantly stabilized, and cobalt(III) complexes can be prepared and isolated.

Since cobalt(III) is not stable in aqueous solution unless a complexing agent is present, it is not possible to prepare a coordination compound by merely adding the complexing agent to a solution of the metal ion as in Choice I. Rather, a cobalt(II) salt is dissolved in water, the complexing agent is added, and the cobalt(II) is oxidized to cobalt(III), *in situ*. In this manner, cobalt(III) can be complexed and stabilized as it is generated. Although most any oxidizing agent might be added to oxidize cobalt(II), it is most convenient to just bubble air (O_2) through the solution.

In this experiment, you will prepare hexamminecobalt(III) chloride. The reaction is

$$4CoCl_2 + 20NH_3 + O_2 + 4NH_4Cl \rightarrow 4[Co(NH_3)_6]Cl_3 + 2H_2O$$

Notice that in $CoCl_2$, cobalt is in the 2+ oxidation state, whereas in the complex cobalt/ammine, cobalt exists in the 3+ oxidation state.

Safety Precautions

- Wear safety glasses at all times while in the laboratory.
- Use glycerine as a lubricant when inserting glass tubing through the rubber stopper. Protect your hands with a towel.
- Cobalt salts are toxic and may irritate the skin. Wash after use.
- Concentrated ammonia solution is a severe respiratory irritant and cardiac stimulant. Work with concentrated ammonia only in the exhaust hood.
- Concentrated hydrochloric acid will burn skin, and its vapor is a severe respiratory irritant. Work in the hood with HCl.
- Ethyl alcohol is volatile and highly flammable. Work with ethyl alcohol in the exhaust hood. No flames are permitted.

Apparatus/Reagents Required

Concentrated aqueous ammonia, cobalt(II) chloride hexahydrate, ammonium chloride, charcoal, concentrated hydrochloric acid, 95% ethyl alcohol

Procedure

Record all data and observations directly in your notebook in ink.

Weigh 4 ± 0.5 g of cobalt(II) chloride hexahydrate into a clean 250-mL beaker. Add approximately 5 mL of water and stir to dissolve the salt.

Weigh out approximately 3 g of ammonium chloride and add to the cobalt chloride solution. Stir to dissolve.

In the exhaust hood, set up the apparatus indicated in Figure 35-1. Use glycerine to lubricate the rubber stoppers when inserting the tubing, and protect your hands with a towel. Make certain that the long glass tube (air inlet) in the suction flask reaches almost to the bottom of the flask.

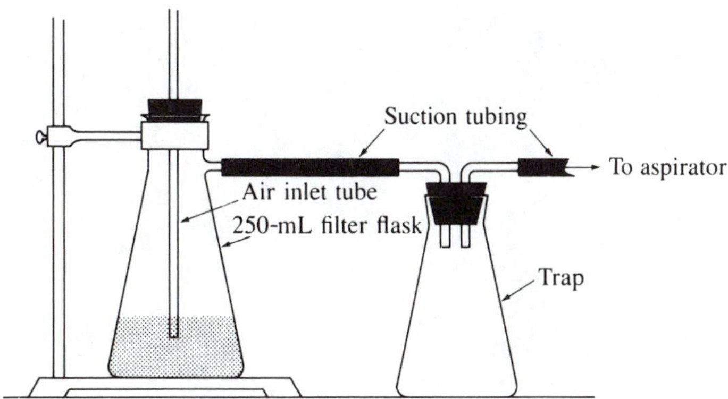

Figure 35-1. Apparatus for preparing a complex of cobalt.

Remove the stopper/air inlet tube from the apparatus, and transfer the cobalt chloride/ammonium chloride mixture to the suction flask. Add a pinch of activated charcoal to the mixture.

In the exhaust hood, with stirring, slowly add 10 mL of concentrated aqueous ammonia (*Caution!*) to the solution in the flask. The solution will be red at this point.

Replace the stopper/air inlet tube in the filter flask. Make certain that the air inlet tube dips beneath the surface of the solution in the flask. Turn on the water aspirator very slowly, and allow air to bubble through the mixture until the red solution has changed color to yellow-brown (1–2 hours). Crude crystals of product will form during this period. Do not allow the bubbling to be so vigorous that solution is lost to the trap. Disconnect the rubber tubing from the aspirator before shutting off the water flow.

Filter the crude crystals of product and discard the liquid filtrate.

Prepare a dilute HCl solution by adding 2–3 mL of concentrated HCl (*Caution!*) to 30–40 mL of distilled water in a 250-mL beaker. Transfer the crude crystals of product to the beaker of dilute acid and stir to dissolve. Test the solution with pH paper. If the solution is not acidic, add concentrated HCl dropwise until the solution becomes acidic.

Heat the solution until all the crystals of product have dissolved. Using tongs or a towel to protect your hands, filter the hot solution into a clean 400-mL beaker to remove charcoal.

In the exhaust hood, slowly add approximately 10 mL of concentrated HCl (*Caution!*) to the filtered solution. Crystals of product should form. If no crystals have formed, transfer the beaker to an ice bath, and scratch the sides of the beaker with a glass rod until crystallization occurs.

Filter the crystals of product under suction on a Büchner funnel.

Mix approximately 5 mL of distilled water with 5 mL of 95% ethyl alcohol. Wash the crystals of product with this water/alcohol mixture, adding the liquid in two small portions under suction.

Wash the crystals with two small portions of 95% ethyl alcohol under suction. Allow the suction to continue until the crystals are dry.

Weigh the crystals of product and calculate the percentage yield. Place the crystals in a labeled sample pack or vial, and turn in to the instructor.

Name ______________________ Section ______________________

Lab Instructor ______________________ Date ______________________

EXPERIMENT 35 Coordination Compounds

PRE-LABORATORY QUESTIONS

CHOICE I. PREPARATION OF A COORDINATION COMPLEX OF COPPER

1. For the synthesis to be performed in this experiment, based on 1.00 g of copper(II) sulfate pentahydrate taken initially, what is the theoretical yield of product tetramminecopper(II) sulfate hydrate?

____

____

____

____

____

2. Using your textbook, write those properties of a molecule or ion that will make the species a good ligand for forming complexes with transition metal ions.

____

____

____

____

____

3. Briefly describe why so many coordination compounds of transition metals are brightly colored.

____

____

____

____

____

CHOICE II. PREPARATION OF A COORDINATION COMPLEX OF COBALT(III)

1. Based on 4.00 g of cobalt(II) chloride hexahydrate starting material, calculate the theoretical yield of the product, $[Co(NH_3)_6]Cl_3$.

2. It was mentioned in the introduction to this experiment that uncomplexed solutions of Co(III) are not stable, and that Co(III) will oxidize water. Write balanced half-reactions and the overall reaction equation for the oxidation of water by Co(III) ion. Calculate the standard potential for this process.

Name ______________________________ Section ______________________

Lab Instructor ______________________________ Date ______________________

EXPERIMENT 35 Coordination Compounds

CHOICE I. PREPARATION OF A COORDINATION COMPLEX OF COPPER

RESULTS/OBSERVATIONS

Weight of copper(II) sulfate pentahydrate taken ______________________________

Theoretical yield of product ______________________________

Weight of copper complex obtained ______________________________

% yield ______________________________

Effect of adding HCl to product ______________________________

Conclusion ______________________________

Balanced equation ______________________________

Observation on heating product ______________________________

Effect on pH paper ______________________________

Odor of vapor from heating product ______________________________

Balanced equation ______________________________

QUESTIONS

1. Suppose rather than reacting copper(II) with ammonia, you had reacted it with an ammonium salt, such as NH_4Cl. Would the same coordination complex have formed? Why or why not? What product would form?

____

____

____

____

2. In this experiment, you replaced water as a ligand (with ammonia) from the coordination sphere of the copper(II) ion. Use your textbook to find what ligands might be capable of replacing ammonia from the complex synthesized. What properties of a ligand make it a strong ligand?

3. Use your textbook or a chemical encyclopedia to describe the structure, shape, and bonding found in the $[Cu(NH_3)_4^{2+}]$ complex ion.

Name ______________________________ Section ______________________

Lab Instructor ______________________________ Date ______________________

EXPERIMENT 35 Coordination Compounds

CHOICE II. PREPARATION OF A COORDINATION COMPLEX OF COBALT(III)

RESULTS/OBSERVATIONS

Mass of $CoCl_2 \cdot 6H_2O$ taken ______________________________

Mass of NH_4Cl taken ______________________________

Color of solution ______________________________

Color on adding concentrated ammonia ______________________________

Color on oxidizing Co^{2+} ______________________________

Color and appearance of filtered product ______________________________

Mass of product ______________________ % of theory ______________________

QUESTIONS

1. Why does the presence of a ligand such as ammonia lower the tendency of Co(III) to be reduced?

____

____

____

____

____

____

____

____

____

2. What purpose did charcoal serve in the synthesis?

3. Why were the crystals of product washed with concentrated HCl, rather than with distilled water?

4. Use your textbook or a chemical encyclopedia to describe the structure, shape, and bonding in the $[Co(NH_3)_6^{3+}]$ complex ion.

EXPERIMENT

36 Inorganic Preparations

Objective

Synthetic chemistry is one of the most important and practical fields of science. The synthetic chemist designs laboratory procedures for the production of new and useful compounds from natural resources or other simpler synthetic substances. In the various choices of this experiment, you will use relatively simple starting reagents to synthesize a more complex and more useful substance. In this experiment you will prepare sodium thiosulfate pentahydrate (Choice I), copper(II) sulfate pentahydrate (Choice II), and sodium bicarbonate (Choice III).

Choice I. Preparation of Sodium Thiosulfate Pentahydrate

Introduction

Sodium thiosulfate pentahydrate, $Na_2S_2O_3 \cdot 5H_2O$, is a substance that finds several important uses, both in the chemistry laboratory and in commercial situations. The thiosulfate ion, $S_2O_3^{2-}$, has a Lewis structure that is comparable to that of the sulfate ion, SO_4^{2-}, with one of the oxygen atoms of sulfate replaced by an additional sulfur atom:

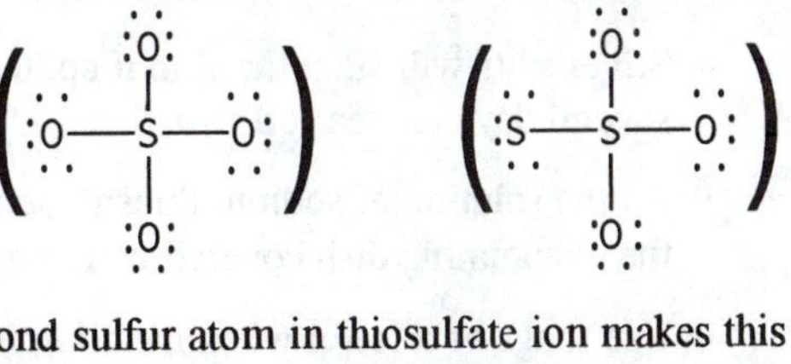

The presence of the second sulfur atom in thiosulfate ion makes this species a better reducing agent than sulfate ion. This ability of thiosulfate to serve as a reducing agent is at the basis of most of its common uses.

Standard solutions of sodium thiosulfate are used as the titrant in analyses of samples containing elemental iodine, I_2,

$$2S_2O_3^{2-} + I_2 \rightarrow S_4O_6^{2-} + 2I^-$$

Such titrations are monitored by watching for the disappearance of the brown color of elemental iodine as the equivalence point is approached. This is usually enhanced by addition of a small amount of starch near the endpoint of the titration, which results in the production of the characteristic blue/black color of the starch/iodine complex. Although not many real samples would be expected to contain elemental iodine itself, oftentimes iodine can be generated in a sample containing an oxidizing agent by addition of an excess amount of potassium iodide

$$2I^- \rightarrow I_2 + 2e^-$$

Naturally, the quantity of elemental iodine produced is directly related to the quantity of oxidizing agent present in the original sample. By titration of the elemental iodine thus produced with standard thiosulfate solution, the quantity of oxidizing agent in the original sample, which resulted in the production of iodine, may be determined.

Commercially, sodium thiosulfate pentahydrate is used in photography, where the substance is known more commonly as hypo. Monochrome photographic film contains a thin coating of silver bromide, AgBr, spread on plastic. Silver bromide is light sensitive: Silver ions are reduced to silver atoms when exposed to light through a camera's lens. The production of elemental silver on the film forms the negative image. When exposed photographic film is developed, any unreacted silver bromide must be removed from the film or it will also be reduced when the film is exposed to further light. Thiosulfate ion complexes and dissolves silver ions, but is not able to dissolve metallic silver

$$AgBr(s) + 2S_2O_3^{2-}(aq) \rightarrow Ag(S_2O_3)_2^{3-}(aq) + Br^-$$

By using sodium thiosulfate to fix the film, only the negative image (elemental silver) remains on the film. All AgBr that had not been reduced by light in forming the image is washed away.

Sodium thiosulfate may be prepared by reaction between aqueous sodium sulfite and elemental sulfur. Since elemental sulfur is not soluble in water, a small amount of detergent is added to the system to promote wetting of the sulfur.

Safety Precautions

- Wear safety glasses at all times while in the laboratory.
- Sodium sulfite is toxic and irritating to the skin. Sodium sulfite evolves toxic, noxious sulfur dioxide gas if acidified or heated.
- Elemental iodine will stain the skin if spilled.
- Silver salts will stain the skin if spilled. The stain is elemental silver and requires several days to wear off.
- As the solution of sodium thiosulfate is concentrated, spattering may occur. Keep the evaporating dish covered with a watchglass while heating.
- Use tongs or a towel when handling hot glassware.

Apparatus/Reagents Required

Sodium sulfite, powdered sulfur, detergent, 0.1 M I_2/KI, 0.1 M silver nitrate, 0.1 M sodium chloride, evaporating dish, crystallizing dish

Procedure

Record all data and observations directly in your notebook in ink.

A. Synthesis of Sodium Thiosulfate

Weigh 15 g of sodium sulfite and 10 g of powdered sulfur (each to the nearest 0.1 g) into a clean 400-mL beaker.

Add approximately 75 mL of water and 5 drops of concentrated laboratory detergent to the beaker. Stir to dissolve the sodium sulfite. Determine the pH of the mixture with pH paper.

Transfer the beaker to a ringstand, cover with a watchglass, and heat the mixture to a *gentle* boil for approximately 30 minutes. The presence of the detergent makes this mixture tend to bubble and foam excessively. Keep the heat as low as possible while still maintaining boiling.

Stir the mixture frequently during the heating period to promote the mixing of the powdered sulfur. During the heating, watch for a subtle change in the appearance and color of the powdered sulfur: the yellow color of the sulfur becomes somewhat lighter as the reaction proceeds.

After boiling for 30 minutes, check the pH of the solution with pH paper. The pH of the solution should have dropped to nearly neutral. If the pH has not dropped to at least pH 8, continue heating for an additional 15 minutes. When the pH has reached the nearly neutral point, remove the heat.

Allow the mixture to cool to the point where it can be handled easily. Filter the mixture through a gravity funnel to remove unreacted sulfur. Collect the filtrate in a clean beaker.

Using the beaker calibration marks, note the approximate volume of the solution at this point.

Transfer the filtrate to an evaporating dish. Cover the evaporating dish with a watchglass, and heat the solution on a ringstand using a very small flame. Do not rapidly boil the solution, or a portion of the product may be lost to spattering.

Continue heating with a small flame until the volume of the solution has been reduced by approximately 50%.

Protect your hands with a towel while transferring the hot contents of the evaporating dish to a clean, dry crystallizing dish. Allow the solution to cool to room temperature. Scratch the bottom and side of the crystallizing dish during this cooling period to assist nucleation of crystal formation.

If no crystals have formed by the time the solution has reached room temperature, return the solution to the evaporating dish, and reduce its volume by an additional 25%. Cool the solution again to room temperature to allow crystallization.

Filter the crystals through a Büchner funnel under suction. Allow the suction to continue for several minutes to promote drying of the crystals.

Remove a small portion of the product for the tests indicated. Allow the remainder of the product to dry in the air for approximately 1 hour. Weigh the product and calculate the percent yield.

B. Tests on Sodium Thiosulfate

Place 5 mL of distilled water in a clean test tube. Add 5–6 drops of 0.1 *M* I_2/KI reagent, followed by a small amount of your product. Describe what happens, and write an equation for the reaction.

Obtain 2–3 mL of 0.1 *M* NaBr in a clean test tube. Add 3–4 drops of 0.1 *M* silver nitrate. Record your observations. Add a small amount of your product and shake the test tube. Record your observations and write balanced equations for the reactions.

Choice II. Preparation of Copper(II) Sulfate Pentahydrate

Introduction

Perhaps the most common copper compound is copper(II) sulfate pentahydrate, $CuSO_4{\cdot}5H_2O$. Of the five water molecules, four are bonded coordinately to the copper(II) ion. Electronic transitions associated with these ligands are responsible for the striking bright blue color of the hydrated salt (the anhydrous salt is white). Copper sulfate, known commercially as blue vitriol, is used as an antibacterial and antifungal agent in agriculture. Copper(II) sulfate pentahydrate will be prepared by dissolving copper(II) oxide in dilute sulfuric acid

$$CuO + H_2SO_4 + 4H_2O \rightarrow CuSO_4{\cdot}5H_2O$$

This synthesis mimics the industrial process. Most copper ores are sulfides, which can be converted to the oxide by roasting.

Some experiments demonstrating various crystallization techniques and their effect on the size and shape of the crystals obtained will be performed. The crystallization techniques differ primarily in how concentrated the solution is when crystallized and in how quickly the crystallization is carried out. If a saturated or supersaturated solution is cooled suddenly, generally very small, fine crystals are obtained. On the other hand, if a relatively dilute solution is allowed to concentrate slowly (for example, by slow evaporation of the solvent), very large, strikingly shaped crystals may be obtained.

Safety Precautions

- Wear safety glasses at all times while in the laboratory.
- Copper salts are toxic if ingested and may be irritating to the skin. Wash after use.
- Sulfuric acid can severely burn the skin. Wash immediately if the acid is spilled.
- Use beaker tongs or a towel to handle the hot beaker of copper sulfate solution during the filtration.
- Stir the mixture of copper oxide and sulfuric acid during the heating period to prevent its boiling over.

Apparatus/Reagents Required

Copper(II) oxide, 6 *M* sulfuric acid, evaporating dish, crystallizing dish, watchglass, plastic wrap

Procedure

Record all data and observations directly in your notebook in ink.

A. Preparation of Copper(II) Sulfate

Weigh out 4 g of copper(II) oxide (to the nearest 0.1 g) and transfer to a clean 400-mL beaker. Record the appearance of the solid. Add 35–40 mL of 6 *M* sulfuric acid (*Caution!*) to the beaker.

Transfer the beaker to a ringstand, and heat the mixture to barely boiling. The mixture has a tendency to boil over: Stir the mixture to prevent this and use as small a flame as possible to barely maintain boiling. Record the color of the solution as the mixture is heated.

As the copper(II) oxide is heated, it will gradually dissolve in the sulfuric acid solution. A small quantity of metallic copper, which will float on the surface, may be produced. Eventually most of this metallic copper will dissolve in the acid.

Continue heating the mixture until all the copper(II) oxide has dissolved and the mixture is homogeneous.

Set up a gravity funnel/filter paper, and have ready a clean 250-mL beaker for catching the filtrate.

Using tongs or a towel to protect your hands, filter the hot solution into the clean 250-mL beaker.

B. Crystallization Techniques

Divide the filtrate of copper(II) sulfate solution into three portions of approximately the same size. The copper sulfate will be crystallized by three different methods, which will lead to very different sizes and shapes for the crystals obtained.

Transfer one portion of the copper(II) sulfate solution to an evaporating dish. Place the evaporating dish over a burner that has been adjusted to a very small flame. Cover the dish with a watchglass and heat the solution very gently to reduce its volume until crystals begin to form in the solution.

Remove the evaporating dish from the heat, allow the mixture to cool, and filter the crystals under suction on a Büchner funnel. Allow air to pass through the crystals for several minutes to aid in drying them.

Transfer the crystals to a clean dry watchglass, and allow them to dry in the air. Examine the crystals with a magnifying glass and sketch their general shape.

Place the dried crystals in a sample vial or pack, label with your name, and turn in to your instructor.

Transfer a second portion of the copper sulfate solution to a small beaker and heat briefly on a ringstand to redissolve any crystals present.

Using tongs or a towel to protect your hands, pour a small portion of the hot solution onto a watchglass. Allow the solution on the watchglass to cool. Examine the crystals that form with a magnifying glass and sketch their general shape.

To the third portion of the copper sulfate solution, add a volume of distilled water approximately equivalent to 50% of the volume of the copper sulfate solution.

Transfer the diluted solution to a labeled crystallizing dish, cover the dish loosely with plastic wrap, and allow it to stand undisturbed until the next laboratory period.

Examine the crystals produced with a magnifying glass and sketch their shape.

Filter the crystals under suction on a Büchner funnel, allowing the suction to continue for several minutes to help dry the crystals. Transfer the dry crystals to a sample vial labeled with your name, and turn in the vial to the instructor.

Choice III. Preparation of Sodium Hydrogen Carbonate

Introduction

Sodium hydrogen carbonate (sodium bicarbonate, $NaHCO_3$) is more commonly known as baking soda and finds many uses, both in the home and in many commercial applications. Sodium hydrogen carbonate, when combined with certain solid acidic substances, is used as a leavening agent in the form of commercial baking powders. When baking powder is mixed with water or milk (in batter, for example), carbon dioxide gas is released

$$HCO_3^- + H^+ \rightarrow H_2O + CO_2$$

The release of carbon dioxide causes the batter to rise and become filled with air bubbles, adding to the texture of the baked good.

Sodium hydrogen carbonate finds several other uses in the home. Since it is weakly basic, baking soda is used in some over-the-counter antacid tablets (bicarb). Finely powdered baking soda can adsorb many molecules on the surface of its crystals and reacts chemically with vapors of acidic substances. For this reason, it is sold as a deodorizer for the home.

The synthesis of sodium hydrogen carbonate in this experiment duplicates some of the steps of the industrial synthesis. Dry ice (solid CO_2) is added to a saturated solution of sodium chloride in concentrated aqueous ammonia

$$NaCl + NH_3 + CO_2 + H_2O \rightarrow NH_4Cl + NaHCO_3$$

In addition to serving as a source of carbon dioxide, dry ice also provides a means for separating the two products of the reaction. Dry ice provides a very low temperature for the reaction. At low temperatures, $NaHCO_3$ is much less soluble in water than is NH_4Cl and crystallizes from the solution much more readily. By filtering the reaction mixture while it is still very cold, the sodium hydrogen carbonate product can be separated.

The yield of sodium hydrogen carbonate from this synthesis is typically much less than the theoretical amount expected (less than 50%). First, although $NaHCO_3$ is less soluble than NH_4Cl at low temperatures, a fair amount of $NaHCO_3$ does remain in solution. Second, a side reaction involving the same reagents is possible, in which significant amounts of ammonium hydrogen carbonate, NH_4HCO_3, are produced. The solid product isolated from the system is typically a mixture of sodium and ammonium hydrogen carbonates. Once the solid has been filtered, however, ammonium hydrogen carbonate can be removed by simple heating of the solid over a steam bath

$$NH_4HCO_3 \rightarrow NH_3 + H_2O + CO_2$$

The products of this reaction are all gases, which are evolved from the solid mixture, leaving pure sodium hydrogen carbonate.

Safety Precautions

- Wear safety glasses at all times while in the laboratory.
- Ammonia is a severe respiratory irritant and a strong cardiac stimulant. Confine all use of ammonia solutions to the exhaust hood.
- Dry ice will cause severe frostbite to the skin if handled with the fingers. Use tongs, a scoop, or insulated gloves to pick up dry ice.
- The mixture of sodium/ammonium hydrogen carbonates will spatter when heated. Use caution.
- Hydrochloric acid is damaging to the skin. Wash if the acid is spilled.

Apparatus/Reagents Required

Sodium chloride, concentrated ammonia solution, crushed dry ice, tongs, 3 *M* hydrochloric acid

Procedure

Record all data and observations directly in your notebook in ink.

Set up a suction filtration apparatus (suction flask, Büchner funnel, filter paper) for use in filtering the product. The product of the reaction must be filtered before it has a chance to warm up too much.

Clean a 250-mL Erlenmeyer flask, and fit with a solid stopper.

Weigh out 15–16 g of NaCl and transfer to the Erlenmeyer flask. This quantity of NaCl is more than is actually needed for the synthesis, but the excess amount will speed up the saturation of the ammoniacal solution.

In the exhaust hood, add 50 mL of concentrated aqueous ammonia solution to the Erlenmeyer flask containing NaCl. Stopper the flask, but keep the flask *in the hood.*

Swirl and shake the flask in the hood for 10–15 minutes to saturate the ammonia solution with NaCl. Do not "cheat" on the time in this step. If the solution is not saturated, no product will be obtained.

After the ammonia solution has been saturated with NaCl, some undissolved NaCl will remain.

Allow the solid to settle, then decant the clear saturated ammonia/NaCl solution into a clean 250-mL beaker, avoiding the transfer of the solid as much as possible. Keep the beaker of solution *in the exhaust hood.*

Using tongs or a plastic scoop to protect your hands from the cold, fill a 150-mL beaker approximately half full of crushed dry ice. This represents 60–70 g of CO_2.

In the hood, over a 15-minute period, add the dry ice in small portions to the ammonia/NaCl solution. Stir the mixture with a glass rod throughout the addition of dry ice. Do not add the dry ice so quickly that the solution freezes. The temperature of the system at this point is 50–60 degrees below 0°C and is beyond the range of the thermometers in your locker.

As the last portions of dry ice are added, a solid will begin to form in the mixture ($NaHCO_3/NH_4HCO_3$). Allow the mixture to stand until the temperature has risen to approximately –5°C and all dry ice has dissolved.

Quickly filter the mixture using suction before the temperature has a chance to rise much above 0°C. Allow suction to continue for several minutes to remove liquid from the solid.

Weigh a clean, dry watchglass to the nearest 0.1 g, and transfer the solid mixture to the watch glass.

Place the watchglass over a beaker of boiling water and heat for 10–15 minutes to decompose ammonium hydrogen carbonate. As the solid on the watchglass is heated, it will begin to melt and bubble as gases are evolved. Beware of spattering.

After the heating period, allow the watchglass and solid to cool; then reweigh the watchglass.

Determine the yield of $NaHCO_3$ in grams. Assuming that approximately 10 g of the initial 15–16 grams of NaCl actually dissolved in the ammonia solution, calculate an approximate percent yield for the synthesis.

Since the sodium hydrogen carbonate product and the sodium chloride starting material are both white powders, perform the following simple test to confirm that the product of the reaction is not merely recrystallized sodium chloride.

Add a small portion of your product to a clean dry test tube and add a small portion of sodium chloride to a second test tube.

Add 5–6 drops of 3 *M* HCl to each test tube and record your observations. Write an equation for the reaction.

Name ______________________ Section ______________

Lab Instructor ______________________ Date ______________

EXPERIMENT 36 Inorganic Preparations

PRE-LABORATORY QUESTIONS

CHOICE I. PREPARATION OF SODIUM THIOSULFATE PENTAHYDRATE

1. Write a balanced chemical equation for the synthesis of sodium thiosulfate pentahydrate, using aqueous sodium sulfite and elemental sulfur as reactants.

____

2. If 15 g of sodium sulfite is reacted with 10 g of powdered sulfur, which substance is present in excess? By how many grams is this substance in excess? Show your reasoning.

____

____

____

____

____

____

3. Write balanced chemical equations for the reaction of thiosulfate ion with each of the following:

Elemental iodine, I_2 ____

Silver bromide, AgBr ____

CHOICE II. PREPARATION OF COPPER(II) SULFATE PENTAHYDRATE

1. In the introduction to this experiment, it is mentioned that four of the five water molecules in copper(II) sulfate pentahydrate are coordinated to the copper(II) ion. Use your textbook or an encyclopedia to determine and describe how the fifth water molecule is bound in this substance.

____

____

____

2. If 10.0 g of copper(II) oxide were treated with an excess amount of aqueous sulfuric acid, what quantity of copper(II) sulfate pentahydrate should be collected? Show your reasoning.

____

____

____

____

____

____

CHOICE III. PREPARATION OF SODIUM HYDROGEN CARBONATE

1. The desired product of this synthesis, sodium hydrogen carbonate, is separated from the other major product, ammonium chloride, on the basis of its solubility at low temperatures. However, it is not possible to separate ammonium hydrogen carbonate from sodium hydrogen carbonate on this basis. Use a handbook of chemical data to find and write information about the solubilities of these three substances.

Solubility	At 0°C	At 20°C
$NaHCO_3$	____________	____________
NH_4Cl	____________	____________
NH_4HCO_3	____________	____________

2. List some commercial/home uses of sodium hydrogen carbonate.

____

____

____

____

____

____

____

____

____

____

Name ______________________ Section ______________

Lab Instructor ______________________ Date ______________

EXPERIMENT 36 Inorganic Preparations

CHOICE I. PREPARATION OF SODIUM THIOSULFATE PENTAHYDRATE

RESULTS/OBSERVATIONS

Mass of sodium sulfite taken ______________________

Mass of sulfur taken ______________________

pH of initial reaction mixture ______________________

Observation on heating mixture

____

pH of mixture after heating ______________________

Explanation of pH change

____

Theoretical yield of product ______________________

Actual yield of product______________ % yield______________

Observation on treating I_2/KI solution with product

____

Observation on mixing NaCl and $AgNO_3$

____

Equation for reaction ______________________

Observation on adding product to NaCl/$AgNO_3$ mixture

____

QUESTIONS

1. How does detergent help sulfur to dissolve better in the sodium sulfite solution?

2. Why is the initial solution of sodium sulfite basic?

3. Sodium thiosulfate pentahydrate is a deliquescent substance. Use a chemical dictionary or encyclopedia to find and write the meaning of deliquescent.

4. Sodium thiosulfate is the salt of a weak acid ($H_2S_2O_3$). What would happen to a solution of sodium thiosulfate if a quantity of strong acid (for example, HCl) were added?

Name ______________________________ Section ______________

Lab Instructor ______________________________ Date ______________

EXPERIMENT 36 Inorganic Preparations

CHOICE II. PREPARATION OF COPPER(II) SULFATE PENTAHYDRATE

RESULTS/OBSERVATIONS

Mass of copper(II) oxide taken ______________________________

Appearance of CuO ______________________________

Observation on dissolving CuO in sulfuric acid

____

Color of CuO/H_2SO_4 solution ______________________________

Appearance of crystals obtained by concentrating the solution

____

____

Appearance of crystals obtained by rapidly cooling the solution on a watchglass

____

____

Appearance of crystals obtained by slow crystallization from diluted solution

____

____

QUESTIONS

1. Copper(II) sulfate may also be prepared by dissolving metallic copper in concentrated sulfuric acid. Write a balanced equation for this reaction. Why do you suppose this alternative synthesis was not used in the laboratory?

2. Describe/explain the great variation in size and shape of the crystals obtained by the various crystallization methods.

Name ______________________ Section ______________________

Lab Instructor ______________________ Date ______________________

EXPERIMENT 36 Inorganic Preparations

CHOICE III. PREPARATION OF SODIUM HYDROGEN CARBONATE

RESULTS/OBSERVATIONS

Observation on saturating ammonia with NaCl

____

Observation on adding dry ice to NaCl/NH_3 solution

____

Approximate temperature at which crystals of product appeared ______________________

Mass of empty watchglass ______________________

Observation on heating product crystals on watchglass

____

Mass of watchglass with $NaHCO_3$ ______________________

Theoretical yield of $NaHCO_3$, based on 10 g of NaCl taken ______________________

Actual yield of $NaHCO_3$ ______________________ % yield ______________________

Observation on treating product with HCl

____

Equation for reaction ______________________

Observation on treating NaCl with HCl

____

QUESTIONS

1. $NaHCO_3$ can also be produced by bubbling CO_2 gas through concentrated sodium hydroxide solutions. Write an equation for this method of synthesis.

2. One of the most important industrial chemical substances is sodium carbonate, Na_2CO_3. This substance can be produced by strong heating of sodium hydrogen carbonate, $NaHCO_3$. Write an equation for this process.

3. The hydrogen carbonate ion is important physiologically as a component of one of the major buffer systems that keep the pH of the human body constant. Use your textbook to find the other component of this buffer system, and describe how the buffer maintains the pH constant.

EXPERIMENT

37 Qualitative Analysis of Organic Compounds

Objective

The elements present in a typical covalent organic or biological compound can be determined by sodium fusion of the compound, followed by qualitative analysis for the ions produced by the fusion.

Introduction

When a new compound is prepared or isolated from some natural source, the techniques of **qualitative analysis** are frequently applied as a first step in identifying the compound. However, the traditional techniques of qualitative analysis generally apply only to inorganic ions in aqueous solution.

When a new compound does not consist of ions, but rather consists of covalently bonded molecules, frequently the molecule can be broken up into ions by controlled decomposition of the compound. The reagent(s) used to cause the decomposition of the original unknown compound are chosen in such a way that the ions produced by the decomposition will reflect clearly those elements that were present initially in the original covalently bonded unknown. In this experiment, a small amount of an unknown organic compound will be dropped into a hot vapor of elemental sodium. This technique is called **sodium fusion**.

The organic compound to be analyzed consists basically of a chain of carbon atoms to which various other atoms are attached. Sodium is an extremely strong reducing agent, that will cause the breakup of the organic compound's carbon atom chain. It also will convert those other atoms that are covalently bonded to the carbon chain to inorganic ions. Once the covalent compound has been fused with sodium, the ionic mixture produced can then be dissolved in water and tested.

Some elements that occur commonly in organic compounds are nitrogen, sulfur, and the halogens (chlorine, bromine, and iodine). When an organic compound undergoes a sodium fusion, the carbon present in the compound is reduced partially to elemental carbon, which is usually quite visible on the walls of the test tube used for the fusion. Any nitrogen present in the original compound is converted by the fusion to the cyanide ion, CN^-. Any sulfur present is converted to the sulfide ion, S^{2-}, and any halogens present are converted to the halide ions, Cl^-, Br^-, and I^-.

The test for cyanide ion involves forming a precipitate of an iron/cyanide complex of characteristic dark blue color (Prussian blue). The test for nitrogen is very sensitive to conditions, so follow the directions given exactly. Sometimes, if the sodium is not heated to a high enough temperature during the fusion, very little cyanide ion is produced. In this case, the dark blue precipitate may not form, but a small amount of a greenish solid may result. The green solid is sometimes taken as a weakly positive test for nitrogen. It is usually helpful to repeat the sodium fusion to confirm the presence of nitrogen if only the greenish solid is obtained.

The presence of sulfide ion from the original organic compound can be confirmed by several techniques. If the solution prepared from the sodium fusion products is acidified, hydrogen sulfide gas, H_2S, will be generated. This may be noted by its reaction with lead(II) ions impregnated on test paper, causing the test paper to darken as black PbS forms.

$$\underset{\textit{colorless}}{Pb^{2+} + S^{2-}} \rightarrow \underset{\textit{black}}{PbS(s)}$$

On the other hand, a solution that contains sulfide ions may be treated with sodium nitroprusside test reagent, in which case a purple/violet color will result.

The presence of halide ions from the original organic compound can be confirmed by treating the aqueous solution of the fusion products with silver ion reagent. Halide ions precipitate with silver ion.

$$\underset{\textit{colorless}}{Ag^{+} + Cl^{-}} \rightarrow \underset{\textit{white}}{AgCl(s)}$$

$$\underset{\textit{colorless}}{Ag^{+} + (Br^{-}, I^{-})} \rightarrow \underset{\textit{yellow}}{AgBr(s), AgI(s)}$$

If more than one halide ion was present in the original compound, the mixed precipitate of silver halides can be analyzed for each of the possible halogens by the usual techniques. Before adding silver ions to the fusion products, however, it may be necessary to treat the sample to remove cyanide ions and sulfide ions (if they were present). Otherwise, these ions also will precipitate with the silver ions, masking the presence of the halogens.

Apparatus/Reagents Required

Clean, dry Pyrex test tube containing a single small pellet of elemental sodium (available at the stockroom), unknown organic compound, methyl alcohol, 0.5 *M* iron(II) sulfate (ferrous sulfate, 0.5 *M* iron(III) chloride (ferric chloride), 3 *M* potassium hydroxide, 3 *M* sulfuric acid, 3 *M* acetic acid, 3 *M* nitric acid, sodium nitroprusside reagent, lead acetate test paper, 0.1 *M* silver nitrate, 3 *M* aqueous ammonia, 0.1 *M* iron(III) nitrate (ferric nitrate), 0.1 *M* potassium permanganate, methylene chloride

Safety Precautions

- Wear safety glasses at all times in the laboratory.
- *Caution!* Elemental sodium is dangerous! It reacts with water, liberating hydrogen gas and enough heat to ignite the hydrogen. Do not at any time add water to the sodium fusion products unless it has been established with certainty that all elemental sodium has been destroyed. Do not touch the piece of sodium provided.
- Occasionally during the sodium fusion, an amount of noxious vapor is released by the reaction test tube. For this reason, *perform the fusion of the unknown in the exhaust hood.*
- When heating liquids in a test tube, use a small flame, hold the test tube at a 45° angle, and move the test tube quickly through the flame. Only a few seconds are required to heat a liquid in a test tube.
- Hydrogen cyanide and hydrogen sulfide are both extremely toxic. Perform those portions of the procedure that generate these gases *in the exhaust hood.*
- Methyl alcohol is poisonous and its vapor is toxic.
- Dilute sulfuric, nitric, and acetic acids are used in the procedure. These acids can be dangerous to skin and clothing if they concentrate by evaporation of water. Wash immediately if these are spilled and inform the instructor.
- Potassium hydroxide is caustic and can burn the skin. Wash immediately if the solution is spilled and inform the instructor.
- Lead compounds are toxic if ingested. Wash after use.
- Silver nitrate solution will stain the skin if spilled. Nitrates are strong oxidizing agents and are toxic.
- Potassium permanganate reagent is a strong oxidizing agent and will stain the skin and clothing if spilled. Wash after use.

Procedure

Record all data and observations directly in your notebook in ink.

Because of the danger of exposing sodium to water, single sodium pellets have been provided for you in clean, dry Pyrex test tubes. Obtain one of these, *take it to the exhaust hood*, and clamp it vertically on a ringstand above a Bunsen burner so that the sodium pellet can be heated quickly and strongly. It is essential that the sodium be *quickly* melted and vaporized before adding the organic compound (to prevent oxidation of the sodium by the air).

Obtain an unknown organic compound and record its identification number. Transfer an amount of the unknown about the size of a pea to the tip of a knife or spatula.

Pull the safety shield of the hood down partway to protect yourself during the sodium fusion.

Light the burner and adjust the flame so that it is as hot as possible. Heat the sodium pellet until it melts and then vaporizes (sodium vapor is dark gray or purple). When the sodium vapor has risen about half an inch in the test tube, carefully add the unknown sample from the spatula or knife. A vigorous reaction will take place, and the test tube should darken as the carbon of the organic compound is deposited on the walls of the test tube.

Continue heating the test tube for 15–20 seconds, then remove the heat and allow the test tube to cool completely to room temperature.

Ask your instructor to examine your sodium fusion to ensure that sufficient destruction of the unknown has taken place.

Before water can be added to the sodium fusion test tube, any excess sodium present must be destroyed. Add about 5 mL of methyl alcohol to the test tube and *allow the test tube to stand in the hood for at least 5 minutes.* Methyl alcohol reacts with sodium, but at a much slower speed than does water.

When it is certain that all sodium has been destroyed, add 15 mL of distilled water to the test tube.

In the exhaust hood, set up a 600-mL beaker half full of water and begin heating for use as a boiling water bath. Place the sodium fusion test tube in the boiling water bath and heat for 5–10 minutes to assist the dissolving of the inorganic ions produced by the fusion.

While the test tube is heating, scrape down the walls of the tube to assist the dissolving. Remember that the black deposit on the test tube is carbon, which does not dissolve in water.

Filter the solution from the sodium fusion through a gravity funnel/filter paper to remove carbon, and divide the filtrate into five portions in separate clean test tubes.

Test for Nitrogen as Cyanide Ion

To one of the five portions of unknown solution filtrate, add 5–6 drops of 0.5 *M* ferrous sulfate solution and 5–6 drops of 3 *M* potassium hydroxide. Stir to mix. A precipitate will form.

Heat the test tube in the boiling water bath for 5 minutes. Cool the solution briefly and stir in 2–3 drops of 0.5 *M* ferric chloride solution.

Dissolve the reddish-brown precipitate of iron hydroxides by adding 3 *M* sulfuric acid dropwise with stirring until the brown precipitate barely dissolves.

Cool the test tube in an ice bath, and then filter the solution through filter paper. If nitrogen was present in the unknown compound tested, a dark blue precipitate of Prussian blue will remain on the filter paper.

A green precipitate remaining on the filter paper is a weakly positive test for nitrogen. To be certain that nitrogen is present, repeat the sodium fusion on an additional sample of the unknown. A brown precipitate on the filter paper indicates that no nitrogen was present in the unknown.

Tests for Sulfur as Sulfide Ion

Two separate tests for sulfur will be performed.

To a second portion of the filtrate from the sodium fusion, add a few drops of sodium nitroprusside reagent. The appearance of a violet/purple color indicates sulfur is present, but this color may not persist for more than a few seconds.

To a third portion of the filtrate from the sodium fusion, add acetic acid dropwise until the solution is acidic to pH paper.

Transfer the test tube to the boiling water bath in the hood. Moisten a piece of lead acetate test paper with a drop of distilled water, and hold the test paper over the open mouth of the test tube heating in the water bath. If hydrogen sulfide is being evolved from the sample as it is being heated, the lead acetate test paper will darken as lead sulfide forms on the paper.

Test for Halogens as Halide Ions

Before the filtrate from the sodium fusion can be tested for the presence of halide ions, any cyanide ion (nitrogen) or sulfide ion (sulfur) must be removed. Transfer one of the filtrate samples to a small beaker, add 10–15 mL of distilled water, and add 3 *M* acetic acid until the solution is acidic to pH paper.

Transfer the beaker to the exhaust hood, and boil the solution for 5 minutes. If the solution threatens to boil away completely during this heating period, add small portions of distilled water.

While the solution is being boiled, test the vapors coming from the beaker with moist lead acetate paper. When the lead acetate paper no longer darkens, you can assume that both cyanide and sulfide ions have been expelled from the solution. Allow the solution to cool to room temperature.

Add a few drops of 0.1 *M* silver nitrate to the solution. If a precipitate of silver halide forms, at least one of the halogens is present in the unknown. Filter the precipitate, and wash it with several drops of distilled water.

Transfer the precipitate to a clean test tube, and add 10 drops of 0.1 *M* silver nitrate, 6–7 drops of 3 *M* ammonia, and 2–3 mL of distilled water.

If the white precipitate *completely* dissolves in the silver nitrate/ammonia solution, chloride is present and bromide and iodide are absent.

If the precipitate does not dissolve, or appears to only *partially* dissolve, then bromide, iodide, and chloride ions may *all* be present.

If the precipitate did not dissolve completely, filter the precipitate, collecting the clear filtrate. Add 3 *M* nitric acid dropwise to the clear filtrate until it is acidic to pH paper.

If a precipitate (AgCl) reappears when the solution is acidified, then chloride is present in the unknown.

Take the remaining portion of the *original* filtered solution from the sodium fusion and transfer it to a small beaker. Add 10–15 mL of distilled water and sufficient 3 *M* acetic acid to make the solution acidic to pH paper.

Transfer the beaker to the exhaust hood and boil the solution for 5 minutes. If the solution threatens to boil away completely during this heating period, add small portions of distilled water to the beaker.

Test the vapors coming from the solution as it boils with moist lead acetate paper to make certain that all cyanide and sulfide ion have been boiled away.

Transfer the solution to a small test tube. Add 10 drops of 3 *M* nitric acid to the solution, followed by 2–3 mL of 0.1 *M* ferric nitrate solution. Add 10 drops of methylene chloride, stopper, and shake. Ferric nitrate is capable of oxidizing iodide ion but not bromide ion.

The appearance of a violet color in the methylene chloride layer at this point indicates the presence of iodine in the original organic sample. Save this sample for determining if bromide ion is present.

Use a disposable pipet and bulb to remove the methylene chloride layer from the test tube used in testing for iodide ion.

Add another 1 mL of ferric nitrate, and 10 drops of methylene chloride. Stopper and shake the test tube. If the methylene chloride layer still shows the violet color of iodine, remove the methylene chloride layer with the disposable pipet, and add another 10 drops of methylene chloride.

Repeat the addition and pipet removal of methylene chloride until no violet color is present (indicating the removal of all iodine from the sample).

When all iodine has been removed, add 5 drops of 3 *M* nitric acid. Then add 0.1 *M* potassium permanganate until the sample retains the pink color of the permanganate. The $KMnO_4$ will oxidize bromide ions to elemental bromine. Add 10 drops of methylene chloride, stopper, and shake. A yellow/orange color in the methylene chloride indicates the presence of bromine in the original organic compound.

Name ______________________ Section ______________________

Lab Instructor ______________________ Date ______________________

EXPERIMENT 37 Qualitative Analysis of Organic Compounds

PRE-LABORATORY QUESTIONS

1. This experiment is concerned with detecting what elements are present in an organic compound. Ultimately for new compounds, the molecular formula of the compound must be determined, followed by a determination of the structure of the compound. What information about a compound (beyond what elements are present) is needed to determine the molecular formula and structure?

2. Sodium nitroprusside is used as a test for the presence of sulfide ion. What is the formula of sodium nitroprusside, and by what other name is this reagent known?

3. Use your textbook or a handbook to find the composition of Prussian blue as produced in the determination of nitrogen.

Name ______________________________ Section ______________

Lab Instructor ______________________________ Date ______________

EXPERIMENT 37 Qualitative Analysis of Organic Compounds

RESULTS/OBSERVATIONS

Identification number of unknown ______________________________

Observation on adding unknown to hot sodium vapor

____

Observation on adding methyl alcohol to sodium fusion products

____

Results/observations on test for nitrogen

____

Results/observations on sodium nitroprusside test for sulfide

____

Results/observations on lead acetate paper test for sulfide

____

Result on adding silver nitrate to fusion product solution

____

Results/observations on adding ammonia followed by nitric acid

____

Results/observations on adding ferric nitrate/methylene chloride

____

Results/observations on adding $KMnO_4$/methylene chloride

____

List those elements present in your unknown: ______________________________

QUESTIONS

1. What errors might have been introduced if the sodium vapor had not been heated to a hot enough temperature before adding the sample to be analyzed?

2. Why was the lead acetate test paper *moistened* before using it to test for hydrogen sulfide gas being evolved from the boiling acidified solution?

3. Why is it necessary to remove sulfide and cyanide ion from the sodium fusion products before testing for the presence of halide ions? How specifically would sulfide and cyanide ions interfere with the halide ion tests?

4. Describe how the three halide ions (Cl^-, Br^-, I^-) were separated from one another and identified.

EXPERIMENT

38 Organic Chemical Compounds

Objective

There are millions of known carbon compounds. In this experiment, several of the major families of carbon compounds will be investigated, which will demonstrate some of the characteristic properties and reactions of each family.

Introduction

Carbon atoms have the correct electronic configuration and size to enable long chains or rings of carbon atoms to be built up. This ability to form long chains is unique among the elements, and is called **catenation**. There are known molecules that contain thousands of carbon atoms, all of which are attached by covalent bonds among the carbon atoms. No other element is able to construct such large molecules with its own kind.

Hydrocarbons

The simplest types of carbon compounds are known as **hydrocarbons**. Such molecules contain only carbon atoms and hydrogen atoms. There are various sorts of molecules among the hydrocarbons, differing in the arrangement of the carbon atoms in the molecule, or in the bonding between the carbon atoms. Hydrocarbons consisting of chainlike arrangements of carbon atoms are called **aliphatic hydrocarbons**, whereas molecules containing the carbon atoms in a closed ring shape are called **cyclic hydrocarbons**. For example, the molecules hexane and cyclohexane represent aliphatic and cyclic molecules, each with six carbon atoms:

$$CH_3-CH_2-CH_2-CH_2-CH_2-CH_3$$

hexane

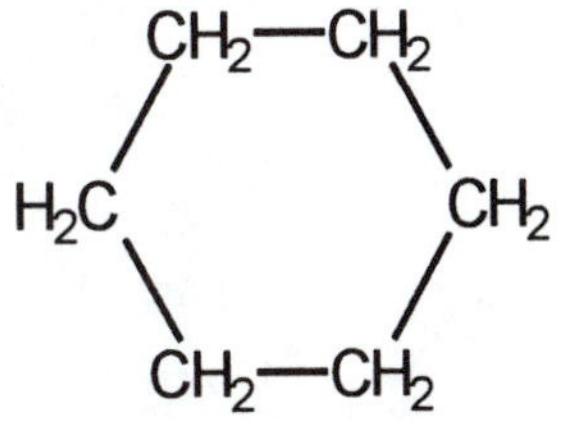

```
      CH2—CH2
     /       \
  H2C         CH2
     \       /
      CH2—CH2
```

cyclohexane

Notice that all the bonds between carbon in hexane and cyclohexane are single bonds. Molecules containing only singly bonded carbon atoms are said to be **saturated**. Other molecules are known in which some of the carbon atoms may be attached by double or even triple bonds and are said to be **unsaturated**. The following are representations of the molecules 1-hexene and cyclohexene, which contain double bonds:

$$CH_2{=}CH{-}CH_2{-}CH_2{-}CH_2{-}CH_3$$

1-hexene

```
      CH=CH
     /     \
  H2C       CH2
     \     /
      CH2—CH2
```

cyclohexene

Hydrocarbons with only single bonds are not very reactive. One reaction all hydrocarbons undergo, however, is combustion. Hydrocarbons react with oxygen, releasing heat and/or light. For example, the simplest hydrocarbon is methane, CH_4, which constitutes the major portion of natural gas. Methane burns in air, producing carbon dioxide, water vapor, and heat:

$$CH_4 + 2O_2 \rightarrow CO_2 + 2H_2O$$

Single-bonded hydrocarbons also will react slowly with the halogen elements (chlorine and bromine). A halogen atom replaces one (or more) of the hydrogen atoms of the hydrocarbon. For example, if methane is treated with bromine, bromomethane is produced, with concurrent release of hydrogen bromide:

$$CH_4 + Br_2 \rightarrow CH_3Br + HBr$$

The hydrogen bromide produced is generally visible as a fog evolving from the reaction container, produced as the HBr mixes with water vapor in the atmosphere.

Molecules with double and triple bonds are much more reactive than single-bonded compounds. Unsaturated hydrocarbons undergo reactions at the site of the double (or triple) bond, in which the fragments of some outside reagent attach themselves to the carbon atoms that had been involved in the double (triple) bond. Such reactions are called **addition reactions**. The orbitals of the carbon atoms that had been used for the extra bond instead are used for attaching to the fragments of the new reagent. For example, the double-bonded compound ethylene (ethene), $CH_2{=}CH_2$, will react with many reagents

$$CH_2{=}CH_2 + H_2 \rightarrow CH_3{-}CH_3$$

$$CH_2{=}CH_2 + HBr \rightarrow CH_3{-}CH_2Br$$

$$CH_2{=}CH_2 + H_2O \rightarrow CH_3{-}CH_2OH$$

$$CH_2{=}CH_2 + Br_2 \rightarrow CH_2Br{-}CH_2Br$$

Notice that in all of these reactions of ethylene, the product is saturated and no longer contains a double bond. Many of the reactions of ethylene indicated only take place in the presence of a catalyst, or under reaction conditions that we will not be able to investigate in this experiment.

Cyclic compounds containing only single or double bonds undergo basically the same reactions indicated earlier for aliphatic hydrocarbons. One class of cyclic compounds, called the **aromatic hydrocarbons**, has unique properties that differ markedly from those we have discussed. The parent compound of this group of hydrocarbons is *benzene:*

benzene

Although we typically draw the structure of benzene indicating double bonds around the ring of carbon atoms, in terms of its reactions, benzene behaves as though it did not have any double bonds. For example, if elemental bromine is added to benzene (with a suitable catalyst), benzene does not undergo an addition reaction as would be expected for a compound with double bonds. Rather, it undergoes a substitution reaction of the sort that would be expected for single-bonded hydrocarbons:

$$C_6H_6 + Br_2 \rightarrow C_6H_5Br + HBr$$

benzene *bromobenzene*

This reaction indicates that perhaps benzene does *not* have any double bonds in reality. Several Lewis dot structures can be drawn for benzene, with the "double bonds" in several different locations in the ring. Benzene exhibits the phenomenon of **resonance**. That is, the structure of benzene shown, as well as any other structure that might be drawn on paper, does not really represent the true electronic structure of benzene. The electrons of the "double bonds" of benzene are, in fact, **delocalized** around the entire ring of the molecule and are not concentrated in any particular bonds between carbon atoms. The ring structure of benzene is part of many organic molecules, including some very common biological compounds.

Functional Groups

While the hydrocarbons just described are of interest, the chemistry of organic compounds containing atoms other than carbon and hydrogen is far more varied. When another atom, or group of atoms, is added to a hydrocarbon framework, the new atom frequently imparts its own very strong characteristic properties to the resulting molecule. An atom or group of atoms that is able to impart new and characteristic properties to an organic molecule is called a **functional group**. With millions of carbon compounds known, it has been helpful to chemists to divide these compounds into **families**, based on what functional group they contain. Generally it is found that all the members of a family containing a particular functional group will have many similar physical properties and will undergo similar chemical reactions.

Alcohols

Simple hydrocarbon chains or rings that contain a hydroxyl group bonded to a carbon atom are called **alcohols**. The following are several such compounds:

$CH_3—OH$ *methanol*

$CH_3—CH_2—OH$ *ethanol*

$CH_3—CH_2—CH_2—OH$ *1-propanol*

$CH_3—CH—CH_3$
OH (on the middle carbon)
2-propanol

The –OH group of alcohols should not be confused with the hydroxide ion, OH^-. Rather than being basic, the hydroxyl group makes the alcohols in many ways like water, H–OH. For example, the smaller alcohols are fully miscible with water in all proportions. Alcohols are similar to water in that they also react with sodium, releasing hydrogen. Alcohols, however, react more slowly with sodium than does water.

Many alcohols can be oxidized by reagents such as permanganate ion or dichromate ion, with the product depending on the structure of the alcohol. For example, ethyl alcohol is oxidized to acetic acid: When wine spoils, vinegar is produced. On the other hand, 2-propanol is oxidized to propanone (acetone) by such oxidizing agents:

$CH_3—CH_2—OH$ (H above and H below the second carbon) → $CH_3—C{=}O$ (OH below the C)

ethanol → *acetic acid*

$CH_3—CH—CH_3$ (OH below the CH) → $CH_3—C—CH_3$ (=O below the C)

2-propanol → *propanone (acetone)*

Alcohols are very important in biochemistry, since many biological molecules contain the hydroxyl group. For example, carbohydrates (sugars) are alcohols, with a particular carbohydrate molecule generally containing several hydroxyl groups.

Aldehydes and Ketones

An oxygen atom that is double bonded to a carbon atom of a hydrocarbon chain is referred to as a **carbonyl group**. Two families of organic compounds contain the carbonyl function, with the difference between the families having to do with where along the chain of carbon atoms the oxygen atom is attached. Molecules having the oxygen atom attached to the first (terminal) carbon atom of the chain are called **aldehydes**, whereas molecules having the oxygen atom attached to an interior carbon atom are called **ketones**.

Aldehydes are intermediates in the oxidation of alcohols. For example, when ethyl alcohol is oxidized, the product is first acetaldehyde, which is then further oxidized to acetic acid:

$$CH_3{-}CH_2{-}OH \rightarrow CH_3{-}CH{=}O \rightarrow CH_3{-}C({=}O){-}OH$$

ethanol *ethanal (acetaldehyde)* *ethanoic acid (acetic acid)*

Ketones are the end product of the oxidation of alcohols in which the hydroxyl group is attached to an interior carbon atom (called secondary alcohols), as was indicated earlier for 2-propanol.

Aldehydes and ketones are important biologically, because all carbohydrates are also either aldehydes or ketones, as well as being alcohols. The common sugars glucose and fructose are shown next: Notice that glucose is an aldehyde, whereas fructose contains the ketone functional group.

$$HC({=}O){-}CH(OH){-}CH(OH){-}CH(OH){-}CH(OH){-}CH_2(OH)$$

glucose

$$H{-}CH(OH){-}C({=}O){-}CH(OH){-}CH(OH){-}CH(OH){-}CH(OH){-}H$$

fructose

Molecules such as glucose and fructose, which contain both the carbonyl and hydroxyl function, are easily oxidized by Benedict's reagent. Benedict's reagent is a specially prepared solution of copper(II) ion that is reduced when added to certain sugars, producing a precipitate of red copper(I) oxide. Benedict's reagent is the basis for the common test for sugars in the urine of diabetics. A strip of paper or tablet containing Benedict's reagent is added to a urine sample. The presence of sugar is confirmed if the color of the test reagent changes to the color of copper(I) oxide. Benedict's reagent reacts only with aldehydes and ketones that also contain a correctly located hydroxyl group, but will not react with simple aldehydes or ketones.

Organic Acids

Molecules that contain the carboxyl group at the end of a carbon atom chain behave as acids. For example, the hydrogen atom of the carboxyl group of acetic acid ionizes as:

$$CH_3{-}C({=}O){-}OH \rightleftharpoons CH_3{-}C({=}O){-}O^- + H^+$$

On paper, the carboxyl group looks like a cross between the carbonyl group of aldehydes and the hydroxyl group of alcohols, but the properties of the carboxyl group are completely different from either of these. The hydrogen atom of the carboxyl group is lost by organic acids because the remaining ion has more than one possible resonance form, which leads to increased stability for this ion, relative to the nonionized molecule.

Organic acids are typically weak acids, with only a few of the molecules in a sample being ionized at any given time. However, organic acids are strong enough to react with bicarbonate ion or to cause an acidic response in pH test papers or indicators.

Organic Bases

There are several types of organic compounds that show basic properties, the most notable of such compounds forming the family of **amines**. Amines are considered to be organic derivatives of the weak base ammonia, NH_3, in which one or more of the hydrogen atoms of ammonia is replaced by chains or rings of carbon atoms. The following are some typical amines:

H—N(H)—H	CH_3—N(H)—H	CH_3—N(H)—CH_3	CH_3—N(CH_2CH_3)—CH_3
ammonia	*methylamine*	*dimethylamine*	*ethyldimethylamine*

Like ammonia, organic amines react with water, releasing hydroxide ion from the water molecules:

$$NH_3 + H_2O \rightarrow NH_4^+ + OH^-$$

$$CH_3\text{–}NH_2 + H_2O \rightarrow CH_3NH_3^+ + OH^-$$

Although amines are weak bases (not fully ionized), they are basic enough to cause a color change in pH test papers and indicators and are able to complex many metal ions in a similar manner to ammonia.

The so-called amino group, $-NH_2$, found in some amines (primary amines) is also present in the class of biological compounds called **amino acids**. Many of the amino acids from which proteins are constructed are remarkably simple, considering the overall complexity of proteins. The amino acids found in human protein contain a carboxyl group, with an amino group attached to the carbon atom next to the carboxyl group in the molecule's carbon chain. This sort of amino acid is referred to as an alpha amino acid. One simple amino acid found in human protein is shown here:

H_2N—C(H)(H)—C(=O)—OH

glycine (2-aminoethanoic acid)

Safety Precautions

- Wear safety glasses at all times while in the laboratory.
- Assume that *all* organic substances are highly flammable. Use only small quantities, and keep them away from open flames. (Use a hot water heating bath if a source of heat is required.)
- Assume that *all* organic substances are toxic and are capable of being absorbed through the skin. Avoid contact, and wash after use. Inform the instructor of any spills.
- Assume that the vapors of *all* organic substances are harmful. Use small quantities and work in the exhaust hood.
- Sodium reacts violently with water, liberating hydrogen and enough heat to ignite the hydrogen. Do not dispose of sodium in the sinks. Add any excess sodium to a small amount of ethyl alcohol, and allow the sodium to react fully with the ethanol.
- Potassium permanganate, potassium dichromate, and Benedict's reagent are toxic. Permanganate and dichromate can burn skin and clothing. Exercise caution and wash after use.
- Bromine causes extremely bad burns to skin, and its vapors are extremely toxic. Work with bromine in the exhaust hood, and hold the bottle with a towel when handling.

Apparatus/Reagents Required

Hexane, 1-hexene, cyclohexane, cyclohexene, toluene, ethyl alcohol, methyl alcohol, isopropyl alcohol, *n*-butyl alcohol, *n*-pentyl alcohol, *n*-octyl alcohol, 10% formaldehyde solution, 10% acetaldehyde solution, acetone, 10% glucose, 10% fructose, acetic acid, propionic acid, butyric acid, ammonia, 10% butylamine, 1% bromine in methylene chloride, 1% aqueous potassium permanganate, Benedict's reagent, sodium pellets, 5% acidified potassium dichromate solution, pH paper, 10% sodium bicarbonate solution, 1 *M* copper sulfate solution

Procedure

Record all data and observations directly in your notebook in ink.

A. Hydrocarbons

Your instructor will ignite small portions of hexane, 1-hexene, cyclohexane, cyclohexene, and toluene to demonstrate that they burn readily (do *not* attempt this yourself). Notice that toluene (an aromatic compound) burns with a much sootier flame than the other hydrocarbons. This is characteristic of aromatics.

Obtain 10-drop portions of hexane, 1-hexene, cyclohexane, cyclohexene, and toluene in separate small test tubes. Take the test tubes to the exhaust hood, and add a few drops of 1% bromine *(Caution!)* to each. Those samples containing double bonds will decolorize the bromine *immediately* (addition reaction).

Those samples that did not decolorize the bromine immediately will *gradually* lose the bromine color (substitution reaction).

Expose those test tubes that did *not* immediately decolorize the bromine to bright sunlight, an ultraviolet lamp, or a high-intensity incandescent lamp. The reaction is sensitive to wavelengths of ultraviolet light.

Blow across the open mouth of the test tubes to see if a fog of hydrogen bromide becomes visible.

Obtain additional samples of hexane, 1-hexene, cyclohexane, cyclohexene, and toluene in separate small test tubes. Add a few drops of 1% potassium permanganate to each test tube; stopper and shake for a few minutes. Those samples containing double bonds will cause a change in the purple color of permanganate.

B. Alcohols

Obtain 10-drop samples of methyl, ethyl, isopropyl, *n*-butyl, *n*-pentyl, and *n*-octyl alcohols in separate clean test tubes. Add 10 drops of distilled water to each test tube; stopper, and shake.

Record the *solubility* of the alcohols in water. In alcohols with only a few carbon atoms, the hydroxyl group contributes a major portion of the molecule's physical properties, making such alcohols miscible with water. As the carbon atom chain of an alcohol gets larger, the alcohol becomes more hydrocarbon in nature and less soluble in water.

Obtain 10-drop samples of methyl, ethyl, isopropyl, *n*-butyl, *n*-pentyl, and *n*-octyl alcohols in separate clean test tubes.

Add 10 drops of acidified potassium dichromate to each sample, transfer to a hot water bath, and heat until a reaction is evident. Dichromate ion oxidizes primary and secondary alcohols.

Obtain 1-mL samples of methyl, ethyl, isopropyl, *n*-butyl, *n*-pentyl, and *n*-octyl alcohols in separate clean test tubes.

Add a *tiny* pellet of sodium to each test tube *(Caution!)*. The vigor of the reaction depends on the structure of the alcohol. Smaller alcohols are more reactive than larger alcohols. If the sodium pellet does not completely dissolve in a particular alcohol sample, add 10 drops additional of that alcohol to the test tube. Continue adding small portions of the alcohol until it is certain that the sodium has completely reacted.

Wait until all the sodium in the test tubes has completely reacted. Then add about 5 mL of distilled water to the test tubes, and test the resulting solution with pH paper. The alkoxide ion that remains from the alcohols after reaction with sodium is strongly basic in aqueous solution.

C. Aldehydes and Ketones

Obtain 10-drop samples of formaldehyde, acetaldehyde, acetone, glucose, and fructose in separate test tubes.

To each test tube add 10 drops of acidified potassium dichromate solution, and transfer the test tubes to a hot water bath. Aldehydes are oxidized by dichromate, being converted into organic acids. Record which samples change color.

Obtain 10-drop samples of formaldehyde, acetaldehyde, acetone, glucose, and fructose in separate test tubes.

To each test tube add 1 mL of Benedict's reagent, and transfer to a hot water bath. Molecules containing a hydroxyl group next to a carbonyl group on a carbon atom chain will cause a color change in Benedict's reagent. This test is applied to urine for the detection of sugar.

D. Organic Acids

Obtain 10-drop samples of acetic acid, propionic acid, and butyric (*Caution:* stench!) acid in separate small test tubes.

Add 10 drops of 10% sodium bicarbonate solution to each test tube: a marked fizzing will result as the acids release carbon dioxide from the bicarbonate.

Obtain 10-drop samples of acetic acid, propionic acid, and butyric acid in separate small test tubes.

Add 5 mL of water to each test tube, and test the solution with pH test paper.

E. Organic Bases

Obtain 10-drop samples of ammonia and of *n*-butylamine (*Caution*: stench!) in separate small test tubes. Add 5 mL of water to each sample. Determine the pH of each sample with pH test paper.

Add a few drops of 1 *M* copper sulfate solution to each test tube and shake. The color of the copper sulfate solution should darken as the metal/amine complex is formed.

Name ______________________ Section ______________

Lab Instructor ______________________ Date ______________

EXPERIMENT 38 Organic Chemical Compounds

PRE-LABORATORY QUESTIONS

Draw structural formulas for each of the following substances. If the common name of the substance is given, write the IUPAC name for the substance.

Hexane IUPAC name ______________

Isopropyl alcohol IUPAC name ______________

1-Hexene IUPAC name ______________

Formaldehyde IUPAC name ______________

Cyclohexane IUPAC name ______________

Acetaldehyde IPUAC name ______________

Cyclohexene IUPAC name ______________________

Acetone IUPAC name ______________________

Toluene IUPAC name ______________________

Acetic acid IUPAC name ______________________

Methyl alcohol IUPAC name ______________________

Butyric acid IUPAC name ______________________

Ethyl alcohol IUPAC name ______________________

n-Butylamine IUPAC name ______________________

Name ______________________________ Section ____________________

Lab Instructor ______________________________ Date ____________________

EXPERIMENT 38 Organic Chemical Compounds

RESULTS/OBSERVATIONS

A. Hydrocarbons

Observation on igniting

____

Which compounds decolorized bromine immediately?

____

Which compounds decolorized bromine on exposure to light?

____

Which compounds produced an HBr fog?

____

Which compounds caused a color change in $KMnO_4$?

____

B. Alcohols

Which alcohols were fully soluble in water?

____

Observation on $K_2Cr_2O_7$ test

____

Odors of ethyl alcohol and *n*-butyl alcohol oxidation products

____

Relative vigor of reaction of alcohols with sodium

____

pH of water solution of alcohol/sodium product

C. Aldehydes and Ketones

Observation on $K_2Cr_2O_7$ test

Odor of acetaldehyde/dichromate product

Observation of Benedict's test

D. Organic Acids

Observation on test with bicarbonate

---

pH of aqueous solution of organic acid samples

E. Organic Bases

pH of aqueous solution of ammonia/*n*-butylamine

Observation on adding $CuSO_4$

QUESTIONS

1. When sodium bicarbonate was added to an organic acid sample, carbon dioxide was evolved. Is this test specific for organic acids, or does it apply to acids in general? Write the balanced equation for the process.

2. When potassium permanganate was shaken with alkenes, the color of the permanganate changed from the purple color of MnO_4^- to the brown/black color of manganese(IV) oxide. What sort of reaction must this be? What is the original alkene converted to in the reaction?

3. When ethyl alcohol was treated with potassium dichromate, the alcohol was oxidized to acetic acid, yet it was indicated that acetaldehyde is an intermediate in this oxidation. How might aldehydes be isolated during such an oxidation, preventing them from being further oxidized to the acid?

4. Why does a primary amine (such as *n*-butyl amine) darken the color of copper(II) sulfate solution?

5. Acetic acid and *n*-butylamine are a typical *weak* acid and base, respectively. Use a handbook of chemical data to write K_a for acetic acid and K_b for *n*-butylamine.

EXPERIMENT

39 The Preparation and and Properties of Esters

Objective

Esters are an important class of organic chemical compounds. The bonding in esters is analogous to that found in some important classes of biological compounds. Several esters will be prepared in this experiment to illustrate the most common methods of synthesis. In Choice I, two esters of salicylic acid with important medicinal properties will be prepared. In Choice II, you will synthesize some esters having characteristic odors.

General Introduction to Esters

When an organic acid, R–COOH, is heated with an alcohol, R′–OH, in the presence of a strong mineral acid, the chief organic product is a member of the family of organic compounds known as **esters**. In Experiment 25 (Chemical Equilibrium) you studied the formation of the ester *n*-propyl acetate, for example.

The general reaction for the esterification of an organic acid with an alcohol is

$$R\text{–}COOH + HO\text{–}R' \rightarrow R\text{–}CO\text{–}OR' + H_2O$$

In this general reaction, R and R′ represent hydrocarbon chains, which may be the same or different. As a specific example, suppose acetic acid, CH_3COOH, is heated with ethyl alcohol, CH_3CH_2OH, in the presence of a mineral acid catalyst. The esterification reaction will be

$$CH_3\text{–}COOH + HO\text{–}CH_2CH_3 \rightarrow CH_3\text{–}COO\text{–}CH_2CH_3 + H_2O$$

The ester product of this reaction ($CH_3\text{–}COO\text{–}CH_2CH_3$) is named ethyl acetate, indicating the acid and alcohol from which it is prepared. Esterification is an equilibrium reaction, which means that the reaction does not go to completion on its own. Frequently, however, the esters produced are extremely volatile and can be removed from the system by distillation. If the ester is not very easily distilled, it may be possible instead to add a desiccant to the equilibrium system, thereby removing water from the system and forcing the equilibrium to the right.

Choice I. Ester Derivatives of Salicylic Acid

Introduction

Two common esters, acetylsalicylic acid and methyl salicylate, are very important over-the-counter drugs. Since ancient times, it has been known that the barks of certain trees, when chewed or brewed as a tea, had *analgesic* (pain-killing) and *antipyretic* (fever-reducing) properties. The active ingredient in such barks was determined to be **salicylic acid**. When *pure* salicylic acid was isolated by chemists, however, it proved to be much too harsh to the linings of the mouth, esophagus, and stomach for direct use as a drug in the pure state. Salicylic acid contains the phenolic (–OH) functional group in addition to the carboxyl (acid) group, and it is the combination of these two groups that leads to the harshness of salicylic acid on the digestive tract.

Because salicylic acid contains both the organic acid group (–COOH) as well as the phenolic (–OH) group, salicylic acid is capable of undergoing two separate esterification reactions, depending on whether it is behaving as an acid (through the –COOH) or as an alcohol analog (through the –OH). Research was conducted in an attempt to modify the salicylic acid molecule in such a manner that its desirable analgesic and antipyretic properties would be preserved, but its harshness to the digestive system would be decreased. The Bayer company of Germany, in the late 1800s, patented an ester of salicylic acid that had been produced by reaction of salicylic acid with acetic acid.

$$C_6H_4(OH)(COOH) + CH_3-\overset{O}{\overset{\|}{C}}-O-\overset{O}{\overset{\|}{C}}-CH_3 \rightarrow C_6H_4(O-\overset{O}{\overset{\|}{C}}-CH_3)(COOH) + CH_3COOH$$

salicylic acid *acetic anhydride* *aspirin* *acetic acid*

The ester, commonly called acetylsalicylic acid, or by its original trade name (aspirin), no longer has the phenolic functional group. The salicylic acid has acted as an alcohol when reacted with acetic acid. Acetylsalicylic acid is much less harsh to the digestive system. When acetylsalicylic acid reaches the intestinal tract, however, the basic environment of the small intestine causes hydrolysis of the ester (the reverse of the esterification reaction) to occur. Acetylsalicylic acid is converted back into salicylic acid in the small intestine and is then absorbed into the bloodstream in that form. Aspirin tablets sold commercially generally contain binders (such as starch) that help keep the tablets dry and prevent the acetylsalicylic acid in the tablets from decomposing into salicylic acid. Since the other component in the production of acetylsalicylic acid is acetic acid, one indication that aspirin tablets have decomposed is an odor of vinegar from the acetic acid released by the hydrolysis.

The second common ester of salicylic acid that is used as a drug is methyl salicylate. When salicylic acid is heated with methyl alcohol, the carboxyl group of salicylic acid is esterified, producing a strong-smelling liquid ester (methyl salicylate).

$$C_6H_4(OH)COOH + CH_3{-}OH \rightarrow C_6H_4(OH){-}C(=O){-}O{-}CH_3 + H_2O$$

salicylic acid *methanol* *methyl salicylate* *water*

The *mint* odor of many common liniments sold for sore muscles and joints is due to this ester. Methyl salicylate is absorbed through the skin when applied topically and may permit the pain-killing properties of salicylic acid to be localized on the irritated area. Methyl salicylate is a skin irritant, however, and causes a sensation of warmth to the area of the skin where it is applied. This is usually considered to be a desirable property of the ester. Methyl salicylate is also used as a flavoring/aroma agent in various products and is referred to commercially as oil of wintergreen.

Safety Precautions

- Wear safety glasses at all times while in the laboratory.
- Most of the organic compounds used or produced in this experiment are highly flammable. All heating will be done using a hotplate, and no flames will be permitted in the laboratory.
- Sulfuric acid is used as a catalyst for the esterification reactions. Sulfuric acid is dangerous and can burn skin very badly. If it is spilled, wash *immediately* before the acid has a chance to cause a burn, and inform the instructor.
- Acetic anhydride is used in the synthesis of aspirin, rather than acetic acid itself. Acetic anhydride can seriously burn the skin, and its vapors are harmful to the respiratory tract. If spilled, wash *immediately* and inform the instructor. Confine the pouring of acetic anhydride to the exhaust hood.
- The vapors of the esters produced in this experiment may be harmful. When determining the odors of the esters produced in this experiment, *do not* deeply inhale the vapors. Merely waft a small amount of vapor from the ester toward your nose.

Apparatus/Reagents Required

Hotplate, suction filtration apparatus, ice, melting point apparatus, salicylic acid, acetic anhydride, methyl alcohol (methanol), 50% sulfuric acid, sodium bicarbonate, 1 *M* iron(III) chloride

Procedure

Record all data and observations directly in your notebook in ink.

A. Preparation of Aspirin

Set a 400-mL beaker about half full of water to warm on a hotplate in the exhaust hood. A water heating bath at approximately 70°C is desired, so use the lowest setting of the hot-plate control. Check the temperature of the water before continuing.

Weigh out approximately 1.5 g of salicylic acid and transfer to a clean 125-mL Erlenmeyer flask.

In the exhaust hood, add approximately 4 mL of acetic anhydride (*Caution!*) and 3–4 drops of 50% sulfuric acid (*Caution!*) to the salicylic acid. Stir until the mixture is homogeneous.

Transfer the Erlenmeyer flask to the beaker of 70°C water (see Figure 39-1) and heat for 15–20 minutes, stirring occasionally. Monitor the temperature of the water bath during this time, and do not let the temperature rise above 70°C.

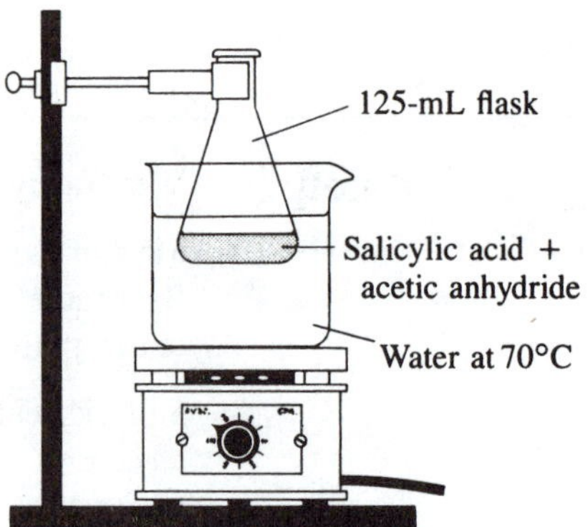

Figure 39-1. Heating bath for synthesis of esters. A hotplate is used because many esters are very volatile and extremely flammable.

At the end of the heating period, cool the Erlenmeyer flask in an ice bath until crystals begin to form. If crystallization does not take place, scratch the walls and bottom of the flask with a stirring rod to promote formation of crystals.

To destroy any excess acetic anhydride that may be present, add 50 mL of cold water, stir, and allow the mixture to stand for at least 15 minutes to permit the hydrolysis to take place.

Remove the liquid from the crystals by filtering under suction in a Büchner funnel (see Figure 39-2). Wash the crystals with two 10-mL portions of cold water to remove any excess reagents. Continue suction through the crystals for several minutes to help dry them.

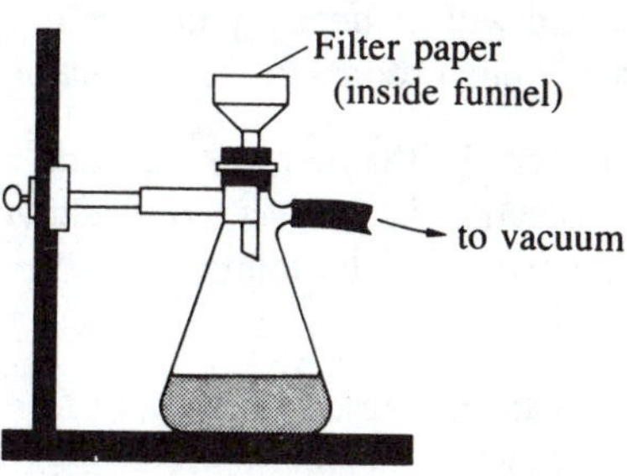

Figure 39-2. Suction filtration apparatus. The filter paper must fit the funnel exactly and must lie flat on the base of the funnel.

You will perform some tests on this *crude* product today, and your instructor may ask you to save a small portion of your crystals for a melting point determination to be performed during the next laboratory period. If your instructor directs, set aside a small portion of your crystals in a beaker to dry until the next lab period.

Dissolve the remaining portion of the aspirin product in 4–5 mL of 95% ethyl alcohol in a 125-mL Erlenmeyer flask. Stir and warm the mixture in the water bath on the hotplate until the crystals have dissolved completely.

Add approximately 15 mL of distilled water to the Erlenmeyer flask, and heat to redissolve any crystals that may have formed.

Remove the Erlenmeyer flask from the hot water bath, and allow it to cool slowly to room temperature. By allowing the solution to cool slowly, relatively large, needlelike crystals of aspirin should form.

Filter the purified product by suction, allowing suction to continue for several minutes to dry the crystals as much as possible.

If your instrucor has indicated that you will be performing the melting point determination for your aspirin in the next lab period, set aside a small portion of the recrystallized product in a labeled test tube or beaker to dry until the next laboratory period.

B. Tests on Aspirin

1. Test with Bicarbonate

Acetylsalicylic acid molecules still contain the organic acid group (carboxyl) and will react with sodium bicarbonate to release carbon dioxide gas:

$$H^+ + HCO_3^- \rightarrow H_2O + CO_2$$

Add a very small portion of your aspirin (crude or purified) to a test tube. Add also a small portion of sodium bicarbonate. Add a small amount of water and note the evolution of carbon dioxide. This test indicates only that aspirin is an acid; it is not a specific test for aspirin.

2. Test with Iron(III)

If the synthesis of aspirin has not been effective, or if aspirin has decomposed with time, then free salicylic acid will be present. This would be harmful if ingested. The standard United States Pharmacopeia test for the presence of salicylic acid is to treat the sample in question with a solution of iron(III). If salicylic acid is present, the phenolic functional group (–OH) of salicylic acid will produce a purple color with iron(III) ions. The *intensity* of the purple color is directly proportional to the amount of salicylic acid present.

Set up four test tubes in a rack. To the first test tube, add a small quantity of pure salicylic acid as a control. To the second and third test tubes, add very small portions of your crude and purified aspirin, respectively. To the fourth test tube, add a commercial aspirin tablet (if available).

Add 5 mL of distilled water to each test tube, and stir to mix. Add 8–10 drops of iron(III) chloride (ferric chloride) solution. The appearance of a pink or purple color in your aspirin or the commercial tablet sample indicates the presence of salicylic acid.

3. Melting Point Determination

After the crude and purified aspirin samples have dried for a week, determine their melting points by the capillary method described in Experiment 5. Compare the observed melting points with the literature value.

C. Preparation of Methyl Salicylate

Place about 1 g (roughly measured) of salicylic acid in a small Erlenmeyer flask and add 5 mL of methyl alcohol. Stir thoroughly to dissolve the salicylic acid.

Add 4–5 drops of 50% sulfuric acid (*Caution!*) and heat the flask in the 70°C water bath on the hot plate in the exhaust hood for approximately 5 minutes.

Pour the contents of the test tube into approximately 50 mL of warm water in a beaker, and cautiously waft some of the vapors toward your nose. There should be a pronounced odor of wintergreen.

Choice II. Preparation of Fragrant Esters

Introduction

The general reaction for the esterification of an organic acid with an alcohol is

$$R{-}COOH + HO{-}R' \rightarrow R{-}CO{-}OR' + H_2O$$

In this general reaction, R and R′ represent hydrocarbon chains, which may be the same or different. Unlike many organic chemical compounds, esters often have very pleasant, fruitlike odors. Many of the odors and flavorings of fruits and flowers are due to the presence of esters in the essential oils of these materials. The table that follows lists some esters with pleasant fragrances, as well as indicating from what alcohol and which acid the ester may be prepared.

Table of Common Esters

Ester	***Aroma***	***Constituents***
n-propyl acetate	pears	*n*-propyl alcohol/acetic acid
methyl butyrate	apples	methyl alcohol/butyric acid
isobutyl propionate	rum	isobutyl alcohol/propionic acid
octyl acetate	oranges	*n*-octyl alcohol/acetic acid
methyl anthranilate	grapes	methyl alcohol/2-aminobenzoic acid
isoamyl acetate	bananas	isoamyl alcohol/acetic acid
ethyl butyrate	pineapples	ethyl alcohol/butyric acid
benzyl acetate	peaches	benzyl alcohol/acetic acid

Generally a fruit or flower may only contain a few drops of ester, giving a very subtle odor. Usually, the ester is part of some complex mixture of substances, which, taken as a whole, have the aroma attributed to the material. When prepared in the laboratory in relatively large amounts, the ester may seem to have a pronounced chemical odor, and it may be difficult to recognize the fruit or flower that has this aroma.

Safety Precautions

- Wear safety glasses at all times while in the laboratory.
- Most of the organic compounds used or produced in this experiment are highly flammable. All heating will be done using a hotplate, and no flames will be permitted in the laboratory.
- Sulfuric acid is used as a catalyst for the esterification reactions. Sulfuric acid is dangerous and can burn skin very badly. If it is spilled, wash *immediately* before the acid has a chance to cause a burn, and inform the instructor.
- The vapors of the esters produced in this experiment may be harmful. When determining the odors of the esters produced in this experiment, *do not* deeply inhale the vapors. Merely waft a small amount of vapor from the ester toward your nose.
- NaOH solution is highly corrosive to eyes and skin. Wash immediately if spilled.

Apparatus/Reagents Required

Hotplate; 50% sulfuric acid; assorted alcohols and organic acids, as provided by the instructor, for the preparation of fruit and flower aromas; methyl salicylate; 20% NaOH

Procedure

Set up a water bath on a hotplate in the exhaust hood. Most of the reactants and products in this choice are highly flammable, and no flames are permitted in the lab during this experiment. Adjust the heating control to maintain a temperature of around 70°C in the water bath.

Some common esters, and the acids/alcohols from which they are synthesized, were indicated in the table in the introduction to this choice. Synthesize at least two of the esters, and note their aromas. Different students might synthesize different esters, as directed by the instructor, and compare the odors of the products.

To synthesize the esters, mix 1–2 mL (or approximately 0.5 g if the acid is a solid) of the appropriate acid with 2–3 mL of the indicated alcohol in a clean dry test tube.

Add 4–5 drops of 50% sulfuric acid to the test tube (*Caution!*), and heat for 10 minutes in the 70°C water bath on the hotplate in the hood.

Pour the resulting ester into a beaker of warm water, and cautiously waft the vapors toward your nose.

Remember that the odor of an ester is very concentrated. Several sniffs may be necessary for you to identify the odor of the ester.

Record which esters you prepared and their aromas.

Esters may be destroyed by reversing the esterification reaction: water and an ester will react with one another (hydrolysis) to give an alcohol and an organic acid. This reaction is often carried out using sodium hydroxide as catalyst (in which case the sodium salt of the organic acid results, rather than the acid itself).

$$R\text{–}COOH + HO\text{–}R' \rightarrow R\text{–}CO\text{–}OR' + H_2O$$

Set up a water bath on the hotplate in the hood. Adjust the heating control to maintain a temperature of approximately 70°C in the water bath.

Obtain 10 drops of methyl salicylate in a clean test tube. Note the odor of the ester by cautiously wafting vapors from the test tube toward your nose.

Add 2–3 mL of water to the test tube, followed by 10–15 drops of 20% NaOH solution (*Caution!*).

Heat the methyl salicylate/NaOH mixture in the water bath for 15–20 minutes. After this heating period, again note the odor of the mixture. If the odor of mint is still noticeable, heat the test tube for an additional 10-minute period.

Write an equation for the hydrolysis of methyl salicylate.

Name ______________________ Section ______________________

Lab Instructor ______________________ Date ______________________

EXPERIMENT 39 The Preparation and Properties of Esters

PRE-LABORATORY QUESTIONS

CHOICE I. ESTER DERIVATIVES OF SALICYLIC ACID

1. From the formulas of salicylic acid and acetylsalicylic acid given in the introduction to this experiment, determine their formula weights.

2. Suppose 5.0 g of salicylic acid is heated with an excess amount of acetic acid. What is the expected yield of acetylsalicylic acid?

CHOICE II. PREPARATION OF FRAGRANT ESTERS

1. Draw the structure of each of the following esters, which are discussed in the introduction to this Choice:

   n-Propyl acetate

   Octyl acetate

   Methyl butyrate

   Isoamyl acetate

   Isobutyl propionate

   Benzyl acetate

2. Write an equation for the hydrolysis of methyl salicylate by aqueous sodium hydroxide solution.

   ____

Name ______________________ Section ______________________

Lab Instructor ______________________ Date ______________________

EXPERIMENT 39 The Preparation and Properties of Esters

CHOICE I. ESTER DERIVATIVES OF SALICYLIC ACID

RESULTS/OBSERVATIONS

Observation on crude aspirin (appearance)

____

Observation of recrystallized aspirin (appearance)

____

Tests on aspirin

Effect of bicarbonate ______________________

Effect of iron(III) on crude aspirin ______________________

Effect of iron(III) on purified aspirin ______________________

Effect of iron(III) on salicylic acid ______________________

Effect of iron(III) on aspirin tablet ______________________

Melting points

Crude aspirin ______________________ Purified aspirin ______________________

Literature value ______________________ Reference ______________________

Observations on methyl salicylate ______________________

Odor? ______________________

QUESTIONS

1. How does sulfuric acid catalyze the preparation of esters? Could some other acid be used?

2. Aspirin was for many years the major household painkiller, but more recently, painkillers such as acetaminophen and ibuprofen have become popular. Find the structures of these substances in a chemical dictionary or handbook (or perhaps your textbook). Are these substances esters?

3. In the aspirin synthesis, excess acetic anhydride was destroyed by the addition of water, which converts the anhydride into acetic acid. Write the balanced chemical equation for the reaction.

Name ______________________________ Section ______________________

Lab Instructor ______________________________ Date ______________________

EXPERIMENT 39 The Preparation and Properties of Esters

CHOICE II. THE PREPARATION OF FRAGRANT ESTERS

RESULTS/OBSERVATIONS

Which esters did you prepare?

Ester 1 ______________________ Ester 2 ______________________

Odor ______________________ Odor ______________________

Formula ______________________ Formula ______________________

Observation on hydrolysis of methyl salicylate

____

____

____

QUESTIONS

1. Ordinarily esterification reactions come to *equilibrium* before the full theoretical yield of ester is realized. Aside from distillation discussed earlier, how might one experimentally shift the equilibrium of the esterification reaction so that a larger amount of ester might be isolated?

____

____

____

____

____

2. Using a handbook or encylopedia of chemistry, list two additional fragrant esters that were not discussed in this experiment.

3. Use your textbook to write a definition for *hydrolysis.*

EXPERIMENT

40 Proteins

Introduction

Proteins are long-chain polymers of the α-amino acids

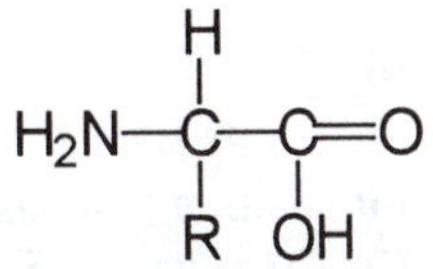

$$H_2N-\underset{R}{\overset{H}{C}}-\underset{OH}{C}=O$$

and make up about 15% of our bodies. Proteins have many functions in the body. Some proteins form the major structural feature of muscles, hair, fingernails. and cartilage. Other proteins help to transport molecules through the body, fight infections, or act as catalysts (enzymes) for biochemical reactions in the cells of the body.

Proteins have several levels of structure, each level of which is very important to a protein's function in the body. The **primary structure** of a protein is the particular sequence of amino acids in the polymer chain: each particular protein has a unique primary structure. The **secondary structure** of a protein describes the basic arrangement in space of the overall chain of amino acids: some protein chains coil into a helical form, while other protein chains may bind together to form a sheet of protein. The **tertiary structure** of a protein describes how the protein chains (whether sheet or helix) fold in space due to the interactions of the side groups of the amino acids: for example, the helical secondary structures of some proteins fold into a globular (spherical) shape to enable them to travel more easily through the bloodstream.

The various levels of structure of a protein are absolutely crucial to the protein's function in the body. For example, if any error occurs in the primary structure of the protein, a genetic disease may result (sickle cell disease and Tay-Sachs disease are both due to errors in protein primary structures). If any change in the environment of a protein occurs, the tertiary and secondary structures of the protein may be changed enough to make the protein lose its ability to act in the body. For example, egg white is basically an aqueous solution of the protein albumin: when an egg is cooked, the increased temperature causes irreversible changes in the tertiary and secondary structures of the albumin, making it coagulate as a solid. When the physical or chemical environment of a protein is modified, and the protein responds by changing its tertiary or secondary structure, the protein is said to have been **denatured**. Denaturation of a protein may be reversible (if the change to the protein's environment is not too drastic) or irreversible (as with the egg white discussed above). Changes in environment that may affect a protein's structure include changes in temperature, changes in the pH (acidity) of the protein's environment, changes in the solvent in which the protein is suspended, or the presence of any other ions or molecules which may chemically interact with the protein (such as ions of the heavy metals).

Safety Precautions

- Wear safety glasses at all times while in the laboratory.
- Sodium hydroxide, nitric acid, and hydrochloric acid solutions are corrosive to eyes and skin. Wash immediately if spilled and clean up all spills on the benchtop.
- Lead acetate and copper(II) sulfate solutions are toxic if ingested. Wash after handling. Dispose of as directed by the instructor.

Apparatus/Reagents Required

1% albumin solution, 1% alanine solution, nonfat milk, gelatin cubes, 10% NaOH solution, 3% $CuSO_4$ solution, 6 *M* nitric acid solution, 6 *M* NaOH solution, 5% lead acetate solution, 3*M* HCl solution, 3 *M* NaOH solution, saturated NaCl solution

Procedure

A. Tests for Proteins

Samples of three proteinaceous materials will be available for testing: egg white (containing the protein albumin); nonfat milk (containing the protein casein), and gelatin. Samples of the free amino acid alanine will also be tested for comparison and contrast.

1. The Biuret Test

The biuret test detects the presence of proteins and polypeptides, but does not detect free amino acids (the test is specific for the peptide linkage).

Place about 10 drops of 1% albumin, nonfat milk, and 1% alanine solutions, and a small cube of gelatin in separate clean small test tubes.

Add 10 drops of 10% sodium hydroxide solution *(Caution!)* to each test tube, followed by 3–4 drops of 3% copper(II) sulfate solution.

Carefully mix the contents of each test tube with a clean stirring rod, being sure to clean the stirring rod when switching between solutions. Make sure to break up the gelatin sample as much as possible when mixing.

The blue color of the copper(II) sulfate solution will change to purple in those samples containing proteinaceous material. Record which samples gave the purple color.

2. The Xanthoproteic Test

Two of the common amino acids found in proteins, tryptophan and tyrosine, contain substituted benzene rings in their side chains. When treated with nitric acid, a nitro group ($-NO_2$) can also be added to the benzene rings, which results in the formation of a yellow color in the protein ("xanthoproteic" means "yellow protein"). For example, if nitric acid is spilled on the skin, the proteins in the skin will turn yellow as the skin is destroyed by the acid.

Set up a 250-mL beaker containing approximately 100 mL of water on a wire gauze/ringstand and heat to boiling. Place about 10 drops of 1% albumin, nonfat milk, and 1% alanine solutions, and a small cube of gelatin in separate clean small test tubes.

Add 8–10 drops of 6 *M* nitric acid *(Caution!)* to each test tube.

Carefully mix the contents of each test tube with a clean stirring rod, being sure to clean the stirring rod when switching between solutions. Make sure to break up the gelatin sample as much as possible when mixing.

Transfer the test tubes to the boiling water bath and heat the samples for 3–4 minutes.

Record which samples produce the yellow color characteristic of a positive xanthoproteic test. Which samples do not give a positive test?

3. The Lead Acetate Test For Sulfur

In strongly basic solutions, the amino acid cysteine from proteins will react with lead acetate solution to produce a black precipitate of lead sulfide, PbS. Cysteine is found in many (but not all) proteins. The strongly basic conditions are necessary to break up polypeptide chains into individual amino acids (hydrolysis).

Set up a 250-mL beaker containing approximately 100 mL of water on a wire gauze/ringstand and heat to boiling. Place about 10 drops of 1% albumin, nonfat milk, and 1% alanine solutions, and a small cube of gelatin in separate clean small test tubes.

If you are willing to make the sacrifice, you may also test small samples of your hair (roll up 2 or 3 strands into a ball which can be fit in a test tube) or some fingernail clippings.

Add 15–20 drops of 6 *M* sodium hydroxide *(Caution!)* to each sample.

Carefully mix the contents of each test tube with a clean stirring rod, being sure to clean the stirring rod when switching between solutions. Make sure to break up the gelatin sample as much as possible when mixing.

Transfer the test tubes to the boiling water bath and heat the samples for 10–15 minutes. If the volume of liquid inside the test tubes begins to decrease during the heating period, add 5–10 drops of water to the test tubes to replace the volume.

After the heating period, allow the samples to cool. Then add 5 drops of 5% lead acetate solution to each test tube.

Record which samples produce a black precipitate of lead sulfide.

B. Denaturation of Proteins

Place 10 drops of 1% albumin solution in each of seven clean small test tubes. Set aside one sample as a control for comparisions.

Heat one albumin sample in a boiling water bath for 5 minutes and describe any change in the appearance of the solution after heating.

To the remaining albumin samples, add 3–4 drops of the following reagents (each to a separate albumin sample): 3 *M* hydrochloric acid; 3 *M* sodium hydroxide; saturated (5.4 *M*) sodium chloride solution; ethyl alcohol; 5% lead acetate solution.

Record any changes in the appearance of each albumin sample after adding the other reagent.

Repeat the tests above, using 10-drop samples of nonfat milk in place of the albumin solution.

Name ______________________________ Section ______________________

Lab Instructor ______________________________ Date ______________________

EXPERIMENT 40 Proteins

PRE-LABORATORY QUESTIONS

1. Use your textbook or an encyclopedia to explain why the amino acids found in animal proteins are referred to as *alpha* (α) amino acids.

____

____

____

____

____

____

2. Use your textbook to draw structural formulas for the following common α-amino acids.

Glycine

Serine

Alanine

Proline

Phenylalanine

Cysteine

3. Explain what is meant by the primary, secondary, and tertiary structures of a protein.

4. What does it mean to say that a protein has been *denatured*?

Name ______________________________ Section ______________

Lab Instructor ______________________________ Date ______________

EXPERIMENT 40 Proteins

RESULTS/OBSERVATIONS

A. Tests for Proteins

Biuret test:

Albumin ______________________________

Milk ______________________________

Alanine ______________________________

Gelatin ______________________________

Xanthoproteic test:

Albumin ______________________________

Milk ______________________________

Alanine ______________________________

Gelatin ______________________________

Lead acetate test:

Albumin ______________________________

Milk ______________________________

Alanine ______________________________

Gelatin ______________________________

Other ______________________________

B. Denaturation of Protein

albumin:

Heat ______________________________

HCl ______________________________

NaOH ______________________________

NaCl ______________________________

Alcohol ______________________________

Lead ______________________________

Milk: Heat____________________

HCl____________________

NaOH____________________

NaCl____________________

Alcohol____________________

Lead____________________

QUESTIONS

1. Why did the 1% alanine solution give a negative result in the biuret test?

2. Use your textbook to explain why a change in pH causes denaturation of a protein.

3. Draw the structure of the simple dipeptides *gly-ala* and *ala-gly*. Circle the peptide bond in each structure.

EXPERIMENT

41 Enzymes

Introduction

Enzymes are proteinaceous materials which function in the body as catalysts to control the speed of biochemical processes. Enzymes are said to be specific since a given enzyme typically acts only on a single type of molecule (or functional group): the molecule upon which an enzyme acts is referred to, in general, as the enzyme's substrate.

Like other proteins, enzymes can be denatured by changes in their environment which cause the tertiary or secondary structures of the protein to deform. The digestion of carbohydrates begins in the mouth, where the enzyme *ptylalin* is present in saliva. Ptyalin is a type of enzyme called an *amylase*, because the substrate it acts upon is the starch amylose (many enzymes are named after the substrate they act upon, with the ending of the substrate's name changed to *-ase* to indicated the enzyme). Ptyalin cleaves the long polysaccharide chain of amylose into smaller units called oligosaccharides. These oligosaccharides are then further digested into individual monosaccharide units in the small intestine.

Enzymes called *protease*s break down proteins into oligopeptides or free amino acids. For example, the protease called *bromelain* is present in fresh pineapple: if fresh pineapple is added to gelatin (a protein), the gelatin will not "set" because the protein is broken down. Meat tenderizers and several contact lens cleaners contain the enzyme *papain*, which is extracted from the papaya plant. Papain breaks down the proteins contained in muscle fibers of meat, making it more "tender," and dissolves the protein deposits which may cloud contact lenses.

Hydrogen peroxide is often used to clean wounds because oxygen is released when H_2O_2 breaks down (this is observed as bubbling)

$$2H_2O_2 \rightarrow 2H_2O + O_2$$

Having an ample supply of oxygen available to a wound is thought to destroy or slow the growth of several dangerous anaerobic microorganisms (e.g., *Clostridium*). Hydrogen peroxide used to clean a wound decomposes due to the action of enzymes called *catalases* (and similar enzymes called *peroxidases*). Such enzymes are also found in abundance in potatos and yeasts.

Safety Precautions

- Wear safety glasses at all times while in the laboratory.
- Lead acetate solution is toxic if ingested. Wash hands after handling. Dispose of as directed by the instructor.
- Hydrogen peroxide solution may be corrosive to the eyes and skin. Wash immediately if spilled. Clean up all spills on the benchtop.

Apparatus/Reagents Required

soda or "oyster" cracker, 5% lead acetate solution, 0.1 *M* I_2/KI solution, gelatin samples, meat tenderizer, contact lens cleaner solution, fresh pineapple, canned/cooked pineapple, 6% hydrogen peroxide solution, yeast suspension, freshly cut potato, cooked potato

Procedure

A. Action Of Ptyalin (Salivary Amylase)

Heat approximately 100 mL of water in a 250-mL beaker to a gentle boil.

Using a small beaker, collect approximately 3 mL of saliva by letting the saliva flow freely from your mouth (do not spit).

Obtain a small soda or "oyster" cracker and crush it finely on a sheet of paper using the bottom of a small beaker.

Label four clean test tubes as 1, 2, 3, and 4. Transfer a small amount (about the size of a matchhead) of the crushed cracker into each test tube.

To the first test tube containing cracker add 2 mL of distilled water as a control.

To the second test tube containing cracker add 1 mL of your saliva and 1 mL of distilled water.

To the third test tube containing cracker add 1 mL of your saliva, followed by 1 mL of 5% lead acetate solution. As a heavy metal, the lead(II) ion should denature the protein of the salivary enzyme.

To the final test tube containing cracker, add 1 mL of your saliva and 1 mL of distilled water, and then place the test tube in the boiling water bath for 5–10 minutes. Heating the saliva sample should denature the protein of the salivary enzyme. Carefully remove the test tube from the boiling water bath and allow it to cool to room temperature.

Set the four test tubes aside on the benchtop for at least 1 hour to allow the salivary enzyme to digest the starches in the cracker. Go on to the other portions of the experiment during the waiting period.

After the test tubes have stood for at least one hour, add 1 drop of 0.1 *M* I_2/KI solution to each of the test tubes. Iodine is a standard laboratory test for the presence of starch: if starch is present even in trace quantities in a sample, addition of I_2/KI solution will cause a deep blue/black color to appear. Record and explain your observations.

B. Action of Proteases.

Your instructor will provide you with samples of prepared gelatin (a protein), either cast in the wells of a spot plate, or as small chunks which you can transfer to small test tubes. Although the gelatin is similar to the type used as a dessert, the gelatin has been made up double-strength to make it drier and more easy to handle.

To one gelatin sample, add a few crystals of meat tenderizer.

To a second gelatin sample, add a few drops of the available commercial contact lens cleaner (record the brand and the name of the enzyme it contains, if available).

To a third gelatin sample, add a small piece of fresh pineapple (or juice extracted from fresh pineapple).

To a fourth gelatin sample, add a piece of cooked or canned pineapple.

Allow the gelatin samples to stand undisturbed for 30–60 minutes. Go on to the last part of the experiment while waiting.

After this time period, examine the gelatin samples for any evidence of breakdown of the gelatin by the proteases. Determine in particular if the gelatin has seemed to melt or partially melt. Record your observations

C. Action of Catalase

Obtain two small slivers of freshly cut potato. Place one of the pieces of potato in a small test tube and transfer to a boiling water bath for 5–10 minutes.

Place 10 drops of 6% hydrogen peroxide *(Caution!)* into each of three small test tubes.

To one test tube, add 8–10 drops of yeast suspension. Record your observations.

To the second test tube, add the small piece of uncooked, freshly cut potato. Record your observations.

To the third test tube, add the small piece of cooked potato. Why does the cooked potato not cause the hydrogen peroxide to decompose?

Name ______________________ Section ______________

Lab Instructor ______________________ Date ______________

EXPERIMENT 41 Enzymes

PRE-LABORATORY QUESTIONS

1. Use your textbook or an encyclopedia to write a specific definition of an *enzyme*.

2. Describe briefly what is meant by the "lock and key" model for the action of enzymes.

3. Use your textbook or an encyclopedia to write the names and functions of five enzymes not mentioned in this experiment.

Name ______________________ Section ______________________

Lab Instructor ______________________ Date ______________________

EXPERIMENT 41 Enzymes

RESULTS/OBSERVATIONS

A. Action of Ptyalin: Results of I_2/KI test

Control ______________________

Saliva ______________________

Saliva/lead ______________________

Saliva/heat ______________________

B. Action of Proteases: Observations on gelatin samples

Meat tenderizer ______________________

Lens cleaner ______________________

Fresh pineapple ______________________

Cooked pineapple ______________________

C. Action of Catalase: Observations on hydrogen peroxide samples

Yeast ______________________

Fresh potato ______________________

Cooked potato ______________________

QUESTIONS

1. Why do the cooked pineapple and cooked potato samples appear to no longer have active enzymes?

____

____

____

____

2. List five ways that an enzyme can be denatured.

3. Some antibiotic drugs work by *inhibiting* the enzymes of microorganisms. Use your textbook or an encyclopedia to explain what is meant by the *inhibition* of enzyme action.

EXPERIMENT

42 Polymeric Substances

Objective

Polymeric substances contain molecules consisting of extremely long chains of repeating simpler units (monomers). In Choice I of this experiment you will prepare an interesting polymeric form of sulfur. In Choice II, a common form of nylon polymer is synthesized.

Choice I. Polymeric Sulfur

Introduction

The ordinary form of sulfur at room conditions, *orthorhombic* sulfur, consists of discrete S_8 molecules. The eight sulfur atoms form a nonplanar ring. Orthorhombic sulfur is brittle, hard, and pale yellow in color. It is insoluble in water but is relatively soluble in carbon disulfide, CS_2. When orthorhombic sulfur is heated to near its boiling point and then quickly quenched in cold water, a *polymeric* allotrope of sulfur, which has very different properties, is formed.

When sulfur is heated, the normal S_8 ring molecules split open, giving free chains of eight sulfur atoms. The terminal sulfur atoms of these chains tend to react with the terminal sulfur atoms of other chains, thereby building up still longer chains. The buildup of polymeric chains is observed during the heating of sulfur as the liquefied sulfur becomes thicker and very viscous (sticky). In contrast with this observation for heating sulfur, most other substances generally become thinner and more free-flowing when heated.

If the thick, sticky, liquefied sulfur were allowed to cool slowly, the original orthorhombic S_8 ring molecules would reform as the temperature drops. If the thick, sticky liquefied sulfur is rapidly quenched by cooling, however, the polymeric allotrope can be "frozen" and examined. The polymeric sulfur is dark brown or black, and its properties are very different from those of orthorhombic sulfur. On standing for several days, the polymeric sulfur slowly converts into orthorhombic sulfur.

Safety Precautions

- Wear safety glasses at all times while in the laboratory.
- Carbon disulfide is very toxic, and its vapor is very volatile and forms explosive mixtures with air. Avoid contact with CS_2. Confine use of CS_2 to the exhaust hood. No flames will be permitted in the room while anyone is using carbon disulfide. *Check with the instructor before lighting any flames.*

Apparatus/Reagents Required

Chunk sulfur, powdered sulfur, carbon disulfide, oil bath/hotplate

Procedure

Record all data and observations directly in your notebook in ink.

A. Properties/Recrystallizaton of Orthorhombic Sulfur

Obtain a large chunk of sulfur and examine its color, hardness, and brittleness. To demonstrate the hardness and brittleness of the sulfur, crush the chunk with the metal base of a ringstand.

Place a very small chunk of the crushed sulfur, or a small amount of powdered sulfur, in a small test tube and bring it to the exhaust hood.

In the exhaust hood, add 10 drops of carbon disulfide (*Caution!*) to the sulfur and stir with a glass rod to dissolve the sulfur. Sulfur is slow to dissolve. If the sulfur has not dissolved after 4–5 minutes of stirring, add 5 drops more of carbon disulfide, and stir to dissolve the remaining sulfur.

While it is still in the exhaust hood, pour the sulfur/carbon disulfide solution into a watchglass, allowing the CS_2 to evaporate slowly. Examine the crystals of orthorhombic sulfur with a magnifying glass and sketch their shape.

B. Preparation of Polymeric Sulfur

Half fill a large test tube with powdered sulfur. Using a clamp, set the test tube up in an oil bath on a hotplate in the exhaust hood.

Begin heating the sulfur, noting changes in color and viscosity as the sulfur is heated. Use a glass rod to stir the sulfur during the heating.

Prepare a beaker of ice-cold water for use in quenching the polymeric sulfur.

When the sulfur has almost begun to boil, rapidly pour the thick molten sulfur into the beaker of cold water.

Using forceps, remove the polymeric sulfur from the cold water. Examine the polymer's hardness and brittleness and compare these properties with those of orthorhombic sulfur.

In the exhaust hood, test the solubility of a tiny piece of the polymeric sulfur in 10 drops of carbon disulfide.

Allow the polymeric sulfur to sit in your laboratory locker until the next lab period. At that time, reexamine the properties of the sulfur, testing for color, hardness, brittleness, and solubility.

Choice II. Preparation of Nylon

Introduction

In general, a polymeric substance contains long-chain molecules, in which a particular unit (monomer) is found *repeating* over and over again along the entire length of the polymer's chain. **Condensation polymers** are formed when two substances that ordinarily would not polymerize on their own are reacted, and the product of the condensation is a monomer that is now capable of polymerizing. A small molecule is also split out during the condensation. Nylon is a condensation polymer of 1,6-hexanedioic acid (adipic acid) or one of its derivatives with 1,6-hexanediamine (hexamethylene diamine). The monomer produced from the condensation of these two substances has a total of 12 carbon atoms, with 6 of these arising from each of the original substances. The nylon resulting from the polymerization of this monomer is sometimes referred to as nylon-66 (other polymers result if other acids or amines are used). The condensation reaction results in the formation of an **amide link** between the acid and the amine, and nylon is therefore a polyamide.

$$n\left(\underset{\text{hexamethylenediamine}}{H{-}N(H){-}(CH_2)_6{-}N(H){-}H} + \underset{\text{adipic acid}}{HO{-}C(=O){-}(CH_2)_4{-}C(=O){-}OH}\right) \longrightarrow$$

$$\left(\underset{\text{nylon-66 repeating unit}}{-N(H){-}(CH_2)_6{-}N(H){-}C(=O){-}(CH_2)_4{-}C(=O)-}\right)_n + nH_2O$$

As the condensation reaction occurs, it is possible to pull the nylon being produced into thin fibers (threads). Because of hydrogen bonding between polymeric molecules, the resulting nylon is very strong and stable.

Safety Precautions

- Wear safety glasses at all times while in the laboratory.
- The reagents used in this synthesis are toxic until mixed to form the nylon. Use caution in dispensing; wash immediately and inform the instructor if spilled.
- Do not handle the nylon resulting from the synthesis until it has been thoroughly washed.
- The reagents used in this experiment are volatile and highly flammable. *Absolutely no flames will be permitted in the laboratory.* Confine work with the reagents to the hood.
- Adipoyl chloride is used rather than adipic acid (1,6-hexanedioic acid). Hydrogen chloride gas is evolved during the reaction. Confine work with the reagents to the hood.

Apparatus/Reagents Required

10% adipoyl chloride in hexane, 20% 1,6-hexanediamine in water, 1 *M* NaOH solution, tongs or forceps

Procedure

Record all data and observations directly in your notebook in ink.

All portions of this experiment up to the washing of the nylon must be performed in the exhaust hood!

Have ready a 250-mL beaker full of cold water for washing the nylon after it is produced.

Place about 5 mL of 1 *M* NaOH on a watchglass. To the NaOH, add approximately 1 mL of 20% 1,6-hexanediamine solution. Stir with a stirring rod to mix the reagents.

Gently add 10–12 drops of the 10% adipoyl chloride/hexane solution to the surface of the aqueous reagents on the watchglass. Try to *float* the adipoyl chloride solution on top of the aqueous layer. Avoid bulk mixing of the two solutions; otherwise, the nylon will congeal into a blob rather than remaining threadlike.

On standing a few seconds, nylon will begin to form at the *interface* between the aqueous and nonaqueous solutions. With forceps or tongs, reach into the liquid on the watchglass and begin to pull on the film of nylon that has formed at the liquid junction. Pull the filament of nylon slowly and evenly from the mixture until you have a thread of nylon about 2 inches long.

Transfer the nylon filament to the beaker of rinsing water and allow it to stand for several minutes. Dispose of the other reagents as directed by your instructor.

Take the beaker with the nylon filament to your bench, rinse the nylon with several additional portions of water, and examine its color, hardness, and strength when pulled.

Name ______________________________ Section ______________

Lab Instructor ______________________________ Date ______________

EXPERIMENT 42 Polymeric Substances

PRE-LABORATORY QUESTIONS

CHOICE I. POLYMERIC SULFUR

1. What, in general terms, is meant by a *polymer*?

____

____

____

____

2. Describe how the polymeric form of sulfur forms from the normal S_8 molecules in which sulfur usually exists.

____

____

____

____

3. Sulfur has several allotropic forms in addition to the polymeric allotrope to be prepared in this experiment. Use your textbook or a chemical encyclopedia to find a representation of the phase diagram of sulfur in which the various forms of sulfur are indicated. Sketch the diagram here.

CHOICE II. PREPARATION OF NYLON

1. Use a handbook of chemical data to find the following physical properties of nylon-66:

   Density____________________ Color____________________

   Melting point____________________ Reference____________________

2. Nylon-66 is a polyamide. What functional group is present in a molecule that makes the molecule an amide?

3. Write a chemical equation showing the formation of a unit of nylon-66 from 1,6-hexanediamine and 1,6-hexanedioic acid.

Name ______________________ Section ______________

Lab Instructor ______________________ Date ______________

EXPERIMENT 42 Polymeric Substances

CHOICE I. POLYMERIC SULFUR

RESULTS/OBSERVATIONS

Observations of color, hardness, brittleness of chunk sulfur

____

____

____

Observation of recrystallized orthorhombic sulfur (from CS_2)

____

____

Shape of crystals ______________________________

Observation of color, hardness, brittleness of plastic sulfur

____

____

____

Solubility in CS_2 ______________________________

Observation of plastic sulfur after standing one week

____

QUESTIONS

1. Why was the polymeric sulfur insoluble in CS_2?

2. It was mentioned that sulfur thickens on heating as the polymeric molecules begin to form whereas most other substances thin out and become more free-flowing when heated. Why would you expect most substances to behave this way when heated?

3. What changes occurred in the polymeric sulfur after standing for a week? Which form of sulfur appeared to be present after the waiting period?

Name ______________________ Section ______________

Lab Instructor ______________________ Date ______________

EXPERIMENT 42 Polymeric Substances

CHOICE II. PREPARATION OF NYLON

RESULTS/OBSERVATIONS

Observation on adding adipoyl chloride solution to 1,6-hexanediamine solution

____

____

Observation on pulling nylon filament from the reaction mixture

____

____

Color, hardness, strength of nylon filament

____

QUESTIONS

1. Why did the adipoyl chloride solution float on the surface of the 1,6-hexanediamine solution?

____

____

____

____

____

____

2. It was mentioned that nylon-66 is not the only nylon that can be synthesized. Use your textbook or a chemical encyclopedia to find the formulas of some other nylons. Draw those formulas here.

3. Use your textbook to draw the structures of the repeating unit of four other common condensation polymers.

Qualitative Analysis: Techniques

Introduction

Qualitative analysis of unknown samples is one of the oldest and most important services performed by chemists. The techniques of **qualitative analysis** determine what elements (or particular groups of elements) are present in a sample, as opposed to how much of each element is present (quantitative analysis of the sort you have performed in earlier experiments). Qualitative analysis also serves two very important purposes in the learning of chemistry by beginning students. First, in performing the various manipulations required for an analysis, you will learn a great deal about laboratory manipulations and techniques. Second, by carefully observing all the chemical interactions that take place during the analysis, you will learn a lot about the properties and reactions of the elements involved.

There are two basic goals in the qualitative analysis of a mixture of cations: (1) separating each cation present in a sample from all the other cations that are present, and (2) identifying positively each cation once it has been separated. With over 100 elements (and with numerous stable polyatomic ions involving those elements), some gross initial separation of these species is necessary. Rather than dealing with the entire group of cations, it is necessary first to subdivide the cations into smaller groups. The subdivision is generally made on the basis of solubility. A reagent is added to the overall mixture of cations that will cause the precipitation of certain groups of the cations. The precipitate is then removed and analyzed for the individual cations it contains, whereas the filtrate is subsequently treated with some other reagent, which will cause precipitation of additional groups of cations.

The overall master assembly of cations is generally divided into five groups based on the precipitation of the cations with various reagents (note that these are *not* the groups or families of the periodic table):

Group I: cations that form chlorides that are insoluble in acid

Group II: cations present in the filtrate that form sulfides that are insoluble in 0.3 *M* acid

Group III: cations present in the filtrate that form insoluble sulfides and hydroxides when base is added

Group IV: cations present in the filtrate that form insoluble carbonates or phosphates

Group V: cations present in the filtrate that do not precipitate

The method of qualitative analysis you will use is especially useful, since it employs only a few drops of the sample and the various reagents needed. Such a method is said to be semimicro in scale. Analyses using small amounts are desirable, both because this scale is less wasteful of possibly expensive reagents, and also because the method requires much less time for analysis. If a macro sample, which produces large volumes of precipitates when reagents are added, were used, a great deal of time would have to be spent filtering and washing these precipitates. In the semimicro method, precipitates do not have to be filtered, but rather are centrifuged. A typical student centrifuge is indicated in Figure 1.

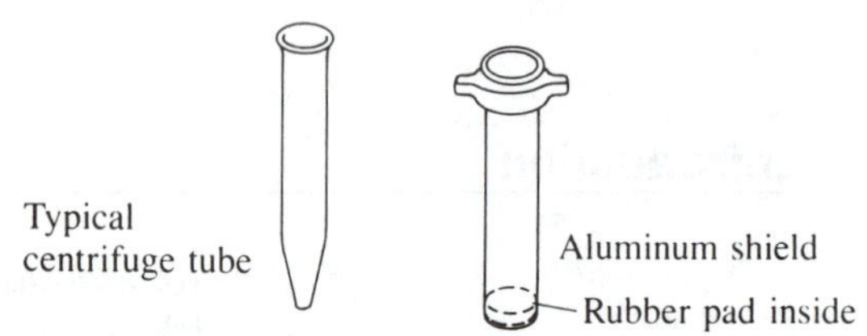

Opposite pairs of tubes should be filled with equal amounts of liquid to prevent excessive vibration.

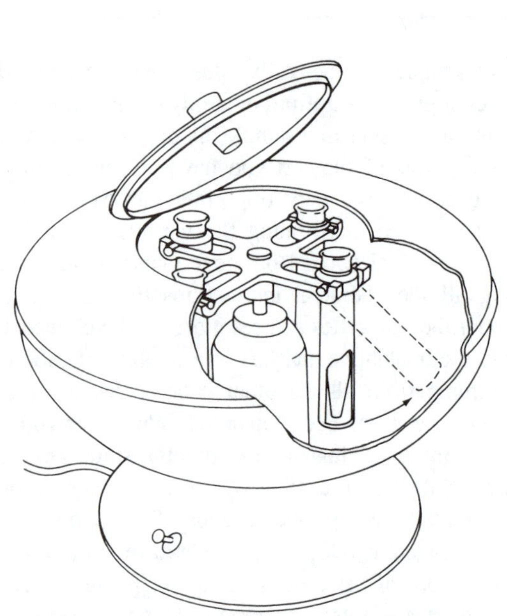

Tubes swing to dotted position when centrifuge is running.

Figure 1. A typical student centrifuge. Precipitates settle rapidly as the tubes spin, avoiding the need for filtration. Make sure the centrifuge is always "balanced."

A sample containing a precipitated substance is transferred to a centrifuge tube and the tube is inserted into the sample holder of the centrifuge. A second centrifuge tube is filled with water (or with another mixture that needs centrifugation) and placed in the sample holder in a position directly opposite to the first centrifuge tube (this is called *balancing* the centrifuge). The centrifuge is then turned on, causing the sample holder and tubes to spin at very high speed. This spinning action causes the precipitate to pack tightly into the narrow bottom of the centrifuge tube (requiring only 2–3 minutes for complete settling of the precipitate).

After centrifugation, the supernatant liquid above the precipitate (frequently referred to as the **centrifugate** rather than filtrate) is removed with a pipet or dropper and transferred to another test tube for further processing. The precipitate remaining in the centrifuge tube is then washed to remove any odd ions that may be adsorbed on the crystals or trapped between the particles. It is then centrifuged again. The wash liquid is removed and discarded.

See Figure 2. To ensure complete removal of odd ions from the precipitate, the washing step may have to be repeated one or more times before the precipitate is treated further.

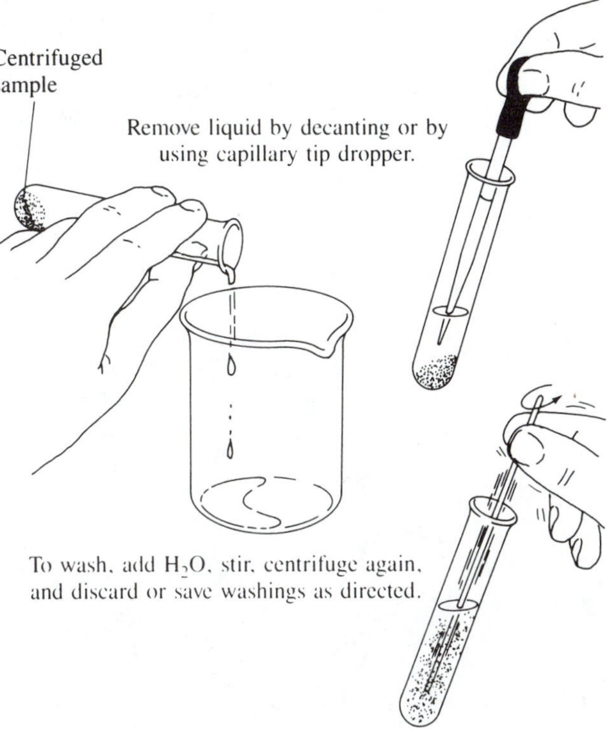

Figure 2. Separation of a centrifugate from a precipitate. Be carefule not to disturb the layer of precipitate when removing liquids with a dropper.

Once the overall sample of cations is divided into separate precipitates or centrifugates representing the five qualitative analysis subgroups, each subgroup is processed in turn to detect what elements are present in each subgroup. This generally involves further separation of the subgroups, followed by a confirming test for the presence of the specific ions being tested for.

In the following four experiments, you will be asked to carry through the analysis of two samples in each experiment. The samples consist of a group known sample (which contains only those cations belonging to the particular group under study in the individual experiment), and a group unknown sample (which may contain any or all of the cations from the group). It is very important that you do not confuse the samples in any portion of the analysis, and that you save all precipitates and centrifugates until you are certain that they have been fully processed. *Label all test tubes* as you go along through the analysis, and *record all observations in detail* in your lab notebook. If you discard a sample when you should not do so, or if you confuse observations or interpretations of tests, you will have to begin the experiments over again. (Keep in mind that there may not be enough time for this.)

In a real qualitative analysis, realize that things would not necessarily be set up so neatly. A real unknown would undoubtedly contain ions from several of the five main subgroups, and would require more extensive pretreatment to separate the main subgroup cations from one another. This is not done in the following experiments in the interest of saving time.

EXPERIMENT

43 Qualitative Analysis of the Group I Cations

Objective

A sample containing only the Group I cations will be analyzed for the presence of silver(I), mercury(I), and lead(II) ions.

Introduction

The Group I cations are those species that form chloride precipitates that are insoluble in acid. The group includes Ag^+, Pb^{2+}, and Hg_2^{2+} (mercurous ion). These cations are precipitated from any other cations that might be present in a sample by addition of 6 M hydrochloric acid. Addition of HCl forms a mixture of AgCl, $PbCl_2$, and Hg_2Cl_2 solids.

$$Ag^+(aq) + Cl^-(aq) \rightarrow AgCl(s)$$

$$Pb^{2+}(aq) + 2Cl^-(aq) \rightarrow PbCl_2(s)$$

$$Hg_2^{2+}(aq) + 2Cl^-(aq) \rightarrow Hg_2Cl_2(s)$$

At this point, the sample is centrifuged and the precipitate of the Group I chlorides isolated (in a real analysis, the centrifugate from above the precipitate is saved for further analysis of the other group cations).

Lead ion is then separated from silver and mercury by taking advantage of the fact that $PbCl_2$ is much more soluble in hot water than in cold water (the solubilities of AgCl and Hg_2Cl_2 do not vary much with temperature). Distilled water is added to the Group I mixed precipitate and the mixture is heated to dissolve $PbCl_2$. The mixture is then centrifuged quickly while still hot, and the centrifugate containing lead ion is removed from the remaining silver/mercury precipitate. The presence of lead ion is then confirmed by addition of chromate ion, CrO_4^{2-}, which forms a characteristic yellow precipitate with lead ion, $PbCrO_4$.

The precipitate containing silver and mercurous ions is then treated with aqueous ammonia. Silver ion is complexed by ammonia; the precipitate of AgCl will dissolve and is removed after centrifugation. A black/gray residue in the centrifuge tube confirms the presence of mercury.

The centrifugate containing complexed silver ion is then treated with acid, which reacts with ammonia allowing the reprecipitation of silver chloride. Alternatively, potassium iodide can be added, which also precipitates the silver (as a creamy yellow-white solid).

In this experiment (and those that follow on the qualitative analysis of Groups II–V) a known sample for only the particular group under study, as well as an unknown sample (containing one or more cations from the specific group) will be analyzed. In real practice, a sample would not be restricted to the members of only one analysis group, but rather would be a general mixture of all possible cations.

Safety Precautions

- Wear safety glasses at all times while in the laboratory.
- The centrifuge spins rapidly and can eject the sample tubes with considerable momentum if it is not correctly balanced. If the centrifuge starts to wobble or creep along the lab bench, immediately disconnect power and balance the centrifuge.
- Lead and mercury compounds are toxic. Wash after use.
- If silver ion or potassium iodide are spilled, they will stain the skin.
- Chromate ion is toxic and will burn the skin if spilled.
- If hydrochloric and nitric acids are spilled, they will burn the skin. Wash immediately if a spill occurs and inform the instructor.
- Ammonia is a strong cardiac stimulant and is irritating to the respiratory tract.

Apparatus/Reagents Required

Group I known sample (containing only Ag^+, Pb^{2+}, Hg_2^{2+}), unknown cation mixture (which contains one or more of the Group I cations), centrifuge and tubes, disposable pipets/droppers, distilled water, 6 *M* HCl, 6 *M* acetic acid, 0.2 *M* potassium chromate, 6 *M* aqueous ammonia, 6 *M* HNO_3, potassium iodide, pH paper

Procedure

Record all data and observations directly in your notebook in ink.

Label all test tubes during the procedure to make certain that samples are not confused or discarded at the wrong point.

Analyze both samples (Group I known, unknown) in parallel, side by side, so that results may be compared. In practice, an unknown would also contain members of later analysis groups, but such cations have been omitted here for simplicity.

A. Precipitation of the Group I Cations

Transfer approximately 1 mL of the sample solution to a test tube and then add 10 drops of 6 *M* HCl to precipitate the Group I cations.

Stir the mixture to mix and then centrifuge until the precipitate is packed firmly in the bottom of the tube (be sure to balance the centrifuge).

To ensure that sufficient HCl has been added to the sample to cause the precipitation of all the Group I cations, add 1 more drop of HCl to the test tube, watching the supernatant liquid in the test tube for the appearance of more precipitate.

If more precipitate forms on the addition of 1 drop of HCl, recentrifuge and retest with additional single drops of HCl until it is certain that all the Group I chlorides have been precipitated. Recentrifuge until the precipitate is packed firmly in the bottom of the centrifuge tube.

For the Group I known [which contains only silver, mercury(I), and lead] and the unknown, you may discard the centrifugate. If a general unknown were used, containing cations from other groups, the centrifugate would be saved for later analysis.

Wash the silver/mercury/lead precipitate in the centrifuge tube by adding 1 mL of ice-cold distilled water and 2 drops of 6 *M* HCl. Stir with a glass rod and centrifuge.

Remove the wash water and discard. Wash a second time, centrifuge, and again discard the wash liquid. The precipitate now contains the chlorides of silver, mercury(I), and lead, separated from all other species.

B. Separation and Confirmation of Lead Ion

To the mixed chloride precipitate from Part A, add 2–3 mL of distilled water.

Heat the mixture in a boiling water bath for 3–4 minutes. Stir the precipitate during the heating period to dissolve the lead chloride.

While the mixture is still hot, centrifuge and remove the centrifugate (which contains most of the lead ion from the sample) to a separate test tube using a dropper.

To the centrifugate (containing lead ion), add 2–3 drops of 6 *M* acetic acid and 5–6 drops of potassium chromate solution. Allow the sample to stand for a few minutes. The appearance of a yellow precipitate of lead chromate confirms the presence of lead ion in the original sample.

C. Separation and Confirmation of Mercury(I) and Silver Ions

To the remaining Group I precipitate, add 2–3 mL of hot water and 1–2 drops of 6 *M* HCl. Stir, and heat in the boiling water bath for 3–4 minutes. Centrifuge the mixture.

Carefully remove and discard the supernatant liquid, which may contain lead ion that had not been completely removed earlier. Add a 2–3-mL portion of hot water and 1–2 drops of 6 *M* HCl.

Stir the mixture, and heat in the hot water bath. Centrifuge, remove, and discard the supernatant liquid. These several washing steps are necessary to remove all traces of lead ion from the silver/mercury precipitate.

Add 2–3 mL of 6 *M* aqueous ammonia to the silver/mercury precipitate.

Stir the mixture, centrifuge, and remove the centrifugate (which contains dissolved silver ion). The presence of a gray-black precipitate in the centrifuge tube at this point confirms the presence of mercury(I) in the original sample.

Divide the centrifugate containing complexed silver ion into two portions in separate clean test tubes.

To one portion, add sufficient 6 *M* nitric acid to make the solution acidic when tested with pH paper. The appearance of a white precipitate of AgCl confirms the presence of silver ion in the original sample.

Add a few crystals of potassium iodide to the second portion of complexed silver ion. The appearance of a creamy yellow precipitate serves as an additional confirmation of silver ion.

Name ______________________ Section ______________________

Lab Instructor ______________________ Date ______________________

EXPERIMENT 43 Qualitative Analysis of the Group I Cations

PRE-LABORATORY QUESTIONS

1. The separation of lead chloride from the chlorides of silver and mercury(I) is based on the differing *solubilities* of these substances in cold and hot water. Use a handbook of chemical data to find the specific solubilities (in g/100 g H_2O) in both cold and hot water for AgCl, $PbCl_2$, and Hg_2Cl_2.

	Cold water	*Hot water*	*Reference*
Silver chloride	______	______	______
Lead(II) chloride	______	______	______
Mercury(I) chloride	______	______	______

2. Because a qualitative analysis may often consist of a rather long series of manipulations, precipitations, centrifugations, and separations, flow charts are often prepared to summarize graphically the steps to be undertaken. For this analysis of Group I, prepare a simple *flow chart* showing what ions are present at each point in the procedure, what reagents are to be added, and the result to be expected at each point.

Name ______________________________ Section ______________

Lab Instructor ______________________________ Date ______________

EXPERIMENT 43 Qualitative Analysis of the Group I Cations

RESULTS/OBSERVATIONS

On adding HCl to the initial sample (Part A)

Group I known ______________________________

Unknown ______________________________

On adding potassium chromate (test for lead, Part B)

Group I known ______________________________

Unknown ______________________________

On adding aqueous ammonia (Part C)

Group I known ______________________________

Unknown ______________________________

On reacidifying with dilute nitric acid

Group I known ______________________________

Unknown ______________________________

On adding potassium iodide

Group I known ______________________________

Unknown ______________________________

Based on your results, which Group I cations are present in your unknown sample?

Ions ______________________ Unknown code number ______________

QUESTIONS

1. The analysis of Group II (which follows in a later experiment) involves precipitation of certain cations as the insoluble sulfides. Why are the Group I cations removed as the insoluble chlorides, rather than as the sulfides?

2. Why is HCl added on a drop-by-drop basis to ensure complete precipitation of Group I chlorides initially, rather than just added all at once?

3. Why is acetic acid added before testing for lead with chromate ion?

4. Why is the silver/mercury precipitate washed to remove lead ion before addition of aqueous ammonia?

5. Why must a centrifuge always be balanced during use?

6. What is the identity of the creamy yellow solid produced when potassium iodide is added to a portion of the final centrifugate?

EXPERIMENT

44 Qualitative Analysis of the Group II Cations

Objective

A known sample containing the Group II cations and an unknown sample will be analyzed for the presence of bismuth(III), copper(II), cadmium(II), and tin(IV).

Introduction

Group II contains those cations forming sulfides that are *insoluble* under *very acidic* conditions (pH approximately 0.5). Among these cations are Bi^{3+}, Cu^{2+}, Cd^{2+}, and Sn^{4+}.

Before precipitation, the pH of a sample containing the Group II cations in solution is adjusted to a pH of approximately 0.5 by the dropwise addition of dilute HCl, which also prevents possible hydrolysis of the ions. Sulfide ion, in the form of hydrogen sulfide, H_2S, is then introduced into the solution, which precipitates the Group II cations as the insoluble sulfide compounds.

$$2Bi^{3+}(aq) + 3S^{2-}(aq) \rightarrow Bi_2S_3(s)$$

$$Cu^{2+}(aq) + S^{2-}(aq) \rightarrow CuS(s)$$

$$Cd^{2+}(aq) + S^{2-}(aq) \rightarrow CdS(s)$$

$$Sn^{4+}(aq) + 2S^{2-}(aq) \rightarrow SnS_2(s)$$

The mixture of Group II sulfides is then centrifuged, and the supernatant liquid (which may contain cations of Groups III–V) is removed and saved for later analysis in a general scheme (in this experiment, the known and unknown contain only Group II cations).

Hydrogen sulfide may be introduced into the mixture of cations by several methods, but most of these methods are objectionable because of the stench and toxicity of H_2S. In the following procedure, hydrogen sulfide is generated directly in the mixture of cations by slow reaction of an organic sulfur compound, thioacetamide, with hot water

$$\underset{\textit{thioacetamide}}{CH_3C(=S)NH_2} + H_2O \rightarrow CH_3C(=O)NH_2 + H_2S$$

The use of thioacetamide should not allow excessive levels of hydrogen sulfide to enter the atmosphere of the laboratory. If the odor becomes objectionable, however, *transfer the mixture to the exhaust hood while H_2S is being evolved.*

Once the precipitate of the Group II sulfides has been isolated, the cations are separated from one another, usually by adjustments of the pH of the mixture. For example, tin(IV) can be separated from the other cations by raising the pH of the solution to basic levels: tin(IV) sulfide dissolves in basic solutions as the $Sn(OH)_6^{2-}$ complex, whereas the other Group II sulfides are insoluble in base.

After tin(IV) is removed, the other Group II sulfides are dissolved in nitric acid. The nitric acid solution is then treated with aqueous ammonia, which precipitates bismuth as $Bi(OH)_3$, leaving Cd(II) and Cu(II) in solution as the ammine complexes.

Copper is then precipitated as the ferrocyanide complex, $Cu_2Fe(CN)_6$. Finally, cadmium is precipitated as the bright yellow sulfide, CdS.

Safety Precautions

- Wear safety glasses at all times while in the laboratory.
- Bismuth, cadmium, and copper compounds are toxic. Wash after use.
- Thioacetamide is *highly toxic*. Wash thoroughly after use. Thioacetamide evolves toxic, noxious hydrogen sulfide gas under the conditions in this experiment. If the odor of hydrogen sulfide is detected during the precipitation of the Group II cations, transfer the mixture to the exhaust hood.
- Sodium dithionite ($Na_2S_2O_4$) is toxic and is a reducing agent. The bulk solid is spontaneously flammable if it becomes wet. Wash after use.
- Nitric, acetic, and hydrochloric acids will burn the skin if spilled. If this occurs, wash immediately and inform the instructor.
- Ammonia is a cardiac stimulant and respiratory irritant. Use in the exhaust hood.
- Potassium hydroxide is caustic if the solution used in this experiment is allowed to concentrate by evaporation. Wash after use and clean up any spills immediately.
- Potassium hexacyanoferrate(II) is toxic. Wash after use.

Apparatus/Reagents Required

Group II known sample (containing all of the ions), Group II unknown (containing one or more of the ions), short-range pH paper (or methyl violet paper), 6 *M* HCl, 6 *M* aqueous ammonia, 1 *M* thioacetamide, 1 *M* ammonium chloride, 3 *M* potassium hydroxide, 6 *M* nitric acid, 0.1 *M* tin(II) chloride, 0.5 *M* potassium hexacyanoferrate(II) (potassium ferrocyanide), sodium dithionite (sodium hydrosulfite, $Na_2S_2O_4$)

Procedure

Record all data and observations directly in your notebook in ink.

A. Precipitation of the Group II Sulfides

Obtain a sample of a Group II known mixture (containing all of Bi^{3+}, Cd^{2+}, Cu^{2+}, and Sn^{4+}) and an unknown (which may contain any or all of the Group II cations). Use about 1 mL of the solutions in the analysis that follows.

The pH of the mixtures must be on the order of pH 0.5 (± 0.3 pH unit). Check this with short-range pH paper, using the color chart on the pH paper dispenser. Withdraw a drop of the mixture with a clean stirring rod, and touch the drop to the pH paper.

If short-range pH paper is not available, a simple low-range pH paper can be made using methyl violet indicator. Place a few drops of methyl violet solution onto a strip of filter paper and allow it to dry. Remove a drop of the solution whose pH is to be tested with a clean stirring rod and touch it to the dried methyl violet spots. A change in color to blue-green confirms that the pH is near the correct level.

If the pH of the known or unknown is too high (pH greater than 0.5), add 6 *M* HCl dropwise to adjust the pH. If the pH is too low (pH less than 0.5), add 6 *M* aqueous ammonia to adjust the pH. The pH should be within 0.5 by ± 0.3 units before continuing.

After the pH has been adjusted, add 20–25 drops (about 1 mL) of 1 *M* thioacetamide solution to the sample.

Heat the sample in a boiling water bath for 5–10 minutes to allow for the hydrolysis of thioacetamide. If the odor of hydrogen sulfide is noticeable in the laboratory, stop heating and transfer to the exhaust hood. During this heating period, the sulfides of the Group II cations will precipitate.

Centrifuge the solution (be sure to balance). In a general analysis, the supernatant liquid (centrifugate) would contain the cations of Groups III–V and would be saved.

Before discarding the supernatant liquid, however, test the liquid to make certain that all the Group II cations have been precipitated. Add 2–3 drops of thioacetamide reagent and allow the mixture to stand for 1–2 minutes.

If no additional precipitate forms, the supernatant liquid can be removed with a pipet or dropper and discarded. If additional precipitate does form, add 5–6 drops of thioacetamide and reheat the test tube in the boiling water bath for a 5–10-minute period. Test again for completeness of precipitation before discarding the supernatant liquid.

Wash the sulfide precipitate with a few drops of 1 *M* ammonium chloride solution (NH_4Cl). Stir the mixture, centrifuge, and remove the supernatant liquid.

B. Separation and Confirmation of Tin

Add 20–25 drops of 3 *M* potassium hydroxide solution, KOH, 2 mL of distilled water, and 1–2 drops of thioacetamide reagent to the Group II sulfide precipitate.

Stir the mixture with a clean glass rod, and heat in the boiling water bath for 2–3 minutes.

Centrifuge the mixture while it is still hot, and decant the supernatant liquid into a clean test tube before it has a chance to cool very much.

The supernatant liquid contains Sn^{4+} as a hydroxide complex, $Sn(OH)_6^{2-}$. The remaining precipitate contains the sulfides of cadmium, copper, and bismuth.

To confirm the presence of tin, treat the supernatant liquid as follows. Acidify the solution with 6 *M* HCl until it is just barely acidic when tested with pH paper (a small amount of precipitate may form upon adding the acid). Add 5–10 drops of thioacetamide reagent, stir, and heat in the boiling water bath for 5 minutes. A yellow precipitate of SnS_2 indicates the presence of tin.

C. Separation and Confirmation of Bismuth(III)

To the precipitate of copper/cadmium/bismuth sulfides, add 15–20 drops of 6 *M* nitric acid. Stir and heat in the boiling water bath for 5 minutes. If any solid remains undissolved in the nitric acid after heating, decant the solution from the solid into a clean test tube.

Add 6 *M* aqueous ammonia to the solution until it is just barely basic when tested with pH paper; then add 2–3 additional drops of ammonia. Stir the contents of the test tube and allow it to cool.

Centrifuge the mixture to collect the precipitate (containing bismuth), and decant the supernatant liquid (containing copper and cadmium) into a clean test tube.

Wash the precipitate (containing bismuth) with 10–15 drops of 6 *M* aqueous ammonia. Stir, centrifuge, and discard the supernatant liquid.

Add 10–12 drops of 3 *M* KOH to the precipitate and stir. Add 5–6 drops of 0.1 *M* tin(II) chloride. $SnCl_2$ will reduce bismuth(III). The formation of a black precipitate of Bi upon addition of tin(II) chloride confirms the presence of bismuth in the original sample.

D. Confirmation of Copper(II) and Cadmium(II)

Divide the liquid containing copper and cadmium into two portions. If copper(II) is present, the solution will be pale blue at this point because of the $Cu(H_2O)_4^{2+}$ ion.

To confirm the presence of copper(II), add 10–12 drops of 6 *M* acetic acid to one of the portions of solution, followed by 5–6 drops of $K_4Fe(CN)_6$ [potassium hexacyanoferrate(II), potassium ferrocyanide]. The appearance of a reddish/brown precipitate of $Cu_2Fe(CN)_6$ confirms the presence of copper in the original sample.

To confirm the presence of cadmium, add a few small crystals of sodium dithionite (sodium hydrosulfite, $Na_2S_2O_4$) to the other portion of solution.

Transfer the test tube to the boiling water bath for 5–10 minutes. Sodium dithionite will remove any copper present, reducing copper(II) to the free metal. If copper(II) is present, the solution will be blue, but the blue color will fade as the reaction with sodium dithionite takes place.

When the solution is no longer blue, and it is certain that all copper ion has been removed, centrifuge the sample and decant the supernatant liquid (containing cadmium) from the reddish/black deposit of metallic copper.

Add 1 mL of 1 *M* thioacetamide to the decanted solution, and heat for 2–3 minutes in the boiling water bath. A yellow precipitate of CdS will form if cadmium is present.

Name ______________________ Section ______________________

Lab Instructor ______________________ Date ______________________

EXPERIMENT 44 Qualitative Analysis of the Group II Cations

PRE-LABORATORY QUESTIONS

1. In addition to the ions to be analyzed for in this experiment (Bi^{3+}, Cu^{2+}, Cd^{2+}, and Sn^{2+}), Group II also formally contains mercury(II), arsenic, and antimony ions. Consult a textbook or handbook of qualitative analysis, and summarize briefly how these additional cations are analyzed in a complete Group II mixture.

____

____

____

____

____

____

____

____

____

____

2. Construct a flow chart that diagrams the separation and confirmation of the four Group II ions to be determined in this experiment.

Name ______________________________ Section ______________

Lab Instructor ______________________________ Date ______________

EXPERIMENT 44 Qualitative Analysis of the Group II Cations

RESULTS/OBSERVATIONS

A. Precipitation of Group II Sulfides

Observation on boiling with thioacetamide

Known ______________________________

Unknown ______________________________

B. Separation and Confirmation of Tin

Observation on adding potassium hydroxide

Known ______________________________

Unknown ______________________________

Observation on acidifying and boiling with thioacetamide

Known ______________________________

Unknown ______________________________

C. Separation and Confirmation of Bismuth

Observation on boiling with nitric acid

Known ______________________________

Unknown ______________________________

Observation on adding ammonia

Known ______________________________

Unknown ______________________________

Observation on adding potassium hydroxide/tin(II) chloride

Known ______________________________

Unknown ______________________________

D. Confirmation of Copper and Cadmium

Observation on adding acetic acid/potassium hexacyanoferrate(II)

Known ______________________________

Unknown ______________________________

Observation on adding sodium dithionite

Known ______________________________

Unknown ______________________________

Observation on adding thioacetamide

Known ______________________________

Unknown ______________________________

Based on your analysis, which Group II cations are present in the unknown sample?

Ions ____________________ Unknown code number ____________

QUESTIONS

1. The initial precipitation of the Group II cations was carried out under rather rigorously controlled conditions of pH 0.5. Why was such a low pH necessary? What other cations (from other groups if present) would have been precipitated if a higher pH were used?

____

____

____

2. Thioacetamide was used as a source of hydrogen sulfide for the precipitation of the Group II cations. Why was gaseous H_2S itself, or some reagent that generates H_2S more rapidly (such as ammonium sulfide), not preferred for the analysis?

____

____

____

3. Write an equation for the reduction of copper(II) to elemental copper by the dithionite ion, $S_2O_4^{2-}$.

____

4. Write an equation for the reduction of bismuth(III) by stannous chloride.

____

EXPERIMENT

45 Qualitative Analysis of the Group III Cations

Objective

A known solution of the Group III cations (Cr^{3+}, Al^{3+}, Fe^{3+}, Ni^{2+}, and Co^{2+}) as well as an unknown sample containing one or more of these cations will be analyzed.

Introduction

Group III contains those cations whose sulfides do not precipitate under highly acidic conditions. In the analysis of a general unknown, the Group III cations are separated from the Group II cations (Experiment 44) on this basis. The general mixture of all cations is acidified and treated with H_2S, which causes the Group II sulfides to precipitate, but leaves the Group III cations still in solution.

Under basic conditions, however, the Group III cations will precipitate as the sulfides. The pH of the solution is raised to basic levels, and thioacetamide is added to generate H_2S *in situ*. At this point, if the sample being treated were a general unknown containing members of Groups IV and V, the mixture would be centrifuged. The precipitate would contain the Group III cations, whereas the centrifugate would contain members of later groups.

$$2Cr^{3+}(aq) + 3S^{2-}(aq) \rightarrow Cr_2S_3(s)$$

$$2Al^{3+}(aq) + 3S^{2-}(aq) \rightarrow Al_2S_3(s)$$

$$2Fe^{3+}(aq) + 3S^{2-}(aq) \rightarrow Fe_2S_3(s)$$

$$Ni^{2+}(aq) + S^{2-}(aq) \rightarrow NiS(aq)$$

$$Co^{2+}(aq) + S^{2-}(aq) \rightarrow CoS(aq)$$

The precipitate of Group III sulfides is then treated with hydrochloric acid, which dissolves the sulfides of chromium, aluminum, and iron. The sulfides of nickel(II) and cobalt(II) do not dissolve very well in HCl at this point, however, and may be removed by centrifugation.

$$Cr_2S_3(s) + 6H^+(aq) \rightarrow 2Cr^{3+}(aq) + 3H_2S(g)$$

$$Al_2S_3(s) + 6H^+(aq) \rightarrow 2Al^{3+}(aq) + 3H_2S(g)$$

$$Fe_2S_3(s) + 6H^+(aq) \rightarrow 2Fe^{3+}(aq) + 3H_2S(g)$$

After the NiS and CoS precipitates are separated from the other members of the group, these sulfides are dissolved in *aqua regia* (a mixture of hydrochloric and nitric acids).

The solution of $Cr^{3+}/Fe^{2+}/Al^{3+}$ species is made basic and is treated with a few drops of household bleach (5% sodium hypochlorite solution). The bleach oxidizes iron(II) to iron(III), which precipitates as the hydroxide under the basic pH conditions. Chromium(III) is also oxidized, to chromium(VI) in the form of the chromate ion, CrO_4^{2-}. Under these conditions, aluminum cannot be oxidized by 5% sodium hypochlorite, but rather is present as the soluble hydroxide complex $Al(OH)_4^-$.

The mixture is centrifuged and the precipitate of $Fe(OH)_3$ is removed and dissolved in acid, with the presence of iron confirmed by the addition of thiocyanate ion (forming the characteristic blood-red $FeNCS^{2+}$ complex).

$$Fe(OH)_3(s) + 3H^+(aq) \rightarrow Fe^{3+}(aq) + 3H_2O$$

$$Fe^{3+}(aq) + SCN^-(aq) \rightarrow [FeSCN]^{2+}(aq)$$

The supernatant solution containing CrO_4^{2-} and $Al(OH)_4^-$ is then buffered at basic pH, under which conditions aluminum forms a gelatinous precipitate of $Al(OH)_3$. Chromate ion, produced by oxidation of Cr^{3+}, remains in solution at this point.

After centrifugation and separation, the presence of aluminum in the precipitate is confirmed with a specific reagent (aluminon) that forms a characteristic "lake" with aluminum hydroxide. The presence of chromium is confirmed in several steps based on the varied colors of chromium compounds having different oxidation states.

The mixed precipitate of nickel and cobalt ions that was separated initially from the other Group III cations is then analyzed for each of these ions. Nickel is precipitated under basic conditions with the organic specific reagent dimethylglyoxime, whereas cobalt is confirmed as the blue thiocyanate complex.

Safety Precautions

- Wear safety glasses at all times while in the laboratory.
- Chromium, cobalt, barium, nickel, and thiocyanate compounds are all toxic. Wash after use.
- Thioacetamide is highly toxic. Wash after use. Thioacetamide evolves toxic, noxious hydrogen sulfide gas under the conditions in this experiment. If the odor of hydrogen sulfide is detected during the precipitation of the Group III cations, transfer the mixture to the exhaust hood.
- Nitric, sulfuric, and hydrochloric acids can all burn skin, especially if they become concentrated through evaporation. Wash immediately if these acids are spilled, and inform the instructor of the spill.
- NaOH is caustic and damaging to the skin if spilled. Wash after use and inform the instructor of any spills.
- Ammonia is a cardiac stimulant and respiratory irritant.
- The dimethylglyoxime is provided as a solution in alcohol. This reagent is flammable!
- Sodium hypochlorite solutions release toxic chlorine gas if oxidizing/reducing agents are added.
- Hydrogen peroxide solutions may burn the skin if spilled.

Apparatus/Reagents Required

Group III known sample (containing chromium, iron, aluminum, cobalt, and nickel ions), Group III unknown (which may contain any or all of the Group III cations), 1 *M* ammonium chloride, 6 *M* aqueous ammonia, 6 *M* HCl, 6 *M* nitric acid, 1 *M* thioacetamide, saturated ammonium thiocyanate solution, 1% dimethylglyoxime (in ethanol), 6 *M* NaOH, 6 *M* sulfuric acid, 5% bleach (sodium hypochlorite), 1 *M* potassium thiocyanate, aluminon reagent, 0.1 *M* barium nitrate, 3% hydrogen peroxide

Procedure

Record all data and observations directly in your notebook in ink.

Obtain a Group III known sample (containing all of Cr^{3+}, Fe^{2+}, Al^{3+}, Co^{2+}, and Ni^{2+}) and a Group III unknown solution (which may contain any or all of these cations).

Place about 8 mL of the sample in a clean casserole or evaporating dish, and reduce the volume of the sample by half. Beware of spattering of the sample during this concentration.

When the volume of the sample has been reduced to about 4 mL, swirl the remaining solution around in the dish to redissolve any solids that have formed, and transfer the solution to a test tube.

A. Separation and Confirmation of Nickel and Cobalt

Add 15–20 drops of 1 *M* ammonium chloride solution to the sample in the test tube and stir.

Add 6 *M* aqueous ammonia to the sample dropwise until the solution is just barely basic (remove a drop of the solution with a stirring rod and touch the drop to a strip of pH test paper). Add 10–12 additional drops of 6 *M* ammonia. The mixture of ammonia/ammonium ion has buffered the solution at a basic pH.

Add 20–25 drops of 1 *M* thioacetamide solution to the sample; stir, and heat the test tube in a boiling water bath for 5–10 minutes. This permits hydrolysis of the thioacetamide and precipitation of the Group III sulfides. If the odor of hydrogen sulfide is noticeable, transfer the heating to the exhaust hood.

Cool the test tube and centrifuge the sample.

Remove the supernatant liquid and test with a few additional drops of thioacetamide, with heating in the boiling water bath, to make certain that all the Group III cations have been precipitated.

If a precipitate forms, add another 15–20 drops of thioacetamide and boil for 5 more minutes.

Centrifuge the mixture, and combine any further precipitate that forms with the first batch of sulfide precipitate. If this were a general analysis of all the groups, the supernatant liquid (centrifugate) at this point would contain the cations of Groups IV and V.

Add 25–30 drops of 6 *M* HCl to the precipitate of the Group III sulfides.

Stir the precipitate/HCl mixture, and heat in the boiling water bath for 10 minutes.

During the heating period, the sulfides of Cr^{3+}, Al^{3+}, and Fe^{2+} will dissolve, whereas the sulfides of Co^{2+} and Ni^{2+} will remain as black solids.

Cool the test tube and add 20–25 drops of distilled water.

Centrifuge the mixture, and transfer the supernatant liquid (containing chromium, aluminum, and iron) to another clean test tube. Save this supernatant for the analysis in Part B.

Wash the black precipitate of CoS/NiS with several small portions of 6 *M* HCl and distilled water; centrifuge and discard the washings.

Add 1 mL of 6 *M* HCl and 1 mL of 6 *M* HNO_3 to the CoS/NiS precipitate. Stir the mixture, and heat the test tube in the boiling water bath for 2–3 minutes.

The CoS/NiS precipitate should dissolve almost completely. If it does not, add 10 drops each of 6 *M* hydrochloric and 6 *M* nitric acids and boil for 2–3 minutes to dissolve the remainder of the Ni/Co precipitate.

Add 2 mL of distilled water to the test tube, and heat the test tube in the boiling water bath for 10 minutes to drive off hydrogen sulfide gas. If the odor of hydrogen sulfide is noticeable, transfer the heating to the exhaust hood.

Divide the solution containing Ni^{2+} and Co^{2+} into two portions in separate clean test tubes.

To confirm the presence of Co^{2+}, slowly add 1 mL of saturated ammonium thiocyanate solution to one of the portions of solution. If Co^{2+} is present, the sample will turn blue because of the production of the blue $Co(NCS)_4^{2-}$ complex ion.

To confirm the presence of Ni^{2+}, add aqueous 6 *M* ammonia to the second portion of sample solution until the sample is just basic to pH paper. Then add 1–2 mL of dimethylglyoxime reagent. The appearance of a fluffy red precipitate of $Ni(DMG)_2$ confirms the presence of nickel(II) in the original sample.

B. Separation and Identification of Iron

To the supernatant solution from Part A containing iron, aluminum, and chromium, add 6 *M* NaOH dropwise until the solution is just barely basic to pH paper. Then add 10–15 additional drops of 6 *M* NaOH. A precipitate is likely to form.

Transfer the test tube to the boiling water bath, stir, and boil for 5 minutes.

Remove the test tube from the boiling water bath, add 2 mL of bleach (sodium hypochlorite solution), and stir with a clean glass rod for 2 minutes.

Return the test tube to the boiling water bath for 2 minutes. Then add 10–15 drops of 6 *M* aqueous ammonia, stir, and boil for 2 more minutes.

At this point, any iron present in the sample has precipitated as $Fe(OH)_3$. Any aluminum and chromium present are present in solution as $Al(OH)_4^-$ and CrO_4^{2-}, respectively.

Centrifuge the mixture, and remove the supernatant liquid containing aluminum and chromium to a separate clean test tube. Save this supernatant liquid for Part C.

Wash the precipitate of $Fe(OH)_3$ with 2 mL of water to which 10 drops of 6 *M* NaOH have been added.

Centrifuge the mixture, and discard the centrifugate. Repeat the washing of the precipitate with a second portion of water/NaOH.

Dissolve the precipitate of $Fe(OH)_3$ by adding 1 mL of distilled water, followed by 6 *M* sulfuric acid dropwise with stirring.

To confirm the presence of iron, add 5–6 drops of 1 *M* potassium thiocyanate. The appearance of the characteristic blood-red $FeNCS^{2+}$ complex confirms the presence of iron in the original sample.

C. Separation and Confirmation of Aluminum and Chromium

To the supernatant solution from Part B that contains $Al(OH)_4^-$ and CrO_4^{2-}, add 6 *M* nitric acid with stirring until the solution is acidic to pH paper.

Add 6 *M* aqueous ammonia dropwise with stirring until the mixture is just barely basic to pH paper (the addition of ammonia buffers the mixture). Then add 5–6 additional drops of 6 *M* ammonia.

At this point, if aluminum is present, a gelatinous white precipitate of $Al(OH)_3$ will be present. If chromium is present, the solution will show the yellow color of chromate ion, CrO_4^{2-}.

Centrifuge the mixture, saving the supernatant liquid that may contain chromium. Wash the precipitate of $Al(OH)_3$ with two or three 2-mL portions of distilled water.

After adding the wash water, heat the mixture in the boiling water bath for 2 minutes before centrifuging and discarding the washings.

Add 10–15 drops of 6 *M* nitric acid to dissolve the precipitate of $Al(OH)_3$, followed by 3–4 drops of aluminon reagent (a specific test reagent for aluminum).

Make the solution just basic to pH paper by adding 6 *M* aqueous ammonia dropwise with stirring. A precipitate of $Al(OH)_3$ will reform under these conditions. Because the precipitate adsorbs the aluminon reagent on its surface, the precipitate of $Al(OH)_3$ appears red. The appearance of the red precipitate confirms the presence of aluminum in the original sample.

If the supernatant solution from which $Al(OH)_3$ was removed is yellow, the sample almost certainly contains chromium. As an initial confirmation of this, add 10–15 drops of 0.1 *M* barium nitrate solution. If chromium is present, a yellow precipitate of $BaCrO_4$ will form.

Transfer the sample to the hot water bath, stir, and boil for 3–4 minutes. Centrifuge the mixture, decant the solution, and wash the precipitate with a small portion of distilled water. Centrifuge and discard the washing liquid.

As a second confirmation of chromium, dissolve the precipitate by adding 10–15 drops of 6 *M* nitric acid, stir the solution, and add several drops of 3% hydrogen peroxide. The transient appearance of a blue color [due to the reduction of Cr(VI) to Cr(II)] confirms the presence of chromium in the original sample.

Name ______________________________ Section ______________________________

Lab Instructor ______________________________ Date ______________________________

EXPERIMENT 45 Qualitative Analysis of the Group III Cations

PRE-LABORATORY QUESTIONS

1. In addition to the cations analyzed in this experiment, Group III also formally includes Mn^{2+} and Zn^{2+} ions. Consult a textbook of qualitative analysis or a handbook and find the means by which these ions are separated from the other members of Group III, as well as the specific confirmatory tests for these ions.

____

____

____

____

____

____

2. For the Group III cations (Cr^{3+}, Al^{3+}, Fe^{3+}, Ni^{2+}, and Co^{2+}) studied in this experiment, construct a flow chart showing how they are separated and identified.

Name ______________________ Section ______________________

Lab Instructor ______________________ Date ______________________

EXPERIMENT 45 Qualitative Analysis of the Group III Cations

RESULTS/OBSERVATIONS

Observation on precipitation of the Group III sulfides

Known ______________________

Unknown ______________________

A. Separation and Confirmation of Nickel and Cobalt

Observation on adding HCl to the sulfides of Group III

Known ______________________

Unknown ______________________

Observation on dissolving CoS/NiS in aqua regia (HCl/HNO_3)

Known ______________________

Unknown ______________________

Observation on adding thiocyanate ion to Co/Ni solution

Known ______________________

Unknown ______________________

Observation on adding DMG to the ammoniacal Co/Ni solution

Known ______________________

Unknown ______________________

B. Separation and Identification of Iron

Observation on adding NaOH to Fe/Al/Cr solution

Known ______________________

Unknown ______________________

Observation on adding bleach to Fe/Al/Cr solution/precipitate

Known ______________________________

Unknown ______________________________

Observation on adding sulfuric acid to $Fe(OH)_3$ precipitate

Known ______________________________

Unknown ______________________________

Observation on adding thiocyanate ion to Fe^{3+} solution

Known ______________________________

Unknown ______________________________

C. Separation and Confirmation of Aluminum and Chromium

Observation as Al/Cr solution is buffered at basic pH

Known ______________________________

Unknown ______________________________

Observation on reprecipitating $Al(OH)_3$/aluminon

Known ______________________________

Unknown ______________________________

Observation on adding Ba^{2+} to the chromate ion solution

Known ______________________________

Unknown ______________________________

Observation on adding hydrogen peroxide to the chromium solution

Known ______________________________

Unknown ______________________________

Based on the tests of your unknown sample, what Group III cations does it contain?

Ions ______________________ Unknown code number ______________

QUESTIONS

1. In several places in the procedure, you were told to acidify the mixture and then to add ammonia (a base) until the solution was basic to pH paper. What purpose did the addition of acid/ammonia serve? Write an equation that shows this.

2. Write balanced equations for the oxidation of Fe^{2+} and Cr^{3+} ions by bleach (hypochlorite ion, OCl^-) .

3. As a second confirmatory test for chromium, you added hydrogen peroxide, which reduced chromium(VI) to chromium(II). You were warned, however, that the blue color of chromium(II) would only be transient. Write a balanced equation for the reduction of chromium(VI). What feature of Cr(II) makes it likely that the color will only persist briefly?

4. In one place in the procedure, aluminum was dissolved by adding NaOH to form the $Al(OH)4^-$ ion, whereas in another place, $Al(OH)_3$ was dissolved by adding acid. What property of aqueous aluminum solutions is being made use of here?

EXPERIMENT

46 Qualitative Analysis of the Group IV and V Cations

Objective

A known sample containing Group IV cations (Ba^{2+}, Ca^{2+}) and the Group V cations (Na^{+}, K^{+}, and NH_4^{+}), as well as an unknown containing any or all of these cations, will be analyzed.

Introduction

In a general mixture of cations, once all insoluble chlorides (Group I) and all insoluble sulfides (Groups II and III) are removed, the sample will contain cations of only the alkali and alkaline earth metals, plus the ammonium ion.

The alkaline earth elements are separated from the other ions as the insoluble carbonate compounds.

$$Ca^{2+}(aq) + CO_3^{2-}(aq) \rightarrow CaCO_3(s)$$

$$Ba^{2+}(aq) + CO_3^{2-}(aq) \rightarrow BaCO_3(s)$$

Once the alkaline earth carbonates are removed by centrifugation, these cations are then brought back into solution by acidifying the mixed carbonate precipitate. Barium is then separated and confirmed as the insoluble yellow chromate, while calcium is identified by precipitation as the white oxalate.

$$Ba^{2+}(aq) + CrO_4^{2-}(aq) \rightarrow BaCrO_4(s)$$

$$Ca^{2+}(aq) + C_2O_4^{2-}(aq) \rightarrow CaC_2O_4(s)$$

The alkali metal ions do not form precipitates with any of the more common reagents, and so these species are determined by flame tests of the supernatant solution remaining after the Group IV carbonates are removed. Sodium imparts its characteristic yellow/orange color to the burner flame, whereas potassium imparts a violet color. A real sample containing potassium is also likely to contain sodium, since the properties and sources of these elements are so similar. Since the intense color imparted to the flame by sodium will mask the weak violet color of potassium, the flame test for potassium is observed through a piece of cobalt blue glass. This glass absorbs the wavelengths of the sodium emissions and permits the violet color of potassium in the flame to be seen.

The test for ammonium ion is performed on a sample of the original unknown, since ammonium ion may have been removed (as NH_3) in other steps of the procedure. A sample of the original mixture is made basic and heated, which generates NH3 gas. The production of ammonia may be confirmed by its odor or by the action of ammonia on pH test paper.

$$NH_4^+(aq) + OH^-(aq) \rightarrow NH_3(g) + H_2O$$

Safety Precautions

- Wear safety glasses at all times while in the laboratory.
- Barium, chromate, and oxalate compounds are toxic. Wash after using them.
- Acetic and hydrochloric acids will burn the skin, especially if the acids are concentrated by evaporation. Use caution. If the acids are spilled, wash immediately and inform the instructor.
- Ammonia is a cardiac stimulant and is severely irritating to the respiratory system.
- Exercise caution in evaporating solutions to dryness. Beware of spattering.

Apparatus/Reagents Required

Group IV/V known sample containing all the ions, Group IV/V unknown sample that may contain any or all of the cations, 6 *M* HCl, 6 *M* aqueous ammonia, 1 *M* ammonium carbonate, 6 *M* acetic acid, 1 *M* potassium chromate, 1 *M* potasssium oxalate, cobalt blue glass, 6 *M* NaOH, nichrome wire

Procedure

Record all data and observations directly in your notebook in ink.

A. Separation and Confirmation of the Group IV Cations

Obtain a known sample containing the Group IV/V cations, as well as an unknown sample that may contain any or all of the cations.

Set aside about half of each sample in separate clean, labeled test tubes for performing the test for the presence of ammonium ion, as well as flame tests for the presence of sodium and potassium.

Reduce the volume of the remaining known and unknown samples to 2–3 mL by evaporating most of the water in clean evaporating dishes or casseroles over a very small burner flame. Beware of spattering as the solutions become more concentrated.

Transfer the reduced samples to separate clean test tubes and add 20–30 drops of 6 *M* HCl.

Add 6 *M* ammonia dropwise to the sample until it is basic to pH paper. (Remove one drop of sample with a clean glass rod and touch the drop to a strip of pH paper.)

After the sample is basic, add 10–15 drops of ammonia to buffer the solution.

Add 20–30 drops of 1 *M* ammonium carbonate to the sample and stir. A precipitate of barium/calcium carbonate should form.

Set up a beaker of water and heat it to 70–80°C. Transfer the test tube containing the precipitated carbonate sample to this warm water bath for 2–3 minutes. (The sample cannot be boiled.)

After the sample has been warmed, centrifuge the mixture, and decant the supernatant liquid (containing Na^+ and K^+) to a separate clean test tube; save this solution for Part B.

Wash the precipitate with 20–30 drops of distilled water, centrifuge, and discard the wash water.

Add 6 *M* acetic acid to the precipitate of barium/calcium carbonates with stirring until the precipitate just dissolves; then add 5 additional drops of acetic acid.

Add 15–20 drops of distilled water and 2–3 drops of 6 *M* aqueous ammonia, followed by 15–20 drops of 1 *M* potassium chromate. The appearance of a yellow precipitate (BaCrO4) confirms the presence of barium in the original sample.

Centrifuge the mixture and remove the supernatant liquid (containing calcium ion). Add 6 *M* ammonia to the solution until it is basic to pH paper. (The solution is orange when removed from the barium precipitate, and becomes yellow when made basic with ammonia.)

Add 10–15 drops of 1 *M* potassium oxalate solution to the solution and allow it to stand for several minutes. The eventual appearance of a white precipitate of calcium oxalate confirms the presence of calcium in the original sample.

B. Confirmation of Sodium and Potassium Ions

Sodium and potassium form no precipitates under normal conditions, and so these ions are confirmed by flame tests. In a real analysis of a general unknown, many of the reagents added in the treatment of the earlier groups themselves are sodium or potassium compounds. For this reason, the flame tests are always performed on a sample of the original unknown, before any of the earlier groups of cations are removed.

Prepare two nichrome flame test wires by forming a small loop in one end of 8-inch strips of wire.

Dip the loop end of the wire into 6 *M* HCl; then heat the wire in the oxidizing portion of a burner flame until no color is imparted to the flame by the wire.

Dip one of the wires into a sample of the original untreated known solution, and transfer a drop of the solution to the burner flame. The appearance of sodium's characteristic yellow/orange color in the flame indicates the presence of sodium in the original known solution.

Clean the wire and repeat the test with a drop of the original untreated unknown solution. Record whether the original unknown gives the yellow/orange flame color of sodium.

Clean the wire and repeat the flame test, using the supernatant solution decanted from the Group IV carbonates in Part A.

The violet color imparted to the burner flame by potassium ion is very faint, very fleeting, and easily masked by the more intense color imparted to the flame by sodium ion. Obtain a piece of cobalt blue glass, and observe the flame tests for potassium through the blue glass. The wavelengths of yellow/orange light emitted by sodium are absorbed by the glass, allowing the violet color of potassium to be observed.

Take 1–2 mL of the original untreated samples and carefully evaporate them practically to dryness to concentrate any potassium present. When the samples have been concentrated, pick up a small amount of the solids remaining on the loop of the second flame test wire. Insert this material into the burner flame and observe through the cobalt blue glass. Potassium is indicated if a violet color is imparted to the flame.

Clean the wire, and repeat the test on a portion of the supernatant liquid (after evaporating most of the water) from which the Group IV carbonates were removed in Part A.

C. Confirmation of Ammonium Ion

As with sodium and potassium, many of the reagents used to separate and confirm the cations of the earlier groups are themselves ammonium compounds. For this reason, the test for the presence of ammonium ion must be performed on a portion of the original, untreated sample.

Transfer a 2–3-mL portion of the original, untreated sample to a small beaker. Suspend a moist piece of pH paper at the top of the beaker, and add 1–2 mL of 6 *M* NaOH to the sample.

Heat the beaker gently over a very small flame without allowing the solution to boil. While the beaker is heated, cautiously waft some of the vapors coming from the sample toward your nose to see if the odor of ammonia can be detected. Observe the pH paper. A change in color to basic indicates the presence of ammonium ion in the original sample.

Name ______________________________ Section ______________________

Lab Instructor ______________________________ Date ______________________

EXPERIMENT 46 Qualitative Analysis of the Group IV and V Cations

PRE-LABORATORY QUESTIONS

1. In addition to the cations determined in this experiment, Group IV also formally includes magnesium, Mg^{2+}. Use a textbook of qualitative analysis or a handbook to determine how magnesium is separated from the other cations of Group IV as well as the specific confirmatory test for the presence of Mg^{2+} ion.

____

____

____

____

____

2. Although the procedure given in the experiment makes use of flame tests to identify the sodium and potassium ions more simply, these cations do each form a characteristic precipitate that is sometimes used in their identification. Use a textbook of qualitative analysis or a handbook to determine the identity of these precipitates.

____

____

____

____

____

3. On the reverse of this page (or on a separate sheet of paper), design a flow chart for the separation of the ions in this experiment.

Name ______________________ Section ______________________

Lab Instructor ______________________ Date ______________________

EXPERIMENT 46 Qualitative Analysis of the Group IV and V Cations

RESULTS/OBSERVATIONS

A. Separation and Confirmation of the Group IV Cations

Observation on adding $(NH_4)_2CO_3$ to the buffered mixture of the Group IV/V cations

Known ______________________

Unknown ______________________

Observation on adding acetic acid to the Ba/Ca precipitate

Known ______________________

Unknown ______________________

Observation on adding potassium chromate to the Ba/Ca solution

Known ______________________

Unknown ______________________

Observation on adding ammonium oxalate to the Ba/Ca solution

Known ______________________

Unknown ______________________

B. Confirmation of Sodium and Potassium Ions

Flame test for sodium

Known ______________________ Unknown ______________________

Flame test for potassium (viewed through cobalt glass)

Known ______________________ Unknown ______________________

C. Confirmation of Ammonium Ion

Observation of pH test paper as basic solution is heated

Known ______________________

Unknown ______________________

Odor of solution as it is heated

Known ______________________________

Unknown ______________________________

Which Group IV/V cations have been shown to be present in your unknown sample?

Ions ______________________ Unknown code number ______________

QUESTIONS

1. Why are the flame tests for the Group V cations performed on a sample of the original unknown, rather than on the supernatant solution from which the earlier groups of cations have been removed?

2. Although sodium emits *yellow/orange* light when excited, a piece of *blue* glass was used to filter the sodium emission to allow the emission of potassium to be seen. Explain.

3. Before the carbonates of barium and calcium are precipitated, the solution is buffered at basic pH. How was the mixture buffered, and why was this necessary?

4. Why was it important that the solution containing the precipitated barium/calcium carbonate mixture *not* be boiled?

5. Write an equation showing how gaseous ammonia is evolved from an ammonium salt when a strong base (such as NaOH) is added.

EXPERIMENT

47 Qualitative Analysis of Selected Anions

Objective

Solid known and unknown samples will be tested for the presence of selected anions. Elimination and confirmation tests will be used to divide the samples tested into groups based on specific chemical properties.

Introduction

In the previous four experiments, methods were outlined for the separation and identification of various cations. The cations were divided into particular groups based on chemical properties shown by the members of each group. For example, cations were placed in Group I if their chloride salts were insoluble in acid. By repeated separation techniques, the cations can be identified and confirmed one by one until all the cations in a mixture have been determined. A similar progressive method can be used to identify the components of a mixture of anions. The anions are separated into smaller groups from the general mixture and then confirmed one by one.

A completely general scheme for the anions is very time-consuming, however, and will not be made use of in this experiment. Rather, individual solid samples containing a single anion will be tested, and the normal confirmatory test for the particular anion will be demonstrated. Then several solid unknown samples will be tested, with each unknown again containing only a single anion. Preliminary screening tests will be performed on the known and unknown samples. On the basis of these preliminary tests, the traditional anion groups are constituted.

Safety Precautions

- Wear safety glasses at all times while in the laboratory.
- All the solid samples in this experiment should be assumed to be toxic. Wash after handling them.
- Barium, nitrite, molybdenum, and permanganate compounds are toxic. Wash after use.
- Sulfuric and nitric acids cause severe burns, especially if the acids are allowed to concentrate through evaporation. Wash immediately if the acids are spilled, and inform the instructor.
- SO_2 and NO_2 gases are toxic and are irritating to the respiratory system. Inhale the smallest possible quantities of these gases in the confirming tests for sulfite and nitrate ions.
- Methylene chloride is toxic and its vapors are harmful.
- Ammonia is a cardiac stimulant and its vapors are irritating to the respiratory tract.
- Silver nitrate will stain the skin if spilled. The stain, which consists of metallic silver, is not harmful. However, it will take several days to wear off.

Apparatus/Reagents Required

Known solid samples of the anions (sulfate, sulfite, carbonate, nitrate, phosphate, chloride, bromide, iodide), two unknown samples (each containing one of the anions), 0.1 *M* barium chloride, 6 *M* nitric acid, 6 *M* sulfuric acid, 0.1 *M* silver nitrate, 0.5 *M* ammonium molybdate reagent, 1 *M* iron(II) sulfate, 1 *M* potassium nitrite, methylene chloride, 0.1 *M* potassium permanganate, 6 *M* aqueous ammonia

Procedure

Record all data and observations directly in your notebook in ink.

The anions may be quickly divided into three subgroups by some simple preliminary tests. Apply the tests to approximately 0.1-g samples of each of the known anion samples and to each of the unknown samples.

The known anion samples are sodium or potassium salts, ensuring that the anion samples will be soluble in water. The anions to be tested are: sulfate, SO_4^{2-}; sulfite, SO_3^{2-}; carbonate, CO_3^{2-}; nitrate, NO_3^-; phosphate, PO_4^{3-}; and the halides (Cl^-, Br^-, and I^-). In addition to the tests on the known anion samples, two unknown samples will be determined.

A. Anions Whose Barium Salts Are Insoluble

Set up a rack of test tubes, and label each of the test tubes for one of the anions to be tested. Also provide labeled test tubes for each of your unknown anion samples.

Place a few crystals of the appropriate anion or unknown sample into its respective test tube. Add about 1 mL of distilled water and swirl each test tube to dissolve the salt.

Add 10 drops of 0.1 *M* barium chloride to each test tube, and record which anions produce insoluble barium salts.

Add 10–15 drops of 6 *M* nitric acid to each test tube, and swirl the test tubes. Record which anions give precipitates with barium ion that are then insoluble in acid. Discard the samples.

B. Anions That Form Volatile Substances When Acidified

Set up a rack of test tubes, and label each of them for one of the anions to be tested. Also provide labeled test tubes for each of your unknown anion samples.

Place a few crystals of the appropriate anion or unknown sample into its respective test tube. Add about 1 mL of distilled water and swirl each test tube to dissolve the salt.

Set up a boiling water bath for use in heating the samples.

Treat each sample of anion/unknown separately to avoid confusing the samples. Add 10–12 drops of 6 *M* sulfuric acid to the sample and note whether any gas is evolved from the sample. Cautiously waft the vapors from the test tube toward your nose to detect any characteristic odor.

Heat each sample one at a time in the boiling water bath to see if any gas/odor is evolved under these conditions. Record which samples produced a volatile product/odor. Discard the samples.

C. Anions That Form Insoluble Silver Compounds

Set up a rack of test tubes, and label each of the test tubes for one of the anions to be tested. Also provide labeled test tubes for each of your unknown anion samples.

Place a few crystals of the appropriate anion or unknown sample into its respective test tube. Add about 1 mL of distilled water and swirl each test tube to dissolve the salt.

Add 10 drops of 0.1 *M* silver nitrate solution to each test tube. Record which samples precipitate with silver ion.

Add 10–15 drops of 6 *M* nitric acid to each sample. Swirl the test tubes and record which samples produce a precipitate with silver ion, that is then insoluble in acid.

D. Confirmation Tests for the Anions

The preliminary tests given in Parts A, B, and C should have indicated to you some general properties of the anions to be determined. Based on these tests, you should have some indication as to what anions your two unknown samples contain. If the results of the preliminary tests are not clear to you, consult with your instructor before proceeding.

1. Sulfate Ion

Sulfate ion is precipitated by barium ion, and the precipitate is insoluble in acid. While other substances may precipitate *initially* with barium ion, such precipitates are subsequently soluble in acid.

Place a small portion of the sulfate known and a small portion of any unknown believed to be sulfate from the preliminary tests into separate clean test tubes. Add 10 drops of water and swirl to dissolve the samples.

Add 10 drops of 0.1 *M* barium chloride solution, followed by 10 drops of 6 *M* nitric acid. If the unknown sample does contain sulfate ion, the precipitate formed with barium ion will not dissolve when acid is added.

2. Phosphate Ion

Phosphate ion is one of the ions that is precipitated by barium ion, but the precipitate is subsequently soluble in acid. Phosphate ion similarly is precipitated by silver ion, but as before, the precipitate subsequently dissolves in acid. Phosphate ion does not generate a volatile product when acidified and boiled. A specific reagent, ammonium molybdate, is used to confirm the presence of phosphate. Phosphate forms a characteristic yellow precipitate with this reagent.

Place a small portion of the phosphate known and of any unknown believed to be phosphate from the preliminary tests into separate clean test tubes. Add 10 drops of water and 5 drops of 6 *M* nitric acid to each test tube. Swirl the test tubes to dissolve the samples.

Add 10 drops of 0.5 *M* ammonium molybdate reagent to each sample, and transfer the test tubes to a boiling water bath. The appearance of a yellow precipitate confirms the presence of phosphate.

The precipitate is sometimes slow in forming if the sample is not saturated, but generally the sample will turn yellow as the ammonium molybdate is added.

Continue heating the samples in the boiling water bath for 5 minutes, then remove and transfer to an ice bath. Check each sample carefully for the yellow precipitate. (There may only be a small amount of precipitate present if the solution was not truly saturated.)

3. Nitrate Ion

Nitrate ion does not precipitate with barium ion or with silver ion and does not generate a volatile product when acidified and heated. A specific test reagent is used for nitrate ion, consisting of acidified iron(II) sulfate. Nitrate ion will oxidize iron(II) to iron(III), and brown nitrogen(IV) oxide gas will be evolved if the mixture is heated.

Place a small portion of the nitrate known and of any unknown that is believed from the preliminary tests to be nitrate into separate clean test tubes. Dissolve the samples in 15–20 drops of 6 *M* sulfuric acid.

Add 10–15 drops of 1 *M* iron(II) sulfate to each sample and transfer to the boiling water bath. The evolution of brown NO_2 gas confirms nitrate in the sample.

4. Sulfite Ion

Sulfite ion precipitates with barium ion, but the precipitate is subsequently soluble in acid. Sulfite ion may also precipitate with silver ion, but the precipitate will dissolve after a few minutes as complexation occurs. Sulfite ion, however, generates sulfur dioxide gas when acidified and heated, and the gas may be identified by its characteristic odor.

Place a small portion of the sulfite known and of any unknown believed to be sulfite ion from the preliminary tests into separate clean test tubes. Add 10–15 drops of distilled water and swirl to dissolve.

Add 10 drops of 6 *M* sulfuric acid and note whether a gas or odor is evolved from the sample. Heat the sample in the boiling water bath for 30 seconds and check again for the characteristic odor of sulfur dioxide.

5. Carbonate Ion

Carbonate precipitates with barium ion and silver ion, but the precipitates are subsequently soluble in acid. Carbonate ion evolves colorless, odorless carbon dioxide gas when acidified. If a more specific test for carbonate is desired, the gas evolved from the carbonate when acidified can be bubbled into limewater, from which insoluble calcium carbonate will precipitate.

Place a small portion of the carbonate known and of any unknown that is believed to be carbonate ion into separate clean test tubes. Add 5–6 drops of 6 *M* sulfuric acid. The vigorous evolution of a colorless, odorless gas can be taken as a positive test for carbonate ion.

6. Halide Ions

Halide ions (chloride, bromide, iodide) do not form precipitates with barium ion and generally do not evolve a volatile product when acidified with a nonoxidizing acid. The halides do precipitate with silver ion, however.

Place a small portion of the chloride, bromide, and iodide known samples into separate clean test tubes (label the test tubes). Also place a small portion of any unknown believed to be one of the halides into a clean test tube. Dissolve each of the samples in 10–15 drops of distilled water.

To the iodide known sample and the unknown sample, add 1 mL of methylene chloride and 20–30 drops of 1 *M* potassium nitrite solution. Stopper the samples and shake. Nitrite ion will oxidize iodide ion to elemental iodine, which is preferentially soluble in the methylene chloride added.

The iodide known sample will develop a purple color in the methylene chloride layer, as will the unknown if it contains iodide. Save the unknown sample if iodide was not present.

Place a small portion of the bromide known sample into a clean test tube. If the unknown tested earlier did not contain iodide ion, it may contain bromide or chloride ion. With a medicine dropper remove the upper (aqueous) layer from the unknown sample that was tested for iodide ion, and transfer the upper layer to a clean test tube.

Add 1 mL of methylene chloride to each test tube, followed by 2 drops of 0.1 *M* potassium permanganate. Stopper the test tubes and shake. Permanganate ion will oxidize bromide ion to elemental bromine, which is preferentially soluble in the methylene chloride added.

The bromide known sample will develop a red color in the methylene chloride layer, as will the unknown if the unknown contains bromide ion. Save the unknown sample if bromide ion was not present.

Place a small sample of the chloride known sample into a clean test tube. If the unknown tested earlier did not contain bromide ion, then remove the upper (aqueous) layer from the sample with a medicine dropper and transfer to a clean test tube.

Add 10–15 drops of 0.1 *M* silver nitrate to each sample. The appearance of a white precipitate is taken as an initial indication that chloride ion is present. Centrifuge the mixture and decant the supernatant liquid. Wash the precipitate with 20–30 drops of distilled water, centrifuge, and discard the rinsings.

To confirm chloride ion, dissolve the precipitate in 15–20 drops of 6 *M* aqueous ammonia. Then add 6 *M* nitric acid dropwise to the solution, which will cause the precipitate of silver chloride to reappear.

Name ______________________ Section ______________________

Lab Instructor ______________________ Date ______________________

EXPERIMENT 47 Qualitative Analysis of Selected Anions

PRE-LABORATORY QUESTION

This experiment provides methods for the detection of a few of the more common anions, but many equally common anions are omitted from the analysis for brevity. Consult a textbook of qualitative analysis or a handbook, and indicate briefly how the presence of the following anions could be detected.

Nitrite ion, NO_2^- ______________________

____

____

____

____

Acetate ion, $C_2H_3O_2^-$ ______________________

____

____

____

____

Sulfide ion, S^{2-} ______________________

____

____

____

____

Name ______________________ Section ______________

Lab Instructor ______________________ Date ______________

EXPERIMENT 47 Qualitative Analysis of Selected Anions

RESULTS/OBSERVATIONS

Identification numbers of unknowns ______________ ______________

A. Which known/unknown samples form precipitates with Ba^{2+}?

____

____

B. Which known/unknown samples evolved a gas when acidified?

____

____

C. Which known/unknown samples form precipitates with Ag^{+}?

____

____

D. Confirmatory Tests: Observations

1. Sulfate ____

____

2. Phosphate ____

____

3. Nitrate ____

____

4. Sulfite ____

____

5. Carbonate

6. Chloride

7. Bromide

8. Iodide

Which anions do your unknowns contain? Give your reasoning.

QUESTIONS

1. Why is nitrite ion used in the oxidation of iodide ion, whereas permanganate is used for bromide ion?

2. The test for carbonate (acidification, with evolution of a colorless, odorless gas) is not truly specific for carbonate ion, since the same result would be obtained if a bicarbonate salt were treated in the same manner. How might you determine whether carbonate or bicarbonate ion were present in a solid sample of an unknown salt?

3. Write an equation showing the reduction of the nitrate ion by acidified iron(II) solution.

EXPERIMENT

48 Identification of an Unknown Salt

Objective

An unknown inorganic ionic compound (salt) will be characterized in experiments planned by the student.

Introduction

Throughout this manual, especially in the previous few experiments on qualitative analysis of cations and anions, you have studied the physical properties of many compounds. Also, you have investigated many of the chemical reactions that can be used to separate and identify substances. In this experiment, you will be given an unknown compound and will be asked to determine as much about the physical and chemical natures of the compound as time permits.

The unknown provided is an ionic salt that is composed of the various cations and anions you have investigated previously. The unknown may be provided as a hydrated solid, as an anhydrous solid, or as a solution. Some of the unknowns will contain only a single cation/anion pair, whereas other unknowns may be double salts, containing two types of cation combined for a single anion species.

You will plan your own analysis for this experiment. Your grade will be based on a correct identification of the cations/anions present in your unknown as well as on the quality of your analysis. *A haphazard approach to the analysis will lead to a low grade, even if by chance you happen to identify the unknown correctly.*

In addition to the determination of which cations/anions are contained in your unknown, you also will be expected to perform as many appropriate physical measurements of your unknown as possible. Measurements that might be appropriate include percentage mass loss of water on heating (for hydrated solids), solubility in water (if the solid is soluble), melting point (for those solids that melt at low enough temperatures), line spectrum, apparent molar mass, and so on. All of these physical measurements have been discussed and explained in earlier experiments.

All of the reagents you have used during the last few weeks for the qualitative analysis experiments will be available today should you need them. In addition, you will be provided with other reagents/apparatus by stockroom personnel, *upon the approval of your instructor*. You should keep a handbook nearby for reference for any additional spot tests or physical data that might help to confirm your unknown. Your instructor will also have several reference works on qualitative analysis available for you to consult.

Record all observations, test results, and numerical data directly in your notebook. Remember that a negative test is sometimes just as significant as a positive test in discerning the identity of an unknown.

As your laboratory report for this experiment, you will be asked to write a detailed summary of the tests and measurements you have made on your unknown, and your conclusions as to the unknown's identity. A well-written report, indicating the organization and findings of the analysis, is expected.

Safety Precautions

- Wear safety glasses at all times while in the laboratory.
- Assume that all acids and alkalis are damaging to skin. Wash immediately if any spills occur, and inform the instructor.
- Assume that all organic reagents/solvents are flammable.
- Assume that the vapors of all volatile substances are harmful, and confine the use of such substances to the exhaust hood.
- Assume that all other compounds are toxic. Wash after use.
- Consult the *Safety Precautions* that were discussed in earlier experiments for any measurements you plan to make.
- Consult with the instructor for specific safety precautions before performing any test that has not been previously performed.
- If you have any questions regarding any substance or procedure, consult with the instructor before attempting to work with it.

Procedure

Record all data and observations directly in your notebook in ink.

Obtain an unknown sample and record its identification number.

Divide the unknown sample into appropriately sized portions for the various tests you have planned. Always keep a spare sample of the unknown in case you lose one of the portions during a test.

A. Physical Tests

Record the results of all tests, whether positive or negative.

If your unknown is a solid, determine the percentage weight loss on heating the solid.

If your unknown is a solution, determine the concentration of the solution by careful, slow evaporation of the water in a measured volume of the original unknown. Use an evaporating dish held over a boiling water bath. Once the solvent has been removed, describe the appearance of the solute.

If your unknown is a solid, determine its solubility in water.

If the unknown is a solid, determine whether it has a melting point that is low enough to measure.

If the unknown is a solid, determine its apparent molar mass by freezing point depression in some appropriate solvent. The molar mass determined will not be the true molar mass (because the substance is ionic).

If the unknown is soluble in water (or is provided as a solution), record the positions of a few of the brightest lines in its emission spectrum.

Perform any other physical measurements you feel appropriate, *checking first with your instructor for permission to perform a specific test.*

B. Chemical Tests

Based on the experiments you have performed in the qualitative analysis of anions and cations, identify the components of your unknown salt.

Remember that there may be more than one cation or more than one anion in some of the unknowns (double salts). Since the unknown only contains a limited number of component species, it may not be necessary to go through the entire progressive scheme of qualitative analysis.

Judicious use of the handbook, which lists the colors and physical properties of common salts, can save you a lot of time and effort. Record the results of all tests performed, whether the results are positive or negative.

Name ______________________ Section ______________________

Lab Instructor ______________________ Date ______________________

EXPERIMENT 48 Identification of an Unknown Salt

RESULTS/OBSERVATIONS

A. Physical Tests

Summarize briefly the physical measurements performed on your unknown sample, along with the numerical values of the determinations made.

B. Chemical Tests

Indicate the specific chemical spot tests you performed on your unknown sample, along with the results (positive or negative) for each test.

Conclusions

Unknown code number ______________________

Substance unknown is believed to be ______________________

Brief summary of reasoning that led to this identification:

Appendix A

Significant Figures

1. Indicating Precision in Measurements

When a measurement is made in the laboratory, generally the measurement is made to the limit of precision permitted by the measuring device used. For example, when a very careful determination of mass is needed, an analytical balance is used, which permits the determination of mass to the nearest 0.0001 g. If such a balance were used in an experiment, it would be silly to record the mass determined for an object as "three grams," if the balance permits a mass determination such as 3.1123 g. It would be just as silly, however, to record your weight as determined on a bathroom scale to the nearest 0.0001 pound. Bathroom scales simply do not justify that much precision.

To indicate the actual precision with which a measurement has been made, scientists are very careful about how many **significant figures** they report in their results and data. The significant figures in a number consist of all the digits in the number whose values are known with complete certainty, plus a final digit that is a best estimate of the reading between the last scale divisions on the instrument used to measure the number.

For example, the thermometer in a student chemistry locker has scale divisions marked on the barrel of the thermometer for every whole degree from –20°C to 110°C. Suppose the mercury level in the thermometer were approximately halfway between the 35- and 36-degree marks. The temperature indicated by the thermometer would be known with complete certainty to be at least 35°. If the student using the thermometer made an estimate that the mercury level was 0.4 of the way from 35° to 36°, then the temperature recorded by the student should be 35.4°C, with the digit in the first decimal place a best estimate of the reading between the smallest scale divisions of the thermometer. The temperature would be reported correctly to three significant figures. Any scientist seeing the temperature 35.4° would realize that a thermometer with a scale divided to the nearest whole degree had been used for the measurement.

To help you, since you are just beginning your study of chemistry, this manual tries in most instances to point out the precision required for an experiment. For example, a procedure may say to "weigh 25 g of salt to the nearest 0.01 g." This statement indicates both how much of the substance to use (25 g) and the precision necessary for the experiment to have meaningful results. (Since the mass is to be measured to the nearest 0.01 g, the mass is needed to four significant-figure precision—two figures before and two figures after the decimal point.)

A point that often confuses beginning chemistry students is the meaning of a zero in a number. Sometimes zeros are significant figures, and sometimes they are not. If a zero is *imbedded* in a number, the zero is always significant and indicates that the reading of the scale of the instrument used for the measurement actually was zero.

Examples:

1.203 g   10.01 mL

When zeros occur at the *beginning* of a number, they are *never significant.* Zeros at the beginning of a number are just indicating the position of the decimal point relative to the first significant digit.

Examples:

0.25 g   0.001 L

One way to show that initial zeros are not significant is to write numbers in scientific notation. For example, the previous numbers could be written as

$$2.5 \times 10^{-1}\ \text{g} \qquad 1 \times 10^{-3}\ \text{L}$$

When numbers with initial zeros are written in this manner, the zeros are not even indicated.

Zeros at the *end* of numbers *may* or *may not be significant,* depending on the actual measurement indicated. For example, in the relationship

1 L = 1000 mL

the three zeros in 1000 mL are significant. They indicate that one liter is defined to be precisely one thousand milliliters. However, if you reported the weight of a beaker as 20.0000 g, your instructor would probably question your result. It is very unlikely that the weight of a beaker would have this precise weight.

To avoid confusion about whether or not zeros at the end of a number are significant, the scientist again resorts to scientific notation. If the 20.0000 g indicated above were exactly precise, this could be indicated by writing the mass as

$$2.00000 \times 10^{1}\ \text{g}$$

When trailing zeros are written in scientific notation, the zeros are assumed to be significant and to represent actual scale readings on the measuring device.

2. Arithmetic and Significant Figures

Although hand-held calculators have taken much of the drudgery out of mathematical manipulations in science, the use of such calculators has led to a whole new sort of problem. Calculators generally are able to handle at least eight digits in their displays; calculators assume that all numbers punched in have this number of significant figures and give all answers with this number of digits. For example, if you enter the simple problem

$$2 \div 3 = ?$$

a calculator will respond with an answer of 0.6666667. Clearly if 2 and 3 represent measurements in the laboratory that were made only to the nearest whole number, then a quotient like 0.6666667 implies too much precision. Not all the digits in the answer indicated by the calculator are significant.

A. Addition and Subtraction

When adding and subtracting numbers that are known to different levels of precision, the answer should be recorded only to the number of decimal places indicated by the number known to the fewest number of decimal places. Consider this addition:

20.2354
1.02
+ 337.114

Although 20.2354 is significant to the fourth decimal place, and 337.114 to the third decimal place, the number 1.02 is known only to the second decimal place. The sum of these numbers then can be reported correctly only to the second decimal place. Perform the arithmetic, but then round the answer so that it extends only to the second decimal place:

20.2354
1.02
+ 337.114
358.3694

which should be reported as 358.37.

B. Multiplication and Division

When multiplying or dividing numbers of different precision, the product or quotient should have only as many digits reported as the least precise number involved in the calculation. For example, consider this multiplication

(3.2795)(4.3302)(2.1)

Both 3.2795 and 4.3302 are known to five significant figures, whereas 2.1 is known with considerably less precision (only to two figures). The product of this multiplication may be reported only to two figures

(3.2795)(4.3302)(2.1) = 29.821871 (calculator) = 30.

Even though a calculator will report 29.821871 as the product, this answer must be rounded to two significant figures to indicate the fact that one of the numbers used in the calculation was known with a lower level of precision.

Appendix B

Exponential Notation

1. Writing Large and Small Numbers in Exponential Notation

Many of the numbers used in science are very large or very small, and are not conveniently written in the sort of notation used for normal-sized numbers. For example, the number of atoms of carbon present in 12.0 g of carbon is

602,000,000,000,000,000,000,000

Clearly, in a number of this magnitude, most of the zeros indicated are merely place-holders, used to locate the decimal point in the correct place.

With very large and very small numbers, it is usually most convenient to express the number in scientific or exponential notation. For example, the number of atoms of carbon could be written in exponential notation as

$$6.02 \times 10^{23}$$

This expression indicates that 6.02 is to be multiplied by 10 twenty-three times. To see how to convert a number to exponential notation, consider the following example.

Example:

Write 125000 in exponential notation.

The standard format for exponential notation uses a factor between 1 and 10 that gives all the appropriate significant digits of the number, multiplied by a power of ten that locates the decimal point and indicates the **order of magnitude** of the number.

The number 125000 can be thought of as 1.25×100000.

100000 is equivalent to $10 \times 10 \times 10 \times 10 \times 10 = 10^5$.

125000 in exponential notation is then 1.25×10^5.

A simple method for converting a large number to exponential notation is as follows. Move the decimal point of the given number to the position after the first significant digit of the number. This gives the multiplying factor for the number when it has been written in exponential notation. Count the number of places the decimal point has been moved from its original position (at the end of the number) to its new position (after the first significant digit). The number of places the decimal point has been moved represents the exponent of the power of ten of the number when expressed in scientific notation.

Example:

Write the following in exponential notation: 1723

Move decimal point to position after the first significant digit: 1.723

Decimal point has been moved three spaces to the left from the original 1723 to give1.723. Exponent will be 3:

$1723 = 1.723 \times 10^3$

Example:

Write the following in exponential notation: 7,230,000

Move decimal point to position after the first significant digit: 7.230000

Decimal point has been moved six spaces to the left from its original position. Exponent will be 6:

$7{,}230{,}000 = 7.23 \times 10^6$

Basically the same methods are used when attempting to express very small numbers in exponential notation. For example, consider this number:

0.0000359

This number is equivalent to 3.59×0.00001, and since 0.00001 is equivalent to 10^{-5}:

$0.0000359 = 3.59 \times 10^{-5}$

A simple method for converting a small number to exponential notation is as follows. Move the decimal point of the given number to the position after the first significant digit of the number. This gives the multiplying factor for the number when it has been written in exponential notation. Count the number of places the decimal point has been moved from its original position (at the beginning of the number) to its new position (after the first significant digit). The number of places the decimal point has been moved represents the exponent of the power of ten of the number when expressed in scientific notation. Because the number is small, the exponent is negative.

Example:

Write the following in exponential notation: 0.0000000072

Move decimal point to position after the first significant digit: 7.2

Decimal point has been moved nine spaces from its original position at the beginning of the number to after the first significant digit. Exponent will thus be –9:

$0.0000000072 = 7.2 \times 10^{-9}$

Example:

Write the following in exponential notation: 0.00498

Move decimal point to position after first significant digit: 4.98

Decimal point has been moved three spaces from its original location. Exponent will be –3:

$0.00498 = 4.98 \times 10^{-3}$

2. Arithmetic with Exponential Numbers

Oftentimes in chemistry problems, it becomes necessary to perform arithmetical operations with numbers written in exponential notation. Such arithmetic is basically no different from arithmetic with normal numbers, but certain methods for handling the exponential portion of the numbers are necessary.

A. Addition and Subtraction of Exponential Numbers

Numbers written in exponential notation can be added together or subtracted only if they have the same power of ten. Consider this problem:

$$1.233 \times 10^3 + 2.67 \times 10^2 + 4.8 \times 10^1$$

It is not possible to just add up the coefficients of these numbers. This problem, if written in normal (nonexponential) notation would be

$$\begin{array}{r} 1233 \\ 267 \\ +\ 48 \\ \hline 1548 \end{array}$$

Before numbers written in exponential notation can be added, they must all be converted to the same power of ten. For example, the numbers could all be written so as to have 10^3 as their power.

$$1233 = 1.233 \times 10^3$$
$$2.67 \times 10^2 = 0.267 \times 10^3$$
$$4.8 \times 10^1 = 0.048 \times 10^3$$

and the sum as $(1.233 + 0.267 + 0.048) \times 10^3 = 1.548 \times 10^3 = 1548$.

B. Multiplication and Division of Exponential Numbers

According to the commutative law of mathematics, the expression $(A \times B)(C \times D)$ could be written validly as $(A \times C)(B \times D)$. When this is applied to numbers written in exponential notation, it becomes very straightforward to find the product of two such numbers.

Example:

Evaluate: $(4.2 \times 10^6)(1.5 \times 10^4)$

$(4.2 \times 10^6)(1.5 \times 10^4) = (4.2 \times 1.5)(10^6 \times 10^4)$

$(4.2 \times 1.5) = 6.3$

$(10^6 \times 10^4) = 10^{(6+4)} = 10^{10}$

$(4.2 \times 10^6)(1.5 \times 10^4) = 6.3 \times 10^{10}$

When dividing numbers written in exponential notation, a similar application of the commutative law of mathematics is made: $(E \times F)/(G \times H) = (E/G) \times (F/H)$

Example:

Evaluate: $(4.8 \times 10^9)/(1.5 \times 10^7)$

$(4.8 \times 10^9)/(1.5 \times 10^7) = (4.8/1.5) \times (10^9/10^7)$

$(4.8/1.5) = 3.2$

$(10^9/10^7) = 10^{(9-7)} = 10^2$

$(4.8 \times 10^9)/(1.5 \times 10^7) = 3.2 \times 10^2$

Appendix C

Plotting Graphs in the General Chemistry Laboratory

Very often the data collected in the chemistry laboratory is best presented pictorially, by means of a simple graph. This section reviews how graphs are most commonly constructed, but you should consult with your instructor as to any special considerations he or she may wish you to include in the graphs you will be plotting in the laboratory.

The use of a graphical presentation allows you to display clearly experimental data, and any relationships that may exist among the data. Graphs also permit interpolation and extrapolation of data, to cover those situations that were not directly investigated in an experiment and to allow predictions to be made for such situations.

In order to be meaningful, your graphs must be well planned. Study the data before attempting to plot it. Recording all data in tabular form in your notebook will be a great help in organizing the data.

The particular sort of graph paper you use for your graphs will differ from experiment to experiment, depending on the precision possible in a particular situation. For example, it would be foolish to use graph paper with half-inch squares for plotting data that had been recorded to 4-significant-figure precision. It is impossible to plot such precise data on paper with such a large grid system. Similarly, it would not be correct to plot very roughly determined data on very fine graph paper. The fine scale would imply too much precision in the determinations.

A major problem in plotting graphs is deciding what each scale division on the axes of the graph should represent. You must learn to scale your data, so that the graphs you construct will fill virtually all of the graph paper page. Scale divisions on graphs should always be spaced at very regular intervals, usually in units of 10, 100, or some other appropriate power of ten. Each grid line on the graph paper should represent some readily evident number (e.g., not 1/3 or 1/4 unit). It is not necessary or desirable to have the intersection of the horizontal and vertical axes on your graph always represent the origin (0, 0). For example, if you were plotting temperatures from 100 to 200 degrees Celsius, it would be silly to start the graph at zero degrees. The axes should be labeled in ink as to what they indicate, and the scale divisions should be clearly marked. By studying your data, you should be able to come up with realistic minimum and maximum limits for the scale of the graph and for what each grid line on the graph paper will represent. Consult the many graphs in your textbook for examples of properly constructed graphs.

By convention, the **horizontal** (x) **axis** of a graph is used for plotting the quantity that the experimenter has varied during an experiment (independent variable). The **vertical** (y) **axis** is used for plotting the quantity that is being measured in the experiment (as a function of the other variable). For example, if you were to perform an experiment in which you measure the pressure of a gas sample as its temperature is varied, you would plot temperature on the horizontal axis, since this is the variable that is being controlled by the experimenter. The pressure, which is measured and which results from the various temperatures used, is then plotted on the vertical axis.

When plotting the actual data points on your graph, use a sharp, hard pencil. Place a single, small, round dot to represent each datum. If more than one set of data is being plotted on the same graph, small squares or triangles may be used for the additional sets of data. If an estimate can be made as to the likely magnitude of error in the experimental measurements, this can be indicated by **error bars** above and below each data point (the size of the error bars on the vertical scale of the graph should indicate the magnitude of the error on the scale). To show the relationship between the data points, draw the best straight line or smooth curve possible through the data points. Straight lines should obviously be inked in with a ruler, whereas curves should be sketched using a drafter's French curve.

If the data plot for an experiment gives rise to a straight line, you may be asked to calculate the **slope** and **intercept** of the line. Although the best way to determine the slope of a straight line is by the techniques of numerical regression, the slope may be approximated by

$$\text{slope} = (y_2 - y_1)/(x_2 - x_1)$$

if the data seem reasonably linear, where (x_1, y_1) and (x_2, y_2) are two points on the line. Once the slope has been determined, either intercept of the line may be determined by setting y_1 or x_1 equal to zero (as appropriate) in the equation of a straight line

$$(y_2 - y_1) = (\text{slope})(x_2 - x_1)$$

If your university or college has microcomputers available for your use, simple BASIC computer programs for determining the slope and intercept of linear data are commonly available. In fact, the microcomputer may even be set up to plot your data for you on a high-speed graphics plotter. Use of computer plots, though, may not be permitted by your instructor. The idea is for *you* to learn how to plot data.

Remember always *why* graphs are plotted. Although it may seem quite a chore when first learning how to construct them, graphs are intended to simplify your understanding of what you have determined in an experiment. Rather than just a list of numbers that may seem meaningless, a well-constructed graph can show you instantaneously if there is consistency and a relationship among your experimental data.

Appendix D

Errors and Error Analysis in General Chemistry Experiments

The likelihood of error in experimental scientific work is a fact of life. No measurement technique is completely perfect. Errors in experiments do happen—no matter how much effort is expended to prevent them. Naturally, however, the scientist must take every reasonable action to exclude errors due to sloppiness, poor preparation, and misinterpretation of results. Other errors, though, are unavoidable.

For such unavoidable errors, an estimation of the magnitude of the error, and how that error will affect the results and conclusions of the experiment, must be included as part of the data treatment for the experiment. Errors in experimental work can generally be classified as either of two types, determinate or indeterminate. **Determinate** errors may include personal errors by the experimenter, errors in the apparatus or instrumentation used for an experiment, and errors in the method used for the experiment. **Indeterminate** errors are introduced because of the fact that no measurement can be made with truly absolute certainty, and that for a series of duplicate measurements, there will always be minor fluctuations in the results obtained.

Personal errors are very common among beginning students, who have not yet mastered the various procedures and techniques used routinely in the chemistry laboratory. Personal errors can be eliminated by close attention to published procedures and to the advice and suggestions of your instructor. Most experiments are repeated several times in an attempt to provide a set of data, so that personal errors can be recognized if the results are not reproducible—that is, if the same measurement does not consistently give the "same" answer.

Errors due to the apparatus or instruments used for an experiment are less easy to detect or correct. However, some major sources of such error can be eliminated. Be sure all glassware you use is clean. This applies especially to volumetric glassware used for transferring or containing specific volumes of liquids or solutions. Oftentimes, volumetric glassware is calibrated before use to ensure that the glassware has been manufactured correctly and has not suffered any loss in precision during previous use. When making mass determinations, always use the same laboratory balance throughout an experiment. Most mass determinations made in the laboratory are done by a difference technique, and many errors due to poorly calibrated balances will cancel out by this method. In many experiments in this manual, you are asked to consult your instructor before starting a procedure, so that the instructor can locate any error in the construction of the apparatus.

Errors in experimental procedures occur because no experimental procedure is perfect and no author of a procedure can anticipate every situation or eventuality among users of the procedure. Even for the most common experimental techniques, procedures are constantly being rewritten and improved. As you perform the experiments in this manual, be aware of suggestions for improving the procedures or for clearing up any ambiguity in them. If you believe you have a better way of doing something, consult with your instructor before trying the new procedure. Be sure you note any changes in procedure in your notebook.

Although errors can be anticipated, and sometimes corrected or allowed for before they occur, they will still occur in virtually all experiments. For numerical data, standard methods have been developed for indicating the magnitude of the error associated with a series of measurements. These methods of error analysis are based on the mathematical science of statistics; these mathematical tools can provide insights about the precision of an experiment or experimental method.

Two words that must be carefully distinguished from one another are used in discussions of error: accuracy and precision. The **accuracy** of a measurement represents how closely the measurement agrees with the true or accepted value for the property being measured. For example, the true boiling point of water under 1 atmosphere barometric pressure is 100.0°C. If the thermometer in your laboratory locker gives a reading of 100°C in boiling water, you would conclude that your thermometer is fairly accurate. The precision of a series of measurements reflects how closely the members of the series agree with each other. If you weighed a series of similar coins on a laboratory balance, and each coin's mass was indicated as 2.65 grams on the balance, you would conclude that the coins had been manufactured to have this precise mass. However, precision in a series of measurements does not necessarily mean that the measurements are accurate. In the preceding example of the coins, even though the balance used indicates the mass of each coin to be 2.65 grams, if the balance is malfunctioning, the mass determined may be completely inaccurate.

Generally, if you have made a valid attempt to exclude personal, apparatus, and procedural errors from an experiment, you may assume that precision in a series of measurements does at least hint that the measurements are also accurate. Expressing the degree of precision in a series of measurements may be done by several standard methods. To demonstrate these methods, consider the following set of data.

Suppose an experiment has been performed to determine the percent by mass of sulfate ion in a sample. To indicate the precision of the method, the experimental determination was repeated four times, with the following results:

Sample	%Sulfate
A	44.02
B	44.11
C	43.98
D	44.09

The **mean** or average value of these determinations is obtained by dividing the sum of the four determinations by the number of determinations.

$$\text{mean} = (44.02 + 44.11 + 43.98 + 44.09)/4 = 44.05$$

The **deviation** of each individual result, without regard to the mathematical sign of the deviation, can then be calculated, and from these, the **average deviation** from the mean.

Sample	% Sulfate	Deviation from Mean
A	44.02	0.03
B	44.11	0.06
C	43.98	0.07
D	44.09	0.04

Mean = 44.05

The average deviation from the mean is then given by

$$(0.03 + 0.06 + 0.07 + 0.04)/4 = 0.05$$

More commonly, average deviations from the mean value of a series of duplicate measurements are expressed on a relative basis, with reference to the mean value. For a mean value of 44.05, and an average deviation from the mean of 0.05, as in the preceding example, the **relative precision** may be expressed as a percentage:

$$(0.05/44.05) \times 100 = 0.1\%$$

The lower the relative deviation from the mean, the more precise a set of measurements is. The precision of an experiment is most commonly expressed by means of the average or relative average deviation from the mean in the situation where only a few measurements have been performed.

Another common method of expressing the precision of a large set of n measurements is the standard deviation, s.

$$s = \sqrt{\frac{(d_1^2 + d_2^2 + d_3^2 + \ldots + d_n^2)}{(n-1)}}$$

where each d represents the deviation of an individual measurement from the mean, and n is the total number of measurements.

For the data given earlier

Sample	% Sulfate	Deviation from Mean	Deviation Squared
A	44.02	0.03	0.0009
B	44.11	0.06	0.0036
C	43.98	0.07	0.0049
D	44.09	0.04	0.0016

Mean = 44.05 Mean deviation = 0.05

the standard deviation, s, would be given by

$$s = \sqrt{\frac{(0.0009 + 0.0036 + 0.0049 + 0.0016)}{(4-1)}} = 0.06$$

and the results of the experiment would be reported as 44.05 ± 0.06 for the percentage of sulfate ion in the sample. As before, the smaller the standard deviation, the more precise is the determination.

Often, one measurement in a series will appear to differ greatly from the other measurements. For example, suppose a fifth determination of the percentage of sulfate was made in the experiment discussed earlier.

Sample	% Sulfate
A	44.02
B	44.11
C	43.98
D	44.09
E	42.15

Obviously, samples A, B, C, and D seem to "agree" with one another, but sample E seems entirely different (maybe some sort of personal error occurred during the experiment). Your first tendency is probably to discard result E; but you need a mathematical method for deciding when a result of a measurement should be rejected as being completely in error. There are various methods for doing this; a very common method is called the *Q-test*, which is applied to the *most deviant result* in a series of measurements. The function Q is defined as follows:

$$Q = \frac{\textit{difference between deviant result and nearest neighbor}}{\textit{range of all measurements in the series}}$$

For the data given, sample E (42.15) is the most deviant, and sample C (43.98) is its nearest neighbor (closest value to the deviant value). The range of the measurements is the difference between the highest value (sample B, 44.11) and the lowest value (sample E, 42.15). Q would then be given by

$$Q = (43.98 - 42.15)/(44.11 - 42.15) = (1.83)/(1.96) = 0.93$$

The calculated value of Q (0.93) is then compared to a table listing standard Q values. If the calculated experimental value of Q is *larger* than the table value of Q, the deviant result should be rejected and not included in subsequent calculations (such as that of the mean or standard deviation). The following short table of standard Q-90 values indicates when a deviant result may be rejected with 90% confidence that the result is in error.

Number of measurements	Q-90
2	—
3	0.94
4	0.76
5	0.64
6	0.56
7	0.51
8	0.47
9	0.44
10	0.41

In the example, there were five measurements made, and the calculated value of Q (0.93) was considerably larger than the table value for five measurements (0.64). This indicates that the results of sample E should be rejected from any subsequent calculations.

There are also other common methods used when considering whether or not to reject a datum in a set of data, and your instructor may prefer that you use one of these methods.

Appendix E

Chemical Nomenclature

Binary compounds—compounds containing only two elements—are named as if they were ionic compounds, giving the name of the positive species first. The name of the negative species is then given, with the ending of the name changed to *-ide*. The simple names of the two elements involved in the binary compound serve as the basis for the two portions of the name.

Example:

$NaBr$

The positive ion is sodium, Na^+.

The negative ion is derived from the element bromine, but the ending of the name is changed to *-ide*: bromide ion.

The compound's name is sodium bromide.

Example:

CaI_2

The positive ion is calcium, Ca^{2+}.

The negative ion is derived from the element iodine, but the ending of the name is changed to *-ide*: iodide ion

The compound's name is calcium iodide.

Example:

Ag_2O

The positive ion is silver, Ag^+.

The negative ion is derived from the element oxygen, but the ending of the name is changed to *-ide*: oxide ion

The compound's name is silver oxide.

For those cases, especially among the transition elements, in which a given element is able to form more than one stable ion, the particular ion involved in a compound must be indicated. This is done by giving the oxidation state (charge) of the element as a Roman numeral in parentheses after the element's name.

Examples:

FeO iron(II) oxide Fe_2O_3 iron(III) oxide

In some cases, the same two elements may react under different conditions to form two (or more) different covalent compounds. For example, if carbon is burned in oxygen, two products are possible, depending on the relative amount of oxygen available.

Examples:

CO carbon monoxide CO_2 carbon dioxide

As in the example, the number of atoms of a particular element in such a covalent compound is indicated by a numerical prefix: mono (1), di (2), tri (3), tetra (4), penta (5), hexa (6), etc.

Examples:

N_2O dinitrogen monoxide

NO nitrogen monoxide

NO_2 nitrogen dioxide

In addition to the simple binary (two-element) compounds already discussed, you will encounter many compounds involving one of the polyatomic ions.

Examples:

NO_3^- nitrate SO_4^{2-} sulfate

Such polyatomic ions are held together as a unit by covalent bonds among the atoms of the ion. However, the species as a whole carries a charge and forms ionic bonds with oppositely charged species. Your textbook gives a list of the more common polyatomic ions, whose names you should memorize. Some examples of the naming of compounds involving polyatomic ions are indicated in the examples that follow. In these names, notice that the polyatomic ion is named as a single entity (rather than in terms of the elements it contains).

Examples:

$NaNO_3$

Positive ion is Na^+, sodium ion.

Negative ion is NO_3^-, nitrate ion.

The compound's name is sodium nitrate.

$CuSO_4$

Positive ion is Cu^{2+}, and since copper is capable of more than one common oxidation state, name is copper(II).

Negative ion is SO_4^{2-}, sulfate ion.

The compound's name is copper(II) sulfate.

$NH_4C_2H_3O_2$

Positive ion is NH_4^+, ammonium ion.

Negative ion is $C_2H_3O_2^-$, acetate ion.

The compound's name is ammonium acetate.

$KMnO_4$

Positive ion is K^+, potassium ion.

Negative ion is MnO_4^-, permanganate ion.

The compound's name is potassium permanganate.

The naming of compounds is confusing to many beginning chemistry students. Use of correct nomenclature, however, is essential to ensure that the correct compound is being discussed or used in a laboratory experiment. For additional examples and practice with nomenclature, consult your textbook.

Appendix F

Vapor Pressure of Water at Various Temperatures

Temperature °C	Pressure mm Hg
0	4.580
5	6.543
10	9.209
15	12.788
16	13.634
17	14.530
18	15.477
19	16.478
20	17.535
21	18.650
22	19.827
23	21.068
24	22.377
25	23.756
26	25.209
27	26.739
28	28.349
29	30.044
30	31.823

Temperature °C	Pressure mm Hg
31	33.696
32	35.663
34	39.899
36	44.563
38	49.692
40	55.324
45	71.882
50	92.511
55	118.03
60	149.37
65	187.55
70	233.71
75	289.10
80	355.11
85	433.62
90	525.77
95	633.91
100	760.00

Appendix G

Concentrated Acid/Base Reagent Data

Acids

Reagent	HCl	HNO_3	H_2SO_4	CH_3COOH
Formula weight of solute	36.46	63.01	98.08	60.05
Density of concentrated reagent	1.19	1.42	1.84	1.05
% concentration	37.2	70.4	96.0	99.8
Molarity	12.1	15.9	18.0	17.4
mL needed for 1 L of 1 *M* solution	82.5	63.0	55.5	57.5

Bases

Reagent	NaOH	NH_3
Formula weight of solute	40.0	35.06
Density of concentrated reagent	1.54	0.90
% concentration	50.5	56.6
Molarity	19.4	14.5
mL needed for 1 L of 1 *M* solution	51.5	69.0

Note: Data as to concentration, density, and amount required to prepare solution will differ slightly from batch to batch of concentrated reagent. Values given here are typical.

Appendix H

Density of Water at Various Temperatures

Temperature, °C	Density, g/mL
0	0.99987
2	0.99997
4	1.0000
6	0.99997
8	0.99988
10	0.99973
12	0.99952
14	0.99927
16	0.99897
18	0.99862
20	0.99823
22	0.99780
24	0.99732
26	0.99681
28	0.99626
30	0.99567
32	0.99505
34	0.99440
36	0.99371
38	0.99299
40	0.99224
45	0.99025
50	0.98807
55	0.98573
60	0.98324

Appendix I

Solubility Information

Simple Rules for Solubility of Ionic Compounds in Water

Nearly all compounds containing Na^+, K^+, and NH_4^+ are readily soluble in water.

Nearly all compounds containing NO_3^- are readily soluble in water.

Most compounds containing Cl^- are soluble in water, with the common exceptions of AgCl, $PbCl_2$, and Hg_2Cl_2.

Most compounds containing SO_4^{2-} are soluble in water, with the common exceptions of $BaSO_4$, $PbSO_4$, and $CaSO_4$.

Most compounds containing OH^- ion are *not* readily soluble in water, with the common exceptions of the hydroxides of the alkali metals (Group 1) and $Ba(OH)_2$.

Most compounds containing other ions not mentioned are *not* appreciably soluble in water.

2.Solubility Products, K_{sp}, for Ionic Substances at 25oC

The molar or gram solubilities of these substances may be calculated from K_{sp} by equilibrium techniques.

Ag_2CO_3	8.1×10^{-12}	$Ba_3(PO4)_2$	$6. \times 10^{-39}$
Ag_2CrO_4	9.0×10^{-12}	$BaCO_3$	1.6×10^{-9}
Ag_2S	1.6×10^{-49}	$BaCrO_4$	8.5×10^{-11}
Ag_2SO_4	1.2×10^{-5}	$BaSO_4$	1.5×10^{-9}
Ag_3PO_4	1.8×10^{-18}	$Ca(OH)_2$	1.3×10^{-6}
AgBr	5.0×10^{-13}	$Ca_3(PO_4)_2$	1.3×10^{-32}
AgCl	1.6×10^{-10}	CaC_2O_4	2.3×10^{-9}
AgI	1.5×10^{-16}	$CaCO_3$	8.7×10^{-9}
AgOH	2.0×10^{-8}	$CaSO_4$	6.1×10^{-5}
$Al(OH)_3$	2.0×10^{-32}	CdS	1.0×10^{-28}
$Ba(OH)_2$	5.0×10^{-3}	CoS	$5. \times 10^{-22}$

$Cr(OH)_3$	6.7×10^{-31}	$Pb(OH)_2$	1.2×10^{-15}
$Cu(OH)_2$	1.6×10^{-19}	$Pb_3(PO4)_2$	$1. \times 10^{-54}$
CuS	8.5×10^{-45}	$PbBr_2$	4.6×10^{-6}
$Fe(OH)_2$	1.8×10^{-15}	$PbCl_2$	1.6×10^{-5}
$Fe(OH)_3$	4.0×10^{-38}	$PbCO_3$	1.5×10^{-15}
FeS	3.7×10^{-19}	$PbCrO_4$	2.0×10^{-16}
Hg_2Cl_2	1.1×10^{-18}	PbI_2	4.6×10^{-6}
HgS	1.6×10^{-54}	PbS	$7.\times 10^{-29}$
$KClO_4$	$1. \times 10^{-2}$	$PbSO_4$	1.3×10^{-8}
$KHC_4H_4O_6$	$3. \times 10^{-4}$	$Sn(OH)_2$	3.0×10^{-27}
$Mg(OH)_2$	8.9×10^{-12}	$Sn(OH)_2$	5×10^{-26}
$MgCO_3$	1.0×10^{-15}	$Sn(OH)_4$	$1. \times 10^{-56}$
$MgNH_4PO_4$	2.5×10^{-13}	SnS	$1. \times 10^{-26}$
MnS	2.3×10^{-13}	$Zn(OH)_2$	4.5×10^{-17}
$Ni(OH)_2$	1.6×10^{-16}	$ZnCO_3$	2.1×10^{-10}
NiS	$3. \times 10^{-21}$	ZnS	2.5×10^{-22}

APPENDIX J

Properties of Substances

Introduction

As a convenience to the user, some properties of the various chemical substances employed in this manual are listed in the table that follows. This table by no means lists all notable properties of the substances mentioned, and should serve only as a quick reference to the substances. *Additional information about the substances to be used in an experiment should be obtained before use of the substances.* Sources of such additional information include the lab instructor or professor, the label on the substance's bottle, or books such as a handbook of dangerous or toxic substances.

In the following table, Corrosiveness and Toxicity risks are indicated as L (low), M (moderate), or H (high).

Name	*Flammability/ Combustibility*	*Corrosivity to skin*	*Toxicity if ingested*	*Notes*
acetaldehyde	Yes	M	M	1, 4
acetic acid	Yes	H	H	1, 4
acetic anhydride	Yes	H	H	
acetone	Yes	M	H	
acetylacetone	Yes	M	H	1, 4
adipoyl chloride	Yes	H	H	1, 4, 7
alanine	No	L	L	
aluminon reagent	No	M	M	
aluminum	Yes*	L	L	
ammonia	No	M	H	1, 4
ammonium carbonate	No	L	M	2
ammonium chloride	No	L	L	
ammonium molybdate	No	M	H	1
ammonium nitrate	No	M	H	3
ammonium sulfate	No	L	M	
ammonium sulfide	No	M	H	4
ammonium thiocyanate	No	L	H	
asparagine	Yes*	L	L	
barium chloride	No	M	H	
barium nitrate	No	M	H	3

Name	*Flammability/ Combustibility*	*Corrosivity to skin*	*Toxicity if ingested*	*Notes*
Benedict's reagent	No	M	M	
benzoic acid, 2-amino-	Yes	M	H	
benzyl alcohol	Yes	L	M	
boric acid	No	L	M	
boron	Yes*	M	H	
bromine	No	H	H	1
butyl alcohol, *n*-	Yes	M	M	
butylamine, *n*-	Yes	M	M	1, 4
butyric acid	Yes	M	M	4
calcium	Yes*	M	L	7
calcium acetate	Yes	H	H	1, 4
calcium carbonate	No	L	L	2
calcium chloride	No	M	M	
calcium hydroxide	No	M	M	
carbon	Yes*	L	M	
carbon disulfide	Yes	H	H	1, 4
chlorine	Yes*	H	H	1, 4
chromium(III) chloride	No	H	H	1
cobalt(II) chloride	No	M	H	
copper	Yes*	L	M	
copper(II) oxide	No	M	M	
copper(II) sulfate	No	M	M	
cyclohexane	Yes	L	H	
cyclohexene	Yes	M	H	4
dichlorobenzene, 1, 4-	Yes	M	H	
diethyl ether	Yes	M	H	
dimethylglyoxime	Yes	M	M	
dry ice	No	H	L	5
Eriochrome Black-T	No	M	H	
ethyl alcohol (ethanol)	Yes	L	H	6
ethyl ether	Yes	M	H	
ethylene diamine	Yes	M	H	1, 4
ethylene diamine tetraacetic acid (EDTA)	Yes*	M	H	
formaldehyde	Yes	M	H	1, 4
fructose	Yes*	L	L	
gelatin	Yes*	L	L	
glucose	Yes*	L	L	
glycine	Yes*	L	L	
graphite	Yes*	M	M	
hexane	Yes	M	H	
hexanediamine	Yes	H	H	4
hexene, 1-	Yes	M	M	
hydrochloric acid	No	H	H	1, 4
hydrogen	Yes	L	M	
hydrogen peroxide, 3%	No	H	M	3
iodine	No	H	H	
iron(II) sulfate	No	L	L	
iron(III) chloride	No	M	M	

Name	***Flammability/ Combustibility***	***Corrosivity to skin***	***Toxicity if ingested***	***Notes***
iron(III) nitrate	No	M	M	3
isoamyl alcohol	Yes	L	M	
isobutyl alcohol	Yes	M	M	
isoleucine	Yes*	L	L	
isopropyl alcohol	Yes	L	M	
lead	Yes*	L	M	
lead acetate	Yes*	M	H	
lithium	Yes	H	H	7
lithium chloride	No	M	H	
magnesium	Yes	L	M	
magnesium chloride	No	L	M	
manganese(II) chloride	No	M	M	3
manganese(IV) oxide	No	M	M	3
mercury	No	H	H	
mercury(II) chloride	No	H	H	
methyl alcohol (methanol)	Yes	H	H	
methyl salicylate	Yes	M	H	
methylene chloride	Yes	H	H	1, 4
n-butyl alcohol	Yes	M	M	
n-butylamine	Yes	M	M	4
n-octyl alcohol	Yes	M	M	
n-pentyl alcohol	Yes	M	M	
n-propyl alcohol	Yes	L	M	
naphthalene	Yes	M	H	
nickel(II) chloride	No	M	H	
nickel(II) sulfate	No	M	H	
ninhydrin	Yes	H	H	
nitric acid	No	H	H	1, 3, 4
nitrogen	No	L	L	
oleic acid	Yes	M	L	
oxygen	Supports	L	L	
pentane	Yes	M	H	
pentanedione, 2,4-	Yes	M	H	1, 4
pentyl alcohol, *n*-	Yes	M	M	
phenanthroline, 1,10-	Yes	M	M	
phosphoric acid	No	H	H	
phosphorus, red	Yes	H	H	
phosphorus, white	Yes	H	H	
potassium	Yes	H	H	7
potassium bromide	No	L	L	
potassium carbonate	No	L	L	
potassium chloride	No	L	L	
potassium chromate	No	H	H	3
potassium dichromate	No	H	H	3
potassium ferrocyanide	No	M	H	
potassium hydrogen phthalate	Yes*	M	M	4
potassium hydroxide	No	H	H	
potassium iodate	No	H	H	3
potassium iodide	No	M	M	
potassium nitrate	No	M	H	3

Name	***Flammability/ Combustibility***	***Corrosivity to skin***	***Toxicity if ingested***	***Notes***
potassium oxalate	Yes*	M	H	
potassium permanganate	No	H	H	3
potassium sulfate	No	L	L	
proline	Yes*	L	L	
propionic acid	Yes	H	H	
propyl alcohol, *n*-	Yes	L	M	
salicylic acid	Yes*	M	M	
serine	Yes*	L	L	
silicon	No	M	M	
silver acetate	Yes*	M	M	
silver chromate	No	H	H	3
silver nitrate	No	M	M	
sodium	Yes	H	H	7
sodium acetate	Yes*	M	M	
sodium bicarbonate	No	L	L	
sodium bromide	No	L	L	
sodium chloride	No	L	L	
sodium chromate	No	H	H	3
sodium hydrogen sulfite	No	M	M	4
sodium hydroxide	No	H	H	
sodium hypochlorite	No	H	H	1, 4
sodium iodide	No	M	M	
sodium nitrate	No	H	H	3
sodium nitrite	No	H	H	3
sodium nitroprusside	No	M	M	
sodium silicate	No	L	L	
sodium sulfate	No	L	L	
sodium sulfite	No	M	M	3, 4
sodium thiosulfate	No	M	M	
starch	Yes*	L	L	
strontium chloride	No	M	M	
sulfur	Yes	L	L	
sulfuric acid	No	H	H	3, 7
tartaric acid	Yes*	M	L	
thioacetamide	Yes*	H	H	4
tin	No	L	L	
tin(II) chloride	No	M	M	3
toluene	Yes	H	H	
zinc	Yes*	L	L	
zinc sulfate	No	L	M	

*Will burn if deliberately ignited, but has very low vapor pressure and is not a fire hazard under normal laboratory conditions.

Notes:

1. This substance is a respiratory irritant. Confine use of the substance to the exhaust hood.

2. This substance vigorously evolves carbon dioxide if acidified.

3. This substance is a strong oxidant/reductant. Although the substance itself may not be flammable, contact with an easily oxidized/reduced material may cause a fire or explosion. Avoid especially contact with any organic substance.

4. This substance possesses or may give rise to a stench. Confine use of this substance to the exhaust hood.

5. Handle dry ice with tongs to avoid frostbite.

6. Ethyl alcohol for use in the laboratory has been denatured. It is not fit to drink and cannot be made fit to drink.

7. This substance reacts vigorously with water and may explode.

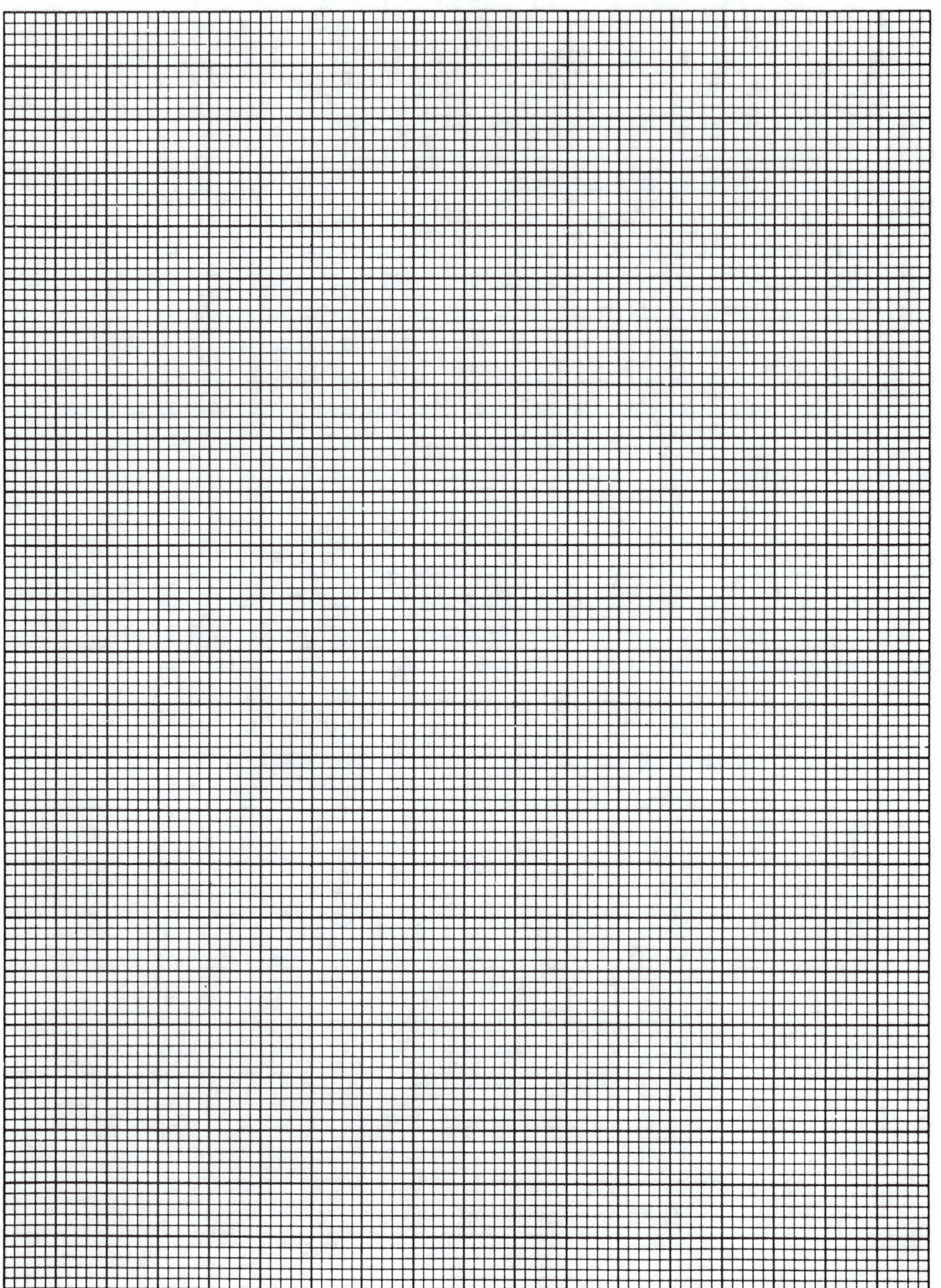

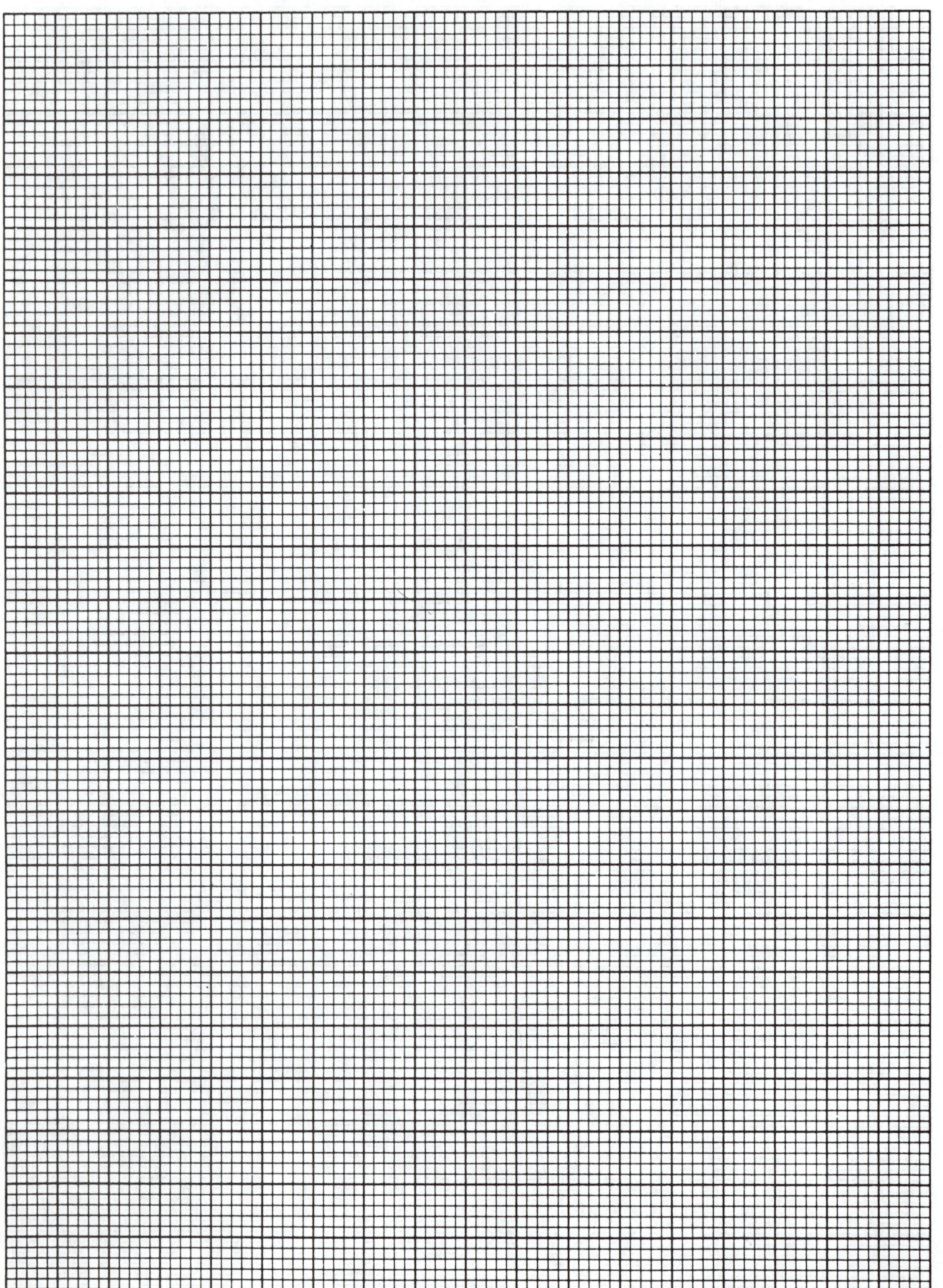

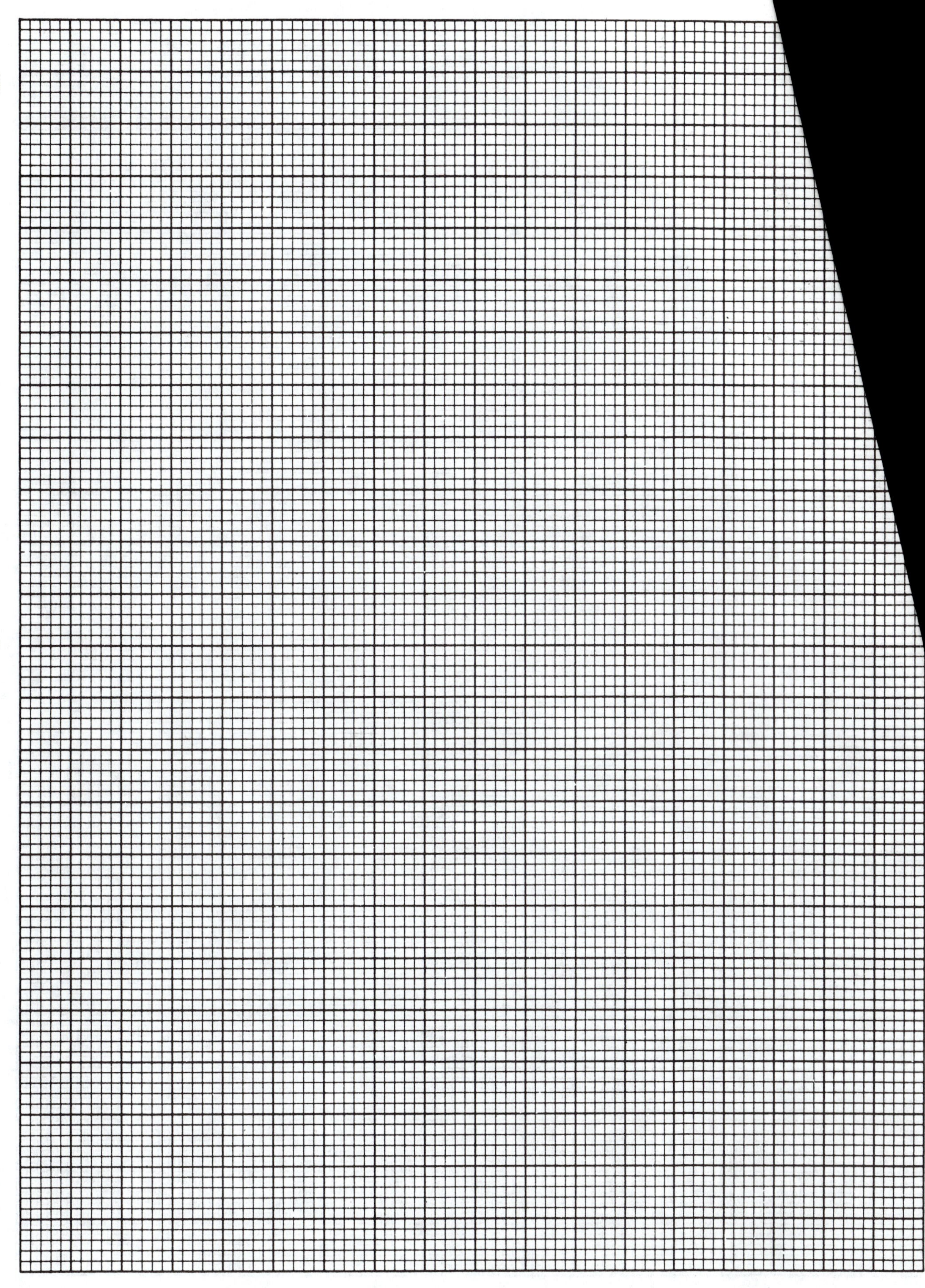

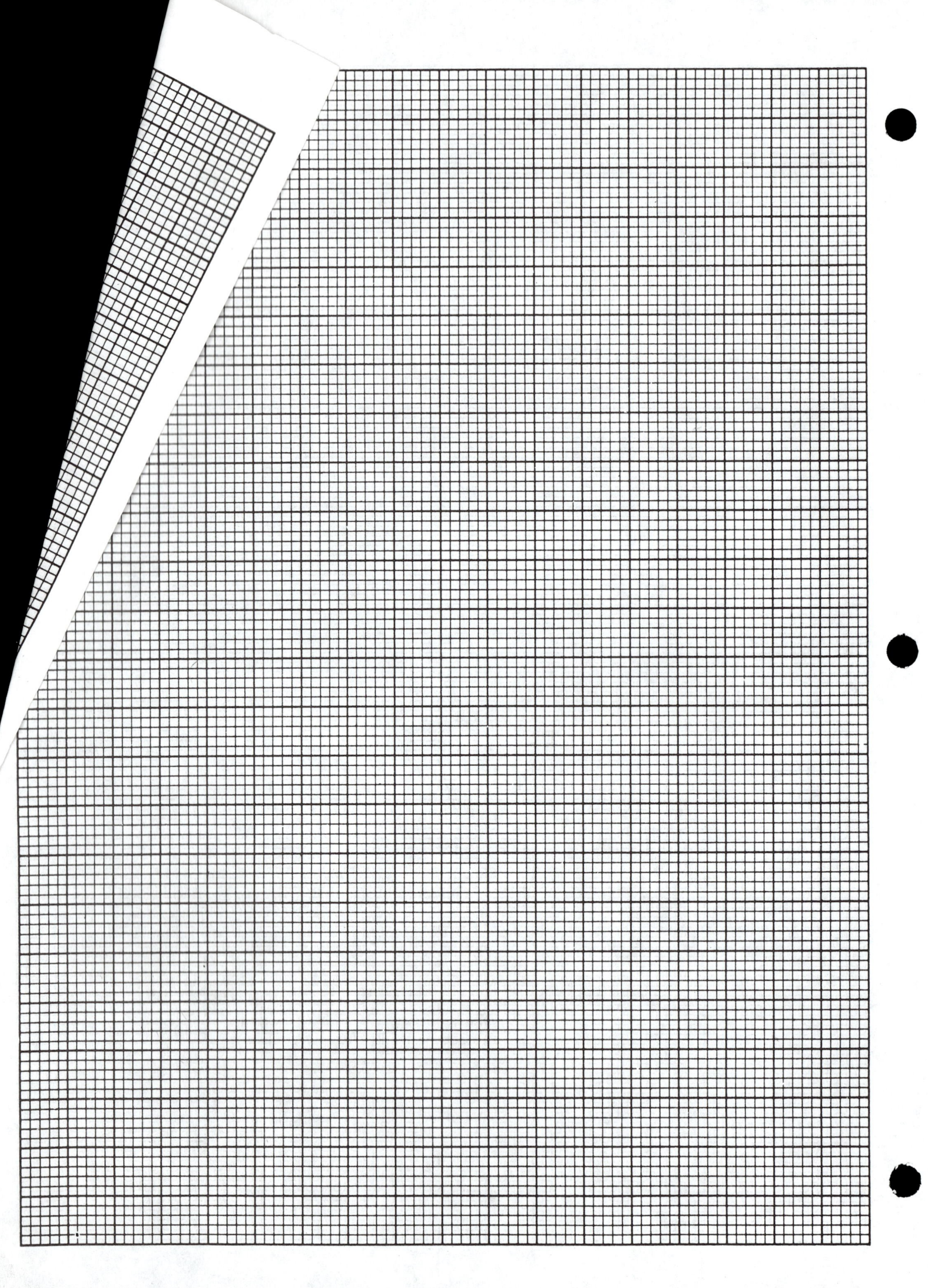